Topics in Current Physics

46

Topics in Current Physics

Founded by Helmut K.V. Lotsch

Volume 39 **Nonequilibrium Vibrational Kinetics**
Editor: M. Capitelli

Volume 40 **Microscopic Methods in Metals**
Editor: U. Gonser

Volume 41 **Structure and Dynamics of Surfaces I**
Editors: W. Schommers and P. von Blanckenhagen

Volume 42 **Metallic Magnetism**
Editor: H. Capellmann

Volume 43 **Structure and Dynamics of Surfaces II**
Phenomena, Models, and Methods
Editors: W. Schommers and P. von Blanckenhagen

Volume 44 **Persistent Spectral Hole-Burning: Science and Applications**
Editor: W.E. Moerner

Volume 45 **Structural Phase Transitions II**
Editors: K.A. Müller and H. Thomas

Volume 46 **Photoacoustic, Photothermal and Photochemical Processes in Gases**
Editor: P. Hess

Volume 47 **Photoacoustic, Photothermal and Photochemical Processes at Surfaces and in Thin Films**
Editor: P. Hess

Volumes 1–38 are listed on the inside back cover

Photoacoustic, Photothermal and Photochemical Processes in Gases

Edited by P. Hess

With Contributions by
R.T. Bailey S. Bernegger D. Bicanic F. Bijnen
C.W.P.M. Blom F.R. Cruickshank G.J. Diebold
M. Fiedler F. Harren P. Hess L. van Hove
H. Jalink M. Kooijman J.B. Mehl P.L. Meyer
M.R. Moldover D. Pugh J. Reuss H. Sauren
M.W. Sigrist J. Snel W. Tonk L. van Hove
L.A.C.J. Voesenek E. Woltering B. Zuidberg

With 115 Figures

Springer-Verlag Berlin Heidelberg New York
London Paris Tokyo Hong Kong

Professor Peter Hess

Physikalisch-Chemisches Institut, Universität, Im Neuenheimer Feld 253,
D-6900 Heidelberg, Fed. Rep. of Germany

ISBN-13:978-3-642-83853-8 e-ISBN-13:978-3-642-83851-4
DOI: 10.1007/978-3-642-83851-4

Softcover reprint of the hardcover 1st edition 1989

2154/3150-543210 – Printed on acid-free paper

Preface

This volume comprises review articles based on talks presented at the 49th WE-Heraeus seminar, which was held at the Physikzentrum, Bad Honnef, November 21–23, 1988. This seminar was truly international and attended by leading scientists in the field. It was the generous support of the WE-Heraeus Foundation that made possible the meeting, and also this book resulting from the seminar. Therefore, I would like to thank the WE-Heraeus Foundation and their representatives involved in this seminar, Werner Buckel and Dieter Röss, directory; Volker Schäfer, manager; and Albrecht Bischoff, organizer, for their help and assistance.

The reviews published in this volume cover the subject of photoacoustic photothermal and photochemical processes in the gas phase, which was treated during the first day of the seminar. The second and third days were devoted to discussion of photoacoustic, photothermal and photochemical processes at surfaces and in thin films. The corresponding reviews will be published in a separate volume of this series.

This is the first book dealing only with gas phase photoacoustics. The goal was to present the subject as broadly as possible, including theory and applications. In addition to giving a snapshot of the present state of research, it also describes the basic principles, the tremendous progress made in recent years in this field and expected trends in future developments.

As editor of this volume, I am indebted to Dr. Lotsch of Springer-Verlag, Heidelberg, for valuable suggestions and assistance during the planning stage. Also, I would like to express my sincere appreciation to Deborah Hollis, of the physics editorial department, for her help and skilful assistance during the final processing of the manuscripts. Finally, I would like to thank Reinhold Michels, of the production department, for his efforts to reduce the time needed for publication of this book.

Heidelberg, April 1989 *Peter Hess*

Contents

List of Contributors

R.T. Bailey
Department of Pure and Applied Chemistry, University of Strathclyde,
Glasgow G1 1XL, Great Britain

S. Bernegger
Institute of Quantum Electronics, ETH, CH-8093 Zürich, Switzerland;
now with "Winterthur", Swiss Insurance Company, CH-8401 Winterthur

D. Bicanic
Department of Agricultural Engineering and Physics, Agricultural University,
Duivendaal 1, 6701 AP Wageningen, The Netherlands

F. Bijnen
Department of Molecular and Laser Physics, Catholic University,
Toernooiveld, 6525 ED Nijmegen, The Netherlands

C.W.P.M. Blom
Department of Experimental Botany, Catholic University,
Toernooiveld, 6525 ED Nijmegen, The Netherlands

F.R. Cruickshank
Department of Pure and Applied Chemistry, University of Strathclyde,
Glasgow G1 1XL, Great Britain

G.J. Diebold
Department of Chemistry, Brown University, Providence, RI 02912, USA

M. Fiedler
Institute of Physical Chemistry, University of Heidelberg,
Im Neuenheimer Feld 253, D-6900 Heidelberg, Fed. Rep. of Germany

F. Harren
Department of Molecular and Laser Physics, Catholic University,
Toernooiveld, 6525 ED Nijmegen, The Netherlands

P. Hess
Institute of Physical Chemistry, University of Heidelberg,
Im Neuenheimer Feld 253, D-6900 Heidelberg, Fed. Rep. of Germany

L. van Hove
Department of Plant Physiology, General Foulkesweg 72,
6703 BW Wageningen, The Netherlands

H. Jalink
Department of Agricultural Engineering and Physics, Agricultural University, Duivendaal 1, 6701 AP Wageningen, The Netherlands

M. Kooijman
Department of Plant Physiology, General Foulkesweg 72, 6703 BW Wageningen, The Netherlands

J.B. Mehl
Department of Physics and Astronomy, University of Delaware, Newark, DE 19716, USA

P.L. Meyer
Institute of Quantum Electronics, ETH, CH-8093 Zürich, Switzerland; now with Swiss Reinsurance Company, CH-8022 Zürich, Switzerland

M.R. Moldover
Thermophysics Division, National Institute for Standards and Technology, Gaithersburg, MD 20899, USA

D. Pugh
Department of Pure and Applied Chemistry, University of Strathclyde, Glasgow G1 1XL, Great Britain

J. Reuss
Department of Molecular and Laser Physics, Catholic University, Toernooiveld, 6525 ED Nijmegen, The Netherlands

H. Sauren
Department of Agricultural Engineering and Physics, Agricultural University, Duivendaal 1, 6701 AP Wageningen, The Netherlands

M.W. Sigrist
Institute of Quantum Electronics, ETH, CH-8093 Zürich, Switzerland

J. Snel
Department of Plant Physiology, General Foulkesweg 72, 6703 BW Wageningen, The Netherlands

W. Tonk
Department of Plant Physiology, General Foulkesweg 72, 6703 BW Wageningen, The Netherlands

L.A.C.J. Voesenek
Department of Experimental Botany, Catholic University, Toernooiveld, 6525 ED Nijmegen, The Netherlands

E. Woltering
Sprenger Institute, Haarsteeg 6, 6701 AA Wageningen, The Netherlands

B. Zuidberg
Department of Physics, Bulak Sumur 10, Gadjad Mada University, Yogyakarta, Indonesia

1. Principles of Photoacoustic and Photothermal Analysis

Peter Hess

With 7 Figures

This chapter provides an elementary introduction to the characteristic features of photothermal and photoacoustic detection and analysis. An overview is presented of the recent advances and developments described in later chapters of this book. These include the development of theoretical models in the time and frequency domains, the investigation of fundamental physical and chemical processes in the gas phase, and applications of the techniques in trace analysis, pollution monitoring and kinetic studies.

1.1 Photoinduced Processes and Detection

1.1.1 Variety of Processes

The absorption of photons by the molecules of a gas sample in accordance with their spectroscopic properties may produce a series of effects. The excited level may lose its energy by radiation processes, such as spontaneous or stimulated emission, and by radiationless deactivation, which channels at least part of the absorbed energy into the translational degrees of freedom. If the photon energy is high enough, direct photochemical decomposition of the excited molecule can be achieved (Fig. 1.1). In the case of vibrational excitation, radiative emission and chemical reactions do not play an important role, because the radiative lifetime of vibrational levels is long compared with the time needed for collisional deactivation at ordinary pressures and the photon energy is too small to induce reactions. However, in the case of electronic excitation, emission of radiation and chemical reaction processes may compete efficiently with collisional deactivation.

Local heating with a short light pulse or a modulated light beam induces local expansion in the irradiated region. This causes a transient pressure change in the case of pulsed excitation or a periodic pressure change and excitation of a sound wave in the case of a modulated excitation source. Branching processes such as thermal radiation and phase transitions in condensed samples limit the transformation of heat into acoustic energy (Fig. 1.1). In principle, all branching processes must be taken into account to achieve a quantitative description of the photoacoustic phenomenon.

1.1.2 Detection Methods

Let us consider the two competing processes radiative and nonradiative decay of excited molecules. The former process can be studied directly by means of laser-

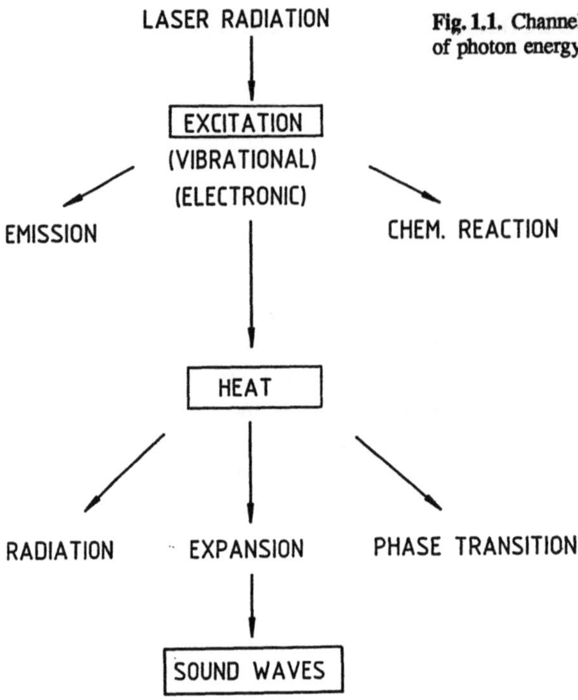

Fig. 1.1. Channels competing with the transformation of photon energy into heat and acoustic energy

induced fluorescence (LIF). In photoacoustics the nonradiative channel generating heat is of primary concern. Thus, the two methods yield supplementary or complementary information. Therefore, it is useful to apply both methods, especially when additional processes such as chemical reactions are induced. Chemical reactions make their own characteristic contribution to the release of heat. If photodissociation occurs, for example, the local increase in the number of moles and the thermalization of the recoil energy of the fragments generate sound waves. Direct photodissociation and subsequent collision-induced chemical reactions occur on different time scales, and therefore it may be possible to separate these processes.

In gas phase photoacoustics two main types of detector are used to monitor the generated thermal or acoustic disturbances, namely refractive index sensors and pressure sensors. Refractive index sensors such as the probe beam deflection technique monitor directly the evolution of heat. The temperature rise in the irradiated region results in a spatial gradient in the refractive index of the gas, which deflects a probe laser beam. These noncontact devices can be applied for remote sensing in aggressive media or the investigation of gas systems at high temperatures. In most gas phase experiments the pressure signal is detected by a sensitive microphone. Commercial condenser microphones with sensitivities up to 100 mV/Pa and cheap miniature electret microphones with typical sensitivities of 10 mV/Pa can be easily employed in photoacoustic experiments and normally possess a higher sensitivity than refractive index sensors.

1.1.3 Literature Review

Since the advent of laser radiation sources, the number of papers published in the field of photoacoustic and photothermal phenomena has increased drastically. There are several books available treating the subject in the form of review articles written by experts or as monographs [1.1–5]. In addition, a series of separate reviews have been published that concentrate on different aspects of the field [1.6–10]. Since the third international topical meeting on photoacoustic and photothermal phenomena, held in 1983, the invited and contributed talks presented at these meetings have been published [1.11–13]. These proceedings volumes give an up-to-date view of current activities and of the progress achieved since the previous conference.

1.2 Principles of Photoacoustics

1.2.1 Basic Considerations

As discussed before, the fraction of the absorbed radiation converted into heat or acoustic energy is detected in a photothermal or photoacoustic experiment. The rate of heat production in the sample plays a central role, and therefore photoacoustics may be considered as a calorimetric technique, where the energy input into the system is controlled by the optical properties and the energy flow into various channels is determined by the kinetic properties.

Photoacoustics is often called photoacoustic spectroscopy (PAS), because energy deposition is governed by the spectroscopic properties. Fig. 1.2 gives a comparison of the characteristic features of photoacoustic spectroscopy and conventional absorption spectroscopy. In photoacoustics the signal produced by the absorbed laser

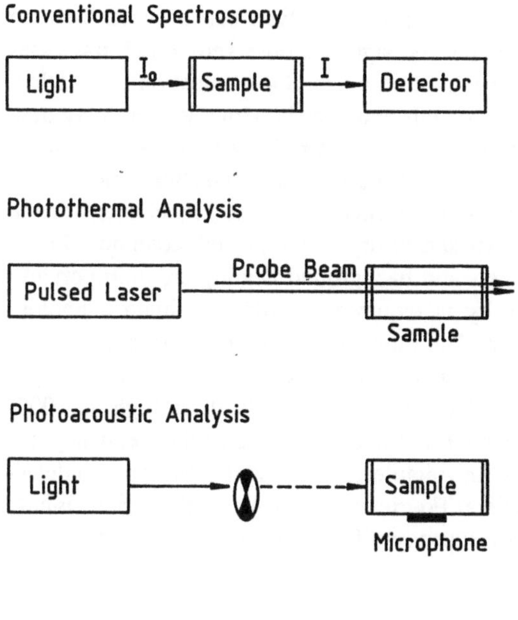

Fig. 1.2. Comparison of direct photothermal and photoacoustic detection schemes with conventional absorption spectroscopy

photons is detected directly because changes induced in the sample by the distribution of the absorbed energy are monitored. The signal amplitude depends on the intensity of the incident radiation as long as saturation effects are avoided. Thus, the use of radiation sources with high light intensities, such as lasers, is advantageous and increases the sensitivity. In photothermal studies the evolution of heat is monitored by a second probe laser, whereas in photoacoustic experiments a microphone is used to detect the sound waves, as illustrated in Fig. 1.2. In conventional absorption spectroscopy the energy absorbed in the sample is determined by measuring the incident light intensity I_0 and the transmitted intensity I (Fig. 1.2). In the case of low optical absorption of the sample and for the higher power levels attainable with lasers, a small difference between two nearly equal signals is obtained. In addition, transparent samples are required and the reflection and scattering of light introduces problems. The requirements on the optical quality of the sample are much less stringent in photoacoustics. Only the signal originating from the absorbed photons is of interest, and therefore measurements can be performed with gaseous, liquid or solid systems, where the latter may be amorphous, opaque, powdered, highly scattering, etc.

1.2.2 Time and Frequency Domain Analysis

Photoacoustic analysis is based on either pulsed or modulated excitation. In the pulsed technique a light pulse of high peak intensity and short duration is employed to generate the acoustic signal. With lasers, peak powers in the large range 10^{-3}–10^6 W may be produced. The goal in this time domain technique is to optimally monitor the thermal signal or the ballistic acoustic wave packet generated by sample absorption. The signal is analyzed in the time domain, allowing a gating technique for noise suppression. Only the first part of the acoustic signal is usually detected because later on window noise, echoes, etc., contribute to the signal, as shown in Fig. 1.3. The pulsed signal consists of many Fourier components which are analyzed at the same time. A boxcar averager or a transient digitizer can be used for time resolved detection. Normally, a strong acoustic signal is observed, which possesses a well-defined delay with respect to the excitation pulse. Short detection times allow effective discrimination against window signals and echoes with a longer delay time and against light-scattering signals with a much shorter delay time. An additional advantage is that the acoustic noise is small during the short detection time.

The efficiency of conversion into acoustic energy may be higher than for the modulation method, resulting in a higher sensitivity of the pulsed technique. In the case of gated detection, boundary effects and background heating are unimportant.

A modulation of the light intensity is required in order to obtain a photoacoustic signal in the frequency domain. The most common technique is amplitude modulation of the radiation beam. With a mechanical chopper ($<$ 15 kHz) or an electro-optic modulation device ($<$ 20 MHz) a modulation depth as high as 100% may be achieved. The photoacoustic signal is usually analyzed in the frequency domain employing a lock-in amplifier. With suitable electronics it is possible to detect the amplitude and phase of one or several Fourier components of the signal. Filters may be used to suppress noise with a frequency different from the signal frequency

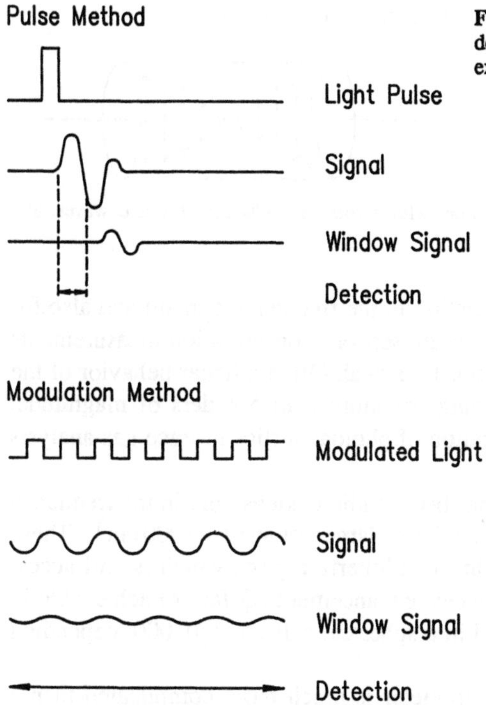

Fig. 1.3. Principles of pulsed excitation with detection in the time domain and modulated excitation with detection in the frequency domain

such as external mechanical noise and electromagnetic pick-up. The window signal, for example, possesses the same frequency, and therefore discrimination against this noise source is very difficult in the continuous wave modulation technique, as shown in Fig. 1.3 schematically.

Because acoustic energy conversion efficiency is low, the cw modulation technique is advantageous in the case of strongly absorbing materials. Additional processes such as boundary effects and background heating are often important and decrease the signal amplitude.

1.2.3 Resonant and Nonresonant Operation

Depending on whether or not there is a coincidence between one of the acoustic resonance frequencies of the photoacoustic cell and the modulation frequency employed, the cell is operated in a resonant or nonresonant mode.

To attain the highest sensitivity in a nonresonant photoacoustic experiment a small cell volume of about 1–10 cm^3 is combined with low modulation frequencies of up to 100 Hz, which are small compared to the resonance frequencies. Such a setup is extremely simple and highly sensitive compared to other techniques. The sensitivity is limited in this case by the relatively large continuum background signal caused by window and wall absorption. The processes occurring in the gas volume and at the surface are too complex to be modelled quantitatively, as indicated in Fig. 1.4. Therefore, it is not possible to develop a quantitative theory of the

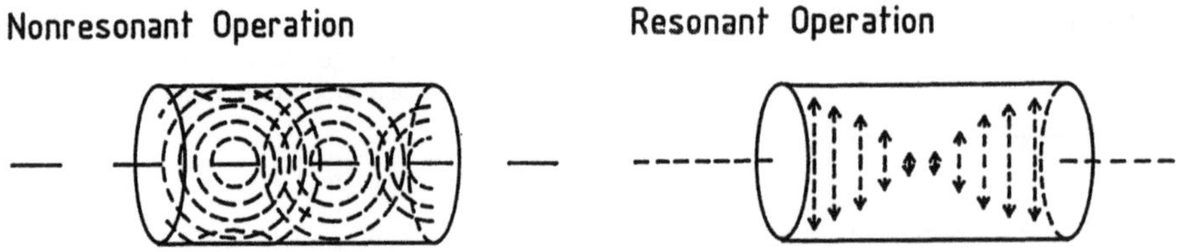

Fig. 1.4. Comparison between nonresonant excitation, with a complex behavior of sound waves, and resonant excitation of a longitudinal mode

photoacoustic signal. This is true for operation in the frequency domain and also for pulsed excitation and detection with a pressure sensor. Concentration measurements are performed by calibrating the photoacoustic signal. Often a linear behavior of the signal is observed in a large dynamic range of more than 5 orders of magnitude. This is an important point for the application of photoacoustics in trace gas analysis and pollution monitoring.

In recent years resonant operation has been studied extensively in the frequency domain using, in particular, cylindrical cavities of large volume ($\sim 1000\,\text{cm}^3$). These resonators possess acoustic resonances in the kilohertz region, which is still accessible by mechanical choppers. The resonance enhancement (Q-factor) achievable in a cylindrical cell may be up to 1000 and in a spherical cell up to 10000, depending on pressure and energy dissipation.

Such a resonant photoacoustic spectrometer is much more complicated than a typical nonresonant setup. However, it allows the precise determination of a series of thermodynamic and kinetic properties of the system. In a computer controlled version the resonance profiles of the acoustic cavity modes can be measured point by point with high accuracy. The position of the acoustic resonance frequencies allows the determination of thermophysical properties, and their dispersion allows the quantitative investigation of kinetic effects. In a similar way the broadening of the resonance profiles by a variety of dissipation processes can be analyzed quantitatively. In a resonance experiment the acoustic energy is channelled into a series of highly symmetric acoustic modes, as shown schematically in Fig. 1.4 for the longitudinal mode in a cylindrical cavity. Therefore, it is not surprising that in this case a quantitative description of the photoacoustic signal is possible. In this analysis the pure photoacoustic signal without resonance enhancement is considered. For this reason it is necessary to measure the Q-factor of each resonance and take it into account in signal analysis.

It has been shown recently that the excitation of acoustic modes in optimized fixed volume resonators is one of the most accurate methods of determining thermophysical and kinetic quantities. Less optimized resonators with low Q-factors are frequently used in trace gas detection.

Acoustic resonances can also be excited with short high-power laser pulses. This is a completely new field with many potential applications in fundamental science and trace gas analysis.

1.2.4 Detection Devices

In photothermal detection, a low-power cw probe laser beam is often employed to monitor the radial refractive index distribution generated with continuous, chopped or pulsed laser radiation. Measurement schemes which detect the thermal effects of incident radiation are collectively referred to as photothermal techniques. The heat distribution develops in the sample as a lens-like element and is called a "thermal lens".

Due to the pulse-to-pulse instability of many pulsed lasers, cw lasers are preferred as pump sources for thermal lens measurements. Pulsed lasers are employed when high peak powers are needed to study nonlinear absorption effects, or in cases where the corresponding spectral properties are of interest.

Optical probing, normally performed with a HeNe laser, measures the influence of the temperature change on the refractive index of the sample. In a complete analysis it should be possible to determine the temperature distribution in the sample, the amount of energy deposited in the sample and the sample absorbance from the measured distribution of the index of refraction.

The lens signal depends, of course, on the density of the sample; the detection limit reported for liquids is about $\alpha l \approx 10^{-6} - 10^{-7}$, where α is the optical absorption coefficient and l is the absorption pathlength. For gases the sensitivity is lower.

In most photoacoustic setups for gas analysis, electret microphones are used as displacement-sensitive detectors to monitor the acoustic waves. Unlike normal condenser microphones, electret microphones need no external dc voltage, they can be manufactured in very small sizes and they are very cheap, due to huge production numbers. This explains the widespread use of these devices despite the somewhat lower sensitivity. As already mentioned, a sensitivity of about 10 mV/Pa is achieved with electret microphones. With the microphones and electronic equipment presently available, a temperature rise of about 10^{-6} K or a power density of about 10^{-9} J/cm^3 s can be detected in a gas. The microphones generally possess a flat frequency response to about 15 kHz and are not sensitive to mechanical vibrations and other distortions. The response to incident pressure waves is linear over many orders of magnitude. Electret microphones also respond well to pressure impulses and thus can be used in time resolved investigations. The response time of a gas–microphone system is on the order of 10^{-4} s and is determined either by the microphone response or by the transit time of the sound wave.

The detection limit of a microphone sensor is about $\alpha l \approx 10^{-8}$ for a gas sample, and thus is much lower than for photothermal detection schemes.

1.3 Recent Advances and Developments

1.3.1 Theory

Considerable progress has been made in recent years in the development of theoretical models for the quantitative description of photothermal and photoacoustic experiments. Figure 1.5 gives a general scheme of the theoretical foundation of pho-

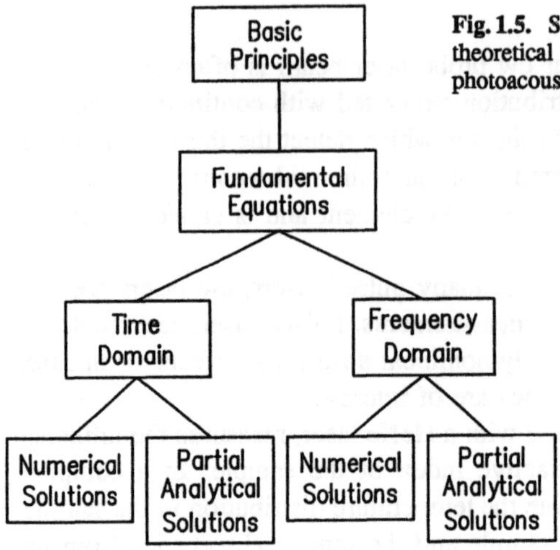

Fig. 1.5. Schematic representation of the structure of theoretical models for the analysis of photothermal and photoacoustic experiments

toacoustics. The fundamental equations describing the corresponding effects can be derived from basic principles in the framework of linear gas dynamics, as shown by *Pugh* in Chap. 2. These fundamental equations provide the starting point for the treatment of specific photothermal or photoacoustic phenomena. The final development of the model is performed either in the time domain, where acoustic waveforms are recorded as a function of time, or in the frequency domain, where the amplitude and phase shift of the acoustic signal are measured as a function of modulation frequency. From a purely mathematical viewpoint, the information obtained from these two different procedures is identical. In Chap. 2 the theoretical modelling is presented for two experimental techniques. One is the pulsed laser excited thermal lens effect, where the radial refractive index distribution is detected by its effect on a second cw probe laser beam in the time domain. It is shown that for certain relationships between important parameters, such as the characteristic times for thermal and diffusional relaxation, the acoustic relaxation time and the relaxation time for energy exchange between internal modes and translation, useful analytical solutions can be found. To include any possible choice of the parameter values or nonlinear disturbances, numerical methods must be applied.

Chapter 2 also introduces the fundamental equations for the investigation of acoustic resonances in fixed volume resonators in the frequency domain. This is one of the most accurate techniques for the determination of thermophysical properties and of kinetic constants presently available. This is due to the existence of a theoretical model not only for the laser excitation of acoustic resonances, but also for fixed volume resonators of high symmetry. The theory of spherical acoustic resonators is presented in Chap. 4 by *Mehl* and *Moldover*. They show that in such a resonance experiment the sound velocity can be measured with an accuracy on the order of 0.01% without sophisticated analysis. A complete theoretical analysis of the resonator behavior improves the accuracy to better than one part per million. This theory takes into account the coupling between acoustic and thermal waves, thermal and viscous effects at the shell boundary, shell motion and imperfect shell geometry. The effects of holes in the resonator wall have also been considered.

A theoretical model for the photoacoustic analysis of a dissociation reaction in a cylindrical resonator is developed in Chap. 5 by *Fiedler* and *Hess*. This treatment is based on the fundamental equations for conservation of mass, momentum and energy, the equation of state, the definition of density and a kinetic equation describing the reaction process. The corresponding Helmholtz equations are solved with the boundary conditions and yield a quantitative description of the experimental results of such a resonance experiment. This includes the resonance frequency dispersion, broadening of the halfwidth of the resonance profile and pressure dependence of the photoacoustic signal.

In Chap. 6 more-complicated chemical reactions are considered by *Diebold*. In this chapter the bases of pulsed and modulated excitation are treated with emphasis on the description of photochemical energy release and gas-phase chemical kinetics. Especially nonlinear effects and the phenomenon of chemical amplification of the signal by chain reactions are discussed on a theoretical basis.

1.3.2 Fundamental Constants and Thermophysical Properties

In photoacoustics, relatively complex dynamic processes must be analyzed. Therefore, it is not surprising that not only thermophysical but also kinetic properties of the system are needed for a quantitative analysis. Figure 1.6 shows schematically some of these properties, which are connected by more or less complicated dependencies. To determine one of these properties experimentally, the best procedure is to select the values of the variables in such a way that the influence of other properties is as small as possible. This will now be discussed in more detail for the determination of the universal gas constant.

The most accurate value of the gas constant was determined recently, as described by *Mehl* and *Moldover* in Chap. 4, by exciting several radial resonances in a

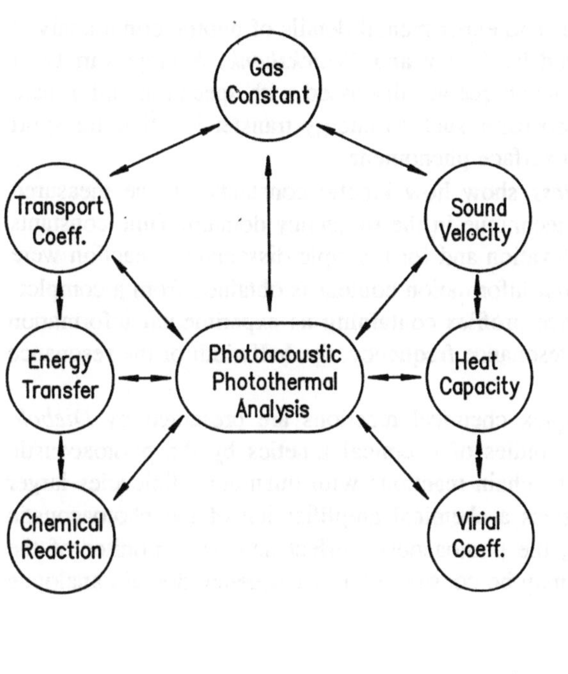

Fig. 1.6. Kinetic processes, fundamental constants and thermophysical properties involved in an analysis of the photoacoustic effect

spherical resonator. Argon was selected as the gas, because with a monoatomic gas relaxation of internal degrees of freedom and chemical relaxation are avoided. The acoustic modes were excited acoustically with a second microphone. The nonselective acoustical excitation method could be used in this case, because the resonances are relatively narrow in argon, especially at higher pressures. In the case of polyatomic molecules with strong broadening of the resonance profile due to dissipation processes, laser excitation of acoustic modes should be preferred, because here the number of excited modes can be reduced drastically by symmetry constraints. In the selected pressure range of 100–500 kPa the influence of the transport properties at the resonator surface and in the bulk was relatively small. The real gas correction was taken into account by extrapolating to zero pressure. The temperature was fixed by the triple point of water. From the corrected eigenfrequencies, considering the perturbation effects mentioned before, a very accurate value for the ratio of the sound velocity to the sphere radius was obtained. Inserting the sphere radius yielded the most accurate value for the sound velocity in ideal argon and finally the gas constant, because the specific heat ratio, the temperature and the mass were known.

A similar procedure can of course be used to determine other thermophysical properties, such as the heat capacities, temperature, virial coefficient, or of course the size of resonators, if the other parameters are known with reasonable accuracy.

1.3.3 Kinetic Processes

The many dynamic processes occurring during a photothermal or photoacoustic analysis allow the accurate investigation of a variety of kinetic effects, for example, transport processes, energy exchange between different degrees of freedom and the kinetics of chemical reactions, as shown in Fig. 1.6. In general, the most accurate analysis of kinetic processes is possible if the different relaxation processes take place on different time scales.

In Chap. 3 different schemes and experimental details of photothermal analysis by optical detection are described by *Bailey* and *Cruickshank*. A large variety of applications of the thermal lens technique are discussed with special attention paid to the investigation of kinetic processes such as energy transfer kinetics, transport phenomena, photochemistry and surface phenomena.

In Chap. 5, *Fiedler* and *Hess* show how kinetic constants can be measured by the photoacoustic resonance technique in the frequency domain. Time constants for rotational and vibrational relaxation and for a simple dissociation reaction were obtained by this method. The high information content is obtained from a complete analysis of the acoustic resonance profiles containing as experimental information the pressure dependence of the resonance frequency, the halfwidth of the resonance and the signal amplitude.

The kinetics of more-complex chemical reactions are presented by *Diebold* in Chap. 6. In his overview of studies of chemical kinetics by the photoacoustic effect, he discusses, in particular, chain reactions with quantum efficiencies larger than 1. This type of reaction gives a chemical amplification of the photoacoustic effect. As stated by the author, the photoacoustic effect acts as a monitor of the concentrations of reactants and may be considered as a frequency domain analogue of flash photolysis. The analytical possibilities and capabilities for in situ analysis are also discussed.

1.3.4 Applications in Trace Analysis and Pollution Monitoring

One of the early applications of the photoacoustic effect was gas detection and trace analysis, owing to the inherent high sensitivity. For a successful analysis of gas mixtures, the selectivity of the method also plays an important role. With tunable lasers a high selectivity can be achieved in the IR spectral region, where molecules can be characterized by specific narrow absorption bands. A third requirement is the capability of in situ detection. Photoacoustic spectoscopy fulfils all three conditions, as shown in Fig. 1.7, and therefore is one of the most valuable analytical methods in gas detection.

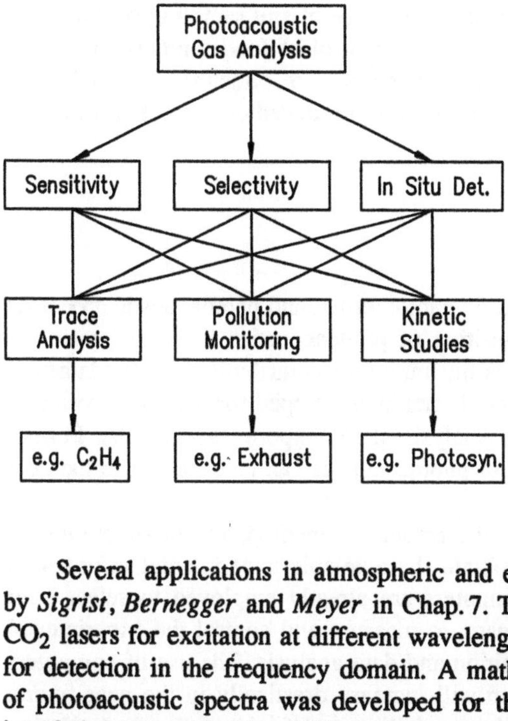

Fig. 1.7. Features of photoacoustic gas analysis with examples considered in more detail in this book

Several applications in atmospheric and exhaust air monitoring are discussed by *Sigrist*, *Bernegger* and *Meyer* in Chap. 7. They used line tunable CO lasers and CO_2 lasers for excitation at different wavelengths, and resonant photoacoustic cells for detection in the frequency domain. A mathematical procedure for the analysis of photoacoustic spectra was developed for this analysis, taking into account the interfering absorption by different components in the gas mixture. Examples are given for air pollution monitoring employing a laboratory-based apparatus and a mobile system for in situ detection of, for example, industrial exhausts at a production site.

The last chapter, written by *Bicanic* et al., discusses several applications of the photoacoustic method in biology and agriculture. The first part presents details of an apparatus optimized for in situ trace gas detection of ethylene. A waveguide CO_2 laser is employed for excitation in the IR fingerprint region and a resonant cell with low Q-factor is operated intracavity under low gas flow conditions. With this setup a detection limit of 6pptv ($\sim 1.5 \times 10^{-10}\mathrm{cm}^{-1}$) was obtained for the detection of ethylene in air. Ethylene is of great interest in biology because it acts as a plant growth regulator. The accumulation of this gas affects its own biosynthesis in a stimulatory or inhibitory way, and therefore the analysis must be performed in a sensitive flow-through system to avoid accumulation of the produced gas. Results on the correlation between ethylene production and senescence of carnation and orchid flowers are presented.

The second part of Chap. 8 describes the application of photoacoustics to the investigation of photosynthesis. The complicated reaction processes are accompanied by radiative and nonradiative losses, which can be detected, for example, by studying chlorophyll fluorescence and heat production by photoacoustic means. The photoacoustic signal is also sensitive to photosynthetic oxygen evolution. The simultaneous observation of fluorescence quenching, thermal dissipation and oxygen evolution yields new information on energy conversion in photosynthesis.

The third part of Chap. 8 is devoted to a more qualitative discussion of the potential use of photoacoustics in agriculture, e.g. the in situ sampling of the atmosphere in agricultural buildings and the quality control of food samples. However, other applications are also considered, e.g. in the field of olfactory psychophysics. In this section a large variety of potential uses are introduced or suggested for future research.

1.4 Summary and Outlook

The different contributions to this book on photothermal and photoacoustic processes in the gas phase demonstrate the transition of photoacoustics from an interesting effect to an established technique making important contributions to fundamental science. It is shown that theoretical models can be developed for several experimental schemes, allowing the quantitative analysis of the experimental results. Further progress in the theoretical foundation of photoacoustics can be expected in the near future.

In advanced techniques such as the resonance method, the incorporation of a computer is necessary in order to obtain the available experimental information in a reasonable time. Laboratory computers are already employed to select laser wavelengths, to modulate the laser beam, to measure and control the experimental parameters and finally for data acquisition and data analysis. The use of computers incorporated in the experimental setup will increase drastically in the near future. In addition, computers will be used for modelling and for the numerical solution of the fundamental equations to investigate the range of applicability of analytical solutions.

In a photothermal or photoacoustic experiment a series of complicated processes is induced by laser radiation. Therefore, it is not surprising that a detailed knowledge of system properties is needed to reach a quantitative analysis. This is a problem if most of the thermophysical and kinetic properties are not accurately known. On the other hand, photoacoustics allows a test of the internal consistency of a large number of system parameters in cases where only one or two quantities must be determined by the current experiment and the others are already available.

It is expected that sophisticated computer controlled setups will make important contributions to the determination of fundamental physical properties and also to the solution of analytical problems by measuring dependences on different experimental parameters in a large range.

References

1.1 Yoh-Han Pao (ed.): *Optoacoustic Spectroscopy and Detection* (Academic, New York 1977)
1.2 A. Rosencwaig: *Photoacoustics and Photoacoustic Spectroscopy* (Wiley, New York 1980)
1.3 D.S. Kliger (ed.): *Ultrasensitive Laser Spectroscopy* (Academic, New York 1983)
1.4 V.P. Zharov, V.S. Letokhov: *Laser Optoacoustic Spectroscopy*, Springer Ser. Opt. Sci., Vol.37 (Springer, Berlin, Heidelberg 1986)
1.5 E. Lüscher, P. Korpiun, H.-J. Coufal, R. Tilgner (eds.): *Photoacoustic Effect: Principles and Applications* (Vieweg, Braunschweig 1984)
1.6 R. Kaiser: Can. J. Phys. **37**, 1499 (1959)
1.7 W.R. Harshbarger, M.B. Robin: Acc. Chem. Res. **6**, 329 (1973)
1.8 M.J. Colles, N.R. Geddes, E. Mehdizadeh: Contemp. Phys. **20**, 11 (1979)
1.9 P. Hess: Top. Curr. Chem. **111**, 1 (1983)
1.10 G.A. West, J.J. Barret, D.R. Siebert, K.V. Reddy: Rev. Sci. Instrum. **54**, 797 (1983)
1.11 J. Badoz, D. Fournier (eds.): *Photoacoustic and Photothermal Spectroscopy*, J. de Phys., Colloque C6 (Les Editions de Physique, Les Ulis 1983)
1.12 L. Bertrand, P. Cielo, R. Leblanc, J.P. Monchalin, B. Mongeau (eds.): Proc. 4th Int. Topical Meeting of Photoacoustics, Thermal and Related Sciences, Can. J. Phys. **64**, 1023-1344 (1985)
1.13 P. Hess, J. Pelzl (eds.): *Photoacoustic and Photothermal Phenomena*, Springer Ser. Opt. Sci., Vol.58 (Springer, Berlin, Heidelberg 1988)

2. Theoretical Foundation of Photoacoustics in the Frequency and Time Domains

David Pugh

With 4 Figures

The first part of this chapter is intended to serve as a brief introduction to the standard, well-established theory of linear gas dynamics, which is the foundation of much of the work reported in subsequent chapters on optoacoustic effects in gases. A form of the theory that systematically includes the effect of a relaxing heat capacity is developed. In later sections, the application of the equations to optoacoustic resonators is discussed and the theory of the time-domain response in the pulsed source coaxial thermal lens experiment is developed. Section 2.7 analyses the theory of the optics of the thermal lens.

2.1 The Equations of Linear Gas Dynamics

For linear gas dynamics [2.1–4], it is useful to think of a small mass m of gas, contained in a small volume δV. This concept is a little blurred on a molecular scale, especially when more than one type of molecule, interdiffusing in concentration gradients, is present, but no errors ensue in the linear theory. Also, in the linear theory, one need not worry about flow line rather than local derivatives, so that the force equation for the mass m is

$$m\frac{\partial \boldsymbol{u}}{\partial t} = (\boldsymbol{f}+\boldsymbol{F})\delta V \quad , \tag{2.1.1}$$

where $\boldsymbol{f}$ and $\boldsymbol{F}$ are the internal and external forces per unit volume; $\boldsymbol{F}$ will be taken as zero until boundary problems arise. In terms of the density $\varrho = m/\delta V$, this becomes

$$\varrho\frac{\partial \boldsymbol{u}}{\partial t} = \boldsymbol{f} \quad . \tag{2.1.2}$$

The continuity equation is

$$\frac{\partial \delta\varrho}{\partial t} = \varrho\nabla\cdot\boldsymbol{u} \quad . \tag{2.1.3}$$

The tensor

$$\varepsilon_{ij} = \frac{\partial u_i}{\partial x_j} \tag{2.1.4}$$

gives the rate of relative displacement of neighboring points of the gas and

$$e_{ij} = \tfrac{1}{2}(\varepsilon_{ij}+\varepsilon_{ji}) \tag{2.1.5}$$

is the rate of strain, which goes correctly to zero for the pure rotation that occurs when $\varepsilon_{ji} = -\varepsilon_{ij}$.

For an anisotropic medium, the general stress–strain relation is

$$\sigma_{ij} = G_{ijkl}e_{kl} - p\delta_{ij} \quad , \tag{2.1.6}$$

where σ_{ij}, e_{ij} and δ_{ij} are tensors, so that G_{ijkl} must be a 4th-order tensor. (The summation convention over repeated indices is implied.) For a viscoelastic medium the stress depends on the history of the strain [2.4],

$$\sigma_{ij}(t) = -p(t)\delta_{ij} + \int_{-\infty}^{t} G_{ijkl}(t-t')\frac{\partial e_{kl}}{\partial t'}dt' \quad . \tag{2.1.7}$$

This is mentioned since one vestige of thermoviscoelasticity [4] still remains in most gases – the relaxing heat capacity function. But in a gas we can assume an instantaneous local equilibrium between pressure, translational temperature and density, governed by the equation of state, so that the nonthermal part of the elastic tensor does not relax. In an isotropic medium, G must be an isotropic tensor – corresponding elements of G must be the same in all Cartesian coordinate systems – and the most general isotropic 4th-order tensor is [2.5]

$$G_{ijkl} = a\delta_{ij}\delta_{kl} + b\delta_{ik}\delta_{jl} + c\delta_{il}\delta_{jk} \quad , \tag{2.1.8}$$

where a, b and c are scalars. This gives

$$\sigma_{ij} = (-p + ae_{ll})\delta_{ij} + ge_{ij} \quad , \tag{2.1.9}$$

where $g = b + c$ and the symmetry of e_{kl} has been used.

Two distinct elastic constants are therefore required to specify the stress–strain relation in an isotropic medium. The shear and bulk viscosity coefficients are defined as

$$\mu = \frac{g}{2} \quad , \quad \eta = a + \frac{2}{3}\mu \quad . \tag{2.1.10}$$

Substituting for a and g gives

$$\begin{aligned}\sigma_{ij} &= (-p + \eta e_{ll}) + 2\mu(e_{ij} - \tfrac{1}{3}e_{ll}) \\ &= (-p + \eta\nabla\cdot\boldsymbol{u})\delta_{ij} + 2\mu(e_{ij} - \tfrac{1}{3}\nabla\cdot\boldsymbol{u}) \quad .\end{aligned} \tag{2.1.11}$$

If the strain is an isotropic dilation, $e_{ij} = e\delta_{ij}$, then

$$\sigma_{ij} = (-p + 3\eta e)\delta_{ij} \quad , \tag{2.1.12}$$

and conversely, if there is no dilation,

$$\nabla\cdot\boldsymbol{u} = \frac{1}{\varrho}\frac{\partial\delta\varrho}{\partial t} = 0$$

and

$$\sigma_{ij} = -p\delta_{ij} + 2\mu e_{ij} \quad . \tag{2.1.13}$$

Effects due to compression and shear are therefore related to separate constants.

Analysis of the forces exerted on a small cube shows that

$$f_i = \frac{\partial\sigma_{ij}}{\partial t} \quad , \tag{2.1.14}$$

and applying this relation to (2.1.11) results in

$$f_i = -\frac{\partial p}{\partial x_i} + \left(\eta - \frac{2}{3}\mu\right)\frac{\partial \nabla \cdot \boldsymbol{u}}{\partial x_i} + 2\mu\frac{\partial e_{ij}}{\partial x_j} \quad . \tag{2.1.15}$$

Using (2.1.3) and combining the components gives

$$\boldsymbol{f} = -\nabla\delta p + (\eta + \tfrac{1}{3}\mu)\nabla(\nabla \cdot \boldsymbol{u}) + \mu\nabla^2\boldsymbol{u} \tag{2.1.16}$$

and application of the vector identity

$$\nabla \times (\nabla \times \boldsymbol{u}) = \nabla(\nabla \cdot \boldsymbol{u}) - \nabla^2\boldsymbol{u} \tag{2.1.17}$$

produces the Navier-Stokes equation

$$\varrho\frac{\partial \boldsymbol{u}}{\partial t} = -\nabla\delta p + \left(\eta + \frac{4}{3}\mu\right)\nabla(\nabla \cdot \boldsymbol{u}) - \mu\nabla \times \nabla \times \boldsymbol{u} \quad . \tag{2.1.18}$$

The velocity $\boldsymbol{u}$ can be separated into longitudinal and transverse parts, such that

$$\boldsymbol{u} = \boldsymbol{u}_\mathrm{l} + \boldsymbol{u}_\mathrm{t} \quad . \tag{2.1.19}$$

Since $\boldsymbol{u}_\mathrm{l}$ and $\boldsymbol{u}_\mathrm{t}$ are orthogonal at each point, the Navier-Stokes equation separates into two:

$$\varrho\frac{\partial \boldsymbol{u}_\mathrm{l}}{\partial t} = -\nabla\delta p + \left(\eta + \frac{4}{3}\mu\right)\nabla(\nabla \cdot \boldsymbol{u}_\mathrm{l}) \quad , \tag{2.1.20}$$

$$\varrho\frac{\partial \boldsymbol{u}_\mathrm{t}}{\partial t} = -\mu\nabla \times (\nabla \times \boldsymbol{u}_\mathrm{t}) = \mu\nabla^2\boldsymbol{u}_\mathrm{t} \quad . \tag{2.1.21}$$

The transverse velocity $\boldsymbol{u}_\mathrm{t}$ is accordingly decoupled from the rest of the problem and need only be introduced again when boundary conditions are treated. The suffix l is therefore omitted for the time being ($\boldsymbol{u}_\mathrm{l} \to \boldsymbol{u}$). The continuity equation (2.1.3) and (2.1.20) are the first two equations of gas dynamics. Taking the divergence of (2.1.20) and using (2.1.3) gives a relation between the pressure and density changes:

$$\frac{\partial^2 \delta\varrho}{\partial t^2} = \nabla^2\delta p + \frac{1}{\varrho}\left(\eta + \frac{4}{3}\mu\right)\frac{\partial}{\partial t}\nabla^2\delta\varrho \quad . \tag{2.1.22}$$

The differential form of the equation of state provides another relation between $\delta\varrho$ and δp, but introduces the temperature change as a third variable,

$$\delta\varrho = \varrho(\kappa_\mathrm{T}\delta p - \alpha\delta T) \tag{2.1.23}$$

where κ_T is the isothermal bulk modulus and α is the coefficient of thermal expansion.

The remaining physical constraint on the system is the energy balance. If the time dependence of excited molecular states or the kinetics of chemical reactions are included, an extended set of equations is obtained for the change in energy. If, on the other hand, we try to treat the gas as a macroscopic continuum, with a composition which, at least in the linear approximation, does not vary in space, we can absorb these effects into the heat capacity memory function. For a nonrelaxing medium (certainly for monatomic gases!) the energy balance equation is

$$\varrho C_V \frac{\partial \delta T}{\partial t} - \frac{T\alpha}{\varrho}\frac{\partial \delta\varrho}{\partial t} - \varrho q - \lambda\nabla^2 \delta T = 0 \quad . \tag{2.1.24}$$

Here C_V is the specific heat (heat capacity of unit mass) at constant volume, λ is the thermal conductivity and q the heat supplied externally to unit volume at the point in question. The term $q(r,t)$ will be described as the source function. In a gas where vibrational relaxation effects occur, or where a chemical equilibrium is perturbed by the disturbance, this equation can be formally replaced by [2.4]

$$\varrho \frac{\partial}{\partial t} \int_{-\infty}^{t} C_V(t-t')\frac{\partial \delta T(t')}{\partial t'}dt' - \frac{T\alpha}{\varrho}\frac{\partial \delta\varrho}{\partial t} - \varrho q - \lambda\nabla^2 \delta T = 0 \quad . \tag{2.1.25}$$

The physical meaning of the relaxing heat capacity can be made more transparent if we consider a system at constant volume with no thermal conduction. Then

$$q = \frac{\partial}{\partial t} \int_{-\infty}^{t} C_V(t-t')\frac{\partial \delta T(t')}{\partial t'}dt' \quad . \tag{2.1.26}$$

If we imagine a small system in good thermal contact with a heat reservoir in which the temperature is instantaneously changed by δT_0 at $t=0$, then we can represent the temperature change in the system by $\delta T_0 h(t)$, where $h(t)$ is the Heaviside unit function, $h(t) = 0$, $t<0$; $h(t)=1$, $t \geq 0$. Then

$$\frac{\partial \delta T(t')}{\partial t'} = \delta T_0 \delta(t') \tag{2.1.27}$$

and

$$q = \frac{\partial}{\partial t} \int_{-\infty}^{t} C_V(t-t')\delta T_0 \delta(t')dt' = \delta T_0 \frac{\partial}{\partial t} C_V(t) \quad . \tag{2.1.28}$$

The total heat transferred at time t is

$$\int_{-\infty}^{t} q\, dt = \delta T_0 C_V(t) \quad . \tag{2.1.29}$$

If the relaxation process is of the first order then [2.6]

$$C_V(t) = C_V(\infty) - \delta C_V e^{-t/\tau} \quad , \tag{2.1.30}$$

where τ is the relaxation time. When the temperature is changed the heat instantaneously transferred to the system is $[C_V(\infty) - \delta C_V]\delta T_0 = C_V(0)\delta T_0$, but as this heat is gradually absorbed into internal modes, more heat is transferred from the heat bath, until the final value of $C_V(\infty)\delta T_0$ is attained.

Solutions of the equations in an unbounded medium can always be expressed in terms of the space-time Fourier transforms. For each variable we write

$$X(\boldsymbol{k},\omega) = \int_{-\infty}^{\infty}\int_{-\infty}^{\infty} X(\boldsymbol{r},t)\exp\left[-\mathrm{i}(\boldsymbol{k}\cdot\boldsymbol{r} - \omega t)\right]d\boldsymbol{r}\, dt \tag{2.1.31}$$

and, in terms of the transformed variables, the equations become

$$-\omega^2\delta\varrho = -k^2\delta p - \frac{1}{\varrho}(\eta + \tfrac{4}{3}\mu)(-\mathrm{i}\omega)k^2\delta\varrho \quad ,$$

$$\delta\varrho = \varrho(\kappa_{\mathrm{T}}\delta p - \alpha\delta T) \quad , \qquad (2.1.32)$$

$$-\mathrm{i}\omega\varrho C_{\mathrm{V}}(\omega)\delta T - \frac{T\alpha}{\varrho}\mathrm{i}\omega\delta\varrho - \varrho q - \lambda k^2\delta T = 0 \quad ,$$

where $C_{\mathrm{V}}(\omega) = \mathrm{i}\omega C_{\mathrm{V}}(t)$ is the usual definition [2.6] of the frequency-dependent specific heat, and where the convolution theorem for Fourier transforms has been used to deal with the integral in (2.1.25). Each Fourier component is a plane wave and, provided complex $\boldsymbol{k}$ values are admitted, (2.1.32a–c) are the basis for the derivation of the usual formulae (see for example [2.1, 2, 6]) for the absorption of energy and changes in velocity and phase of longitudinal ultrasonic waves in gases.

Equation (2.1.25) can also provide the starting point for the treatment of effects in the time domain when relaxation of the heat capacity occurs.

In the following we apply the general equations given in this section to two special cases. In the first, q is a periodic function of time and the steady-state solutions at the same frequency in a cylindrical resonator are treated; in the second, q is a pulse of Gaussian spatial cross section, decaying exponentially with time. More detailed accounts of recent investigations in both these regimes will be given in later chapters in this volume.

2.2 Theory of the Cylindrical Optoacoustic Resonator

We begin our consideration of the cylindrical optoacoustic resonator [2.1, 7–9] by taking the frequency/time Fourier transform of (2.1.22, 23 and 25), replacing each variable X by

$$X(\omega) = \int_{-\infty}^{\infty} X(t)\mathrm{e}^{-\mathrm{i}\omega t}dt \quad . \qquad (2.2.1)$$

The integral in (2.1.25) resolves itself into a product of the transforms of $C_{\mathrm{V}}(t)$ and $\partial\delta T/\partial t$, by the convolution theorem for Fourier transforms, and the equations become

$$\nabla^2\delta p = [-\omega^2 - (\mathrm{i}\omega\eta'/\varrho)\nabla^2]\delta\varrho \quad , \qquad (2.2.2)$$

$$\delta\varrho = \varrho(\kappa_{\mathrm{T}}\delta p - \alpha\delta T) \quad , \qquad (2.2.3)$$

$$\mathrm{i}\omega\varrho C_{\mathrm{V}}\delta T - \mathrm{i}\omega(T\alpha/\varrho)\delta\varrho - \varrho q - \lambda\nabla^2\delta T = 0 \quad , \qquad (2.2.4)$$

where $\eta' = \eta + \frac{4}{3}\mu$.

Following the usual procedure [2.1], $\delta\varrho$ is removed by substitution from (2.2.3) and the characteristic lengths, $l_{\mathrm{v}}' = \eta'/\varrho c$ and $l_{\mathrm{h}} = \lambda/\varrho C_p c$ are introduced, where

$$c = (\varrho\kappa_{\mathrm{S}})^{-1/2} = (\gamma/\varrho\kappa_{\mathrm{T}})^{1/2}.$$

In (2.2.2–4) the only frequency-dependent "constant" is C_V. After the introduction of l'_V and l_h this frequency dependence is also present in $\gamma = C_P/C_V$ and c. The two equations for δT and δp become

$$\nabla^2 \delta p = (\gamma/c^2)(-\omega^2 - i\omega l'_V c \nabla^2)(\delta p - \alpha \delta T) \quad , \tag{2.2.5}$$

$$l_h c \nabla^2 \delta T = i\omega\{\delta T - [(\gamma - 1)/\gamma\alpha]\delta p\} + q/C_P \quad , \tag{2.2.6}$$

where the thermodynamic identity $C_P = C_V + T\alpha\beta/\varrho$ has also been used. A fourth-order equation for δT can be obtained by substituting for δp from (2.2.6) into (2.2.5). There is no term in ∇^3 so that this equation can be factorized to give

$$(\nabla^2 + k_p^2)(\nabla^2 + k_t^2)\delta T = q' \quad , \tag{2.2.7}$$

where k_p and k_t are complex combinations of the constants in (2.2.5, 6) and q' is a function of q, which can be worked out explicitly if necessary. An outline of how (2.2.7) can be treated in such a way as to obtain exact solutions for the realistic boundary conditions will be given [2.10–12], before we introduce the approximations which underlie the more usual treatment of acoustic resonators.

We consider only the case of radial solutions for cylindrical symmetry, where the appropriate expansion functions are the Bessel functions $J_0(k_n r)$, the k_n being determined by the boundary conditions. The homogeneous form of (2.2.7), $q' = 0$, is satisfied by any function of the form

$$A J_0(k_p r) + B J_0(k_t r) \quad . \tag{2.2.8}$$

If the radius of the cell is a, then the ratio A/B can always be chosen to make

$$A J_0(k_p a) + B J_0(k_t a) = 0 \quad , \tag{2.2.9}$$

which satisfies the boundary condition $\delta T = 0$. A second condition involves u_{rad}, the radial component of the velocity. If the walls are rigid we must have $u_{\mathrm{rad}}(a) = 0$. The quantities k_p and k_t are functions of frequency, and the second boundary condition therefore restricts the allowed solutions to a partcular discrete set of frequencies. The functions of type (2.2.8), with k_p and k_t chosen to satisfy the boundary conditions, constitute a complete set of functions, in terms of which the solutions of the inhomogeneous equation can be expanded. This analysis is still not general enough. It is also necessary to introduce boundary conditions on the tangential velocity components at the interface. Normally these are taken as the nonslip conditions, that u_t must vanish at $r = a$. To satisfy the additional conditions, solutions of the equation for the transverse components of the velocity field, from (2.1.21), must also be added. Equation (2.1.21) has the form of a diffusion equation and the spatial decay constant is very large, so that these terms, introduced to satisfy the detailed boundary conditions, decay rapidly as we move away from the boundary. In effect, they produce a viscous loss in a boundary layer of the gas, which, in their absence, would be apparently attributable to friction with the walls. The presence of the transverse terms has an effect on the allowed values of k_p and k_t. The second term in (2.2.8) is similarly confined to a boundary layer, and we can demonstrate the origins of these boundary effects by examining this equation in a little more detail.

If the viscous effect is absent and the thermal conduction is much slower than sound propagation, the equations (2.2.5, 6) separate into the wave equation for δp and the thermal conduction equation for δT. If these equations are compared with (2.2.7) it is apparent that k_p is real and k_t is imaginary in the limit that sound propagation and thermal conduction are completely separable. Coupling between the two modes is small and therefore k_p and k_t retain, respectively, a large element of real and imaginary character even when the interaction between the two modes is fully included. Now,

$$J_0(\mathrm{i}x) = I_0(x)$$

where $I_0(x)$ is the modified Bessel function, which is nonoscillatory and increases rapidly with increasing x. In the normalized eigenfunctions the contribution from $J_0(k_t r)$ is accordingly confined to a thin layer near the wall. The solution throughout the bulk of the gas is represented to a good approximation by $J_0(k_p r)$, which is the solution of the simple wave equation

$$(\nabla^2 + k_p^2)\delta p = 0 \quad . \tag{2.2.10}$$

If such an approximate solution is used, then only one boundary condition can be explicitly introduced, and this must be taken as

$$u_{\mathrm{rad}}(a) \approx \nabla_r \delta p = J_0'(k_p a) = 0 \quad , \tag{2.2.11}$$

which gives a set of real eigenvalues k_n [2.13]. In making this approximation, all surface effects have been lost and the surface losses have to be added back into the result in a rather arbitrary manner. Effects of the relaxing heat capacity and bulk viscous and thermal losses are still retained in the equations. The procedure that uses the $J_0(k_n r)$, solutions of the "unperturbed" wave equation (2.2.10), as approximate eigenfunctions but calculates the eigenvalues, the frequencies, to a higher-order approximation is a form of perturbation theory.

To carry out the approximate treatment, we return to (2.2.5 and 6) and neglect terms that are nonlinear in l_v' and l_h. If terms in l_h^2 are neglected, (2.2.6) can be solved recursively for δT:

$$\begin{aligned}
\delta T &= \frac{\gamma-1}{\gamma\alpha}\delta p + \frac{l_\mathrm{h} c}{\mathrm{i}\omega}\nabla^2 \delta T - \frac{q}{\mathrm{i}\omega C_\mathrm{P}} \\
&= \left(\frac{\gamma-1}{\gamma\alpha}\right)\left(1 + \frac{l_\mathrm{h} c}{\mathrm{i}\omega}\nabla^2\right)\delta p - \frac{1}{\mathrm{i}\omega C_\mathrm{P}}\left(1 + \frac{l_\mathrm{h} c}{\mathrm{i}\omega}\nabla^2\right) q \\
&= \left(1 + \frac{l_\mathrm{h} c}{\mathrm{i}\omega}\nabla^2\right)\left[\left(\frac{\gamma-1}{\gamma\alpha}\right)\delta p - \frac{q}{\mathrm{i}\omega C_\mathrm{P}}\right] \quad .
\end{aligned} \tag{2.2.12}$$

Substitution of this expression into (2.2.5) leads eventually to

$$\nabla^2 \delta p = -\left(\frac{\omega}{c}\right)^2 \left(1 - \frac{\mathrm{i}\omega}{c}[l_\mathrm{v}' + (\gamma-1) l_\mathrm{h}]\right)\delta p + \mathrm{i}\omega H(q) \quad , \tag{2.2.13}$$

where

$$H(q) = \frac{\mathrm{i}\gamma\alpha}{C_\mathrm{P} c^2}\left(1 - \mathrm{i}\frac{\omega}{c}[l_\mathrm{v}' + (\gamma-1) l_\mathrm{h} - (l_\mathrm{v}' - l_\mathrm{h})\nabla^2]\right) q \quad . \tag{2.2.14}$$

Writing (2.2.13) as

$$\nabla^2 \delta p + k_\omega^2 \delta p = \mathrm{i}\omega H(q) \tag{2.2.15}$$

and expanding

$$\delta p = \sum_n A_n J_0(k_n r) \tag{2.2.16}$$

leads to

$$\delta p = \sum_n \frac{\mathrm{i}\omega H_n}{k_\omega^2 - k_n^2} J_0(k_n r) \quad , \tag{2.2.17}$$

where

$$H_n = 2\pi \int_0^\infty H(q) J_0(k_n r) r\, dr \quad . \tag{2.2.18}$$

If we let $\omega_{0,n} = k_n c_0$, where c_0 is the low-frequency sound velocity, the denominator in (2.2.17) can be written as

$$\omega^2 - \omega_{0,n}^2 \left(\frac{c}{c_0}\right)^2 - \frac{\mathrm{i}\omega^3}{c}[l_\mathrm{v}' + (\gamma - 1) l_\mathrm{h}] \quad . \tag{2.2.19}$$

Since c is complex, the first two terms produce both a shift in the resonant frequency and a broadening associated with the dispersion of C_V, while the final term, which is purely imaginary, leads to an additional broadening from bulk viscous and thermal effects. The surface losses are added to the imaginary part of the denominator.

Details of the development of these equations for practical calculations for resonators and a much more extensive treatment of the exact theory for the spherical resonator can be found in [2.7–11] and also in Chaps. 4 and 5 of this book.

2.3 The Pulsed Source Thermal Lens Effect

A schematic description of the experimental arrangement for the pulsed source thermal lens effect [2.12–20] is given here. It is described in more detail in Chap. 3. In outline, a laser pulse of Gaussian radial cross section and of very short duration excites some of the molecules of the gas. The disturbance leads, eventually, to the development of a nonuniform radial refractive index distribution, which is detected by its effect on a second continuous mode probe beam, which is not itself absorbed by the gas. Typically, the pulsed beam would have a frequency in the infrared and excite vibrational modes, while the probe beam is often a HeNe laser.

In all the work to be described, a cylindrical gas cell is used and axial effects are neglected, which implies that the change in intensity of the pulse beam along the axis of the cell, as a consequence of absorption, is small. The optics of the thermal lens is discussed in Sect. 2.7.

It is possible to solve (2.1.22–24) in the time domain by numerical techniques for a given source function $q(r,t)$ with the inclusion of all terms and boundary conditions appropriate for the cell containing the gas. Extensive numerical work has, in fact, been carried out recently by *Barker* et al. [2.21–23] and by *Jacobs* [2.24]. This work will be reviewed in Sect. 2.6. In many cases the time scales associated with the various terms in the equations are so different that the problem can be separated into simpler parts. An additional simplification arises from the fact that the ratio of the laser beam radius to the cell radius is small (1/50) and for most purposes it is valid to treat the gas as an unbounded medium. Only when the pressure wave echoes from the cell walls return to the central region sampled by the probe beam need the effects of the finite radial extension of the medium be included in the analysis. In the following sections we present an analysis based on the above assumptions, which leads to reasonably explicit analytical solutions for the time-dependent thermal lens signal. The analysis incorporates as special cases the earlier results of *Siebert* [2.14] and of *Twarowski* and *Kliger* [2.15, 16]. An approach similar to that developed here was originally attempted by *Longaker* and *Litvak* [2.25], but in their work the two-dimensional nature of the cylindrical gas dynamics, at least in the first approximation, was not established. Graphical results fo the density response, presumably obtained numerically, were presented by *Brueck* et al. [2.26] with no explicit mathematical analysis. Results for the pressure and density profiles in a laser cavity, which are essentially the same as those given here, were obtained by *Tough* and *Willets* [2.27]. The analysis described here is based on [2.20].

The source term, $q(r,t)$ in (2.1.25) must be specified in more detail. In the work reported to date, it is only through the time dependence of q that the effect of internal molecular energetics has been studied, although the method is also potentially useful for investigating effects of the relaxing heat capacity. Energy from the absorbed laser pulse creates "instantaneously" a distribution of vibrationally excited molecules $\varrho^*(r,t)$, which spatially replicates the shape of the laser pulse. The release of translational energy from the excited molecules constitutes the source term q. One complication that must be considered here, as in the resonator theory where it may affect the overlap integral, is diffusion of the excited molecules before they give up their excess energy. If the diffusion process is much slower than the decay, it is permissible to write

$$q(\boldsymbol{r},t) = q(\boldsymbol{r})f(t) \quad , \tag{2.3.1}$$

where, usually, one would assume that

$$q(r) = \exp(-r^2/r_G^2) \quad , \quad f(t) = \exp(-t/\tau) \quad . \tag{2.3.2}$$

The spatial distribution ϱ^* remains fixed, as determined by the laser pulse radius, during the entire decay process. To investigate the validity of this approximation, the equation governing the time dependence of ϱ^* in the presence of diffusion can be solved:

$$\frac{\partial \varrho^*}{\partial t} = D^*\nabla^2\varrho^* - \omega_s\varrho^* \quad , \tag{2.3.3}$$

where $\omega_s = 1/\tau_s$; τ_s is the source relaxation time.

The solution of this type of equation when the initial spatial distribution is Gaussian is particularly simple, because the Green's function for the equation is also a Gaussian. A pulse $\delta(r,0)$ at $t = 0$ gives a Gaussian distribution

$$\frac{1}{4D^*t}\exp(-r^2/4D^*t)$$

at time t. It is known that at time $t = 0$ the distribution is $\varrho^* = \exp(-r^2/r_G^2)$, which can be thought of as arising from a δ-function pulse at an earlier time $-\tau_D$, such that

$$\frac{1}{4D^*(t+\tau_D)}\exp[-r^2/4D^*(t+\tau_D)]|_{t=0} = \exp(-r^2/r_G^2) \quad . \tag{2.3.4}$$

The characteristic relaxation time τ_D must therefore have the value $\tau_D = r_G^2/4D^*$, and the required solution of (2.3.3) is

$$\varrho^*(r,t) = \varrho^*(0,0)\frac{r_G^2}{4D^*(t+\tau_D)}\exp[-r^2/4D^*(t+\tau_D)] \quad . \tag{2.3.5}$$

An estimate of the diffusional relaxation time can be obtained from the simple kinetic theory value of D^*:

$$D^* = \frac{\sqrt{3}\pi}{8}\frac{\bar{c}}{N_0\pi\sigma^2(\varrho/m)} \quad , \tag{2.3.6}$$

where $\bar{c}$ is the r.m.s. velocity $\sqrt{3/\gamma}c$.

In the same way, if the thermal and propagating waves are separated in (2.2.6), the usual thermal conduction equation leads to a characteristic time for thermal relaxation:

$$\tau_\kappa = r_G^2/4\kappa \tag{2.3.7}$$

where κ is the thermal diffusivity, $\kappa = \lambda/\varrho C_p$. It follows from kinetic theory that

$$\tau_\kappa/\tau_D = \gamma \quad . \tag{2.3.8}$$

Two other relaxation times are of interest in the problem. The acoustic, or pressure wave, relaxation time is the time for a pressure pulse to cross the radius of the exciting laser beam:

$$\tau_p = r_G/c \tag{2.3.9}$$

and, finally, the V-T (vibration-translation) relaxation of the source, τ_s. These relaxation times are listed in Table 1.1.

A preliminary survey of the magnitudes of the relaxation times enables one to decide which parts of the system of (2.2.5, 6) must be retained for a particular study. Some examples of physically realistic sets of values are exhibited in Table 2.2.

The V-T relaxation time is inversely proportional to the pressure, and at lower pressures it is always likely to be longer than the acoustic time and may even approach τ_κ.

Table 2.1. Relaxation times

τ_s	Source, internal modes to translation
$\tau_p = r_G/c$	Pressure wave
$\tau_\kappa = r_G^2/4\kappa$	Thermal
$\tau_D = r_G^2/4D$	Diffusional

Table 2.2. Physically realistic values of relaxation times for SF_6/cyclopropane. (r_G = 0.718 mm, c = $2.2 \times 10^2\,ms^{-1}$, T = 354 K)

p [torr]	τ_p [s]	τ_s [s]	τ_K [s]
50	3×10^{-6}	5×10^{-6}	1.2×10^{-3}
350	3×10^{-6}	1×10^{-6}	7.9×10^{-3}

Since τ_D is almost always very much greater than τ_s, the excited distribution ϱ^* decays before it diffuses to any appreciable extent and the diffusion process has no effect on the time development. The decay of the thermal lens signal strength at long time is therefore effected through the dissipation of the excess energy in the central core region by thermal conduction, rather than by diffusion of energetic molecules. The chracteristic time for this decay is therefore τ_K. The most frequently occurring cases are discussed below. In all cases it is assumed that the diffusion time is long in comparison with τ_s.

Case 1: $\tau_K \gg \tau_p > \tau_s$. the excess energy stored in the excited molecules is very rapidly converted to translational energy – in a time much less than that required for a pressure pulse to cross the distribution. On release of the vibrational energy to translation, the temperature of the gas is changed, essentially instantaneously, and so is the pressure, to the value required by the equation of state at constant density. At this time there is no change in density and there is no thermal lens signal, since the refractive index in a perfect gas (and to a very good approximation in other gases) is a function only of the density. As the pressure disturbance moves away with the acoustic pulse, the density begins to change and the signal builds up to a maximum in a time $\sim \tau_p$. Characteristic of this case is an "overshoot" associated with the rarefaction following the positive pressure pulse. If τ_K is much longer than τ_p, the signal then becomes virtually constant on a short time scale. In this plateau region, the pressure has regained its unperturbed value and all the excess energy deposited by the laser is still in the core region, leading to temperature and density distributions still following the spatial profile of the laser pulse. On a longer time scale, usually milliseconds, the thermal conduction process restores the signal strength to zero. Echoes, produced by the pressure pulses reflected from the walls of the cell, are observed as small features on the decay curve. A typical system of this kind, for example, is SF_6-doped cyclopropane [2.20].

Case 2. $\tau_K \gg \tau_s > \tau_p$. In this case, on the time scale of the acoustic transit, the energy leaks out into translational motion slowly. No appreciable pressure perturbation is produced at any stage, $\delta\varrho$ and δT are always related through the gas equation at the equilibrium pressure, and the characteristics of the rising lens signal are determined by τ_s alone. No overshoot is observed in this case. Recovery is by thermal conduction as before.

A mathematical analysis of the above cases is given in the following sections [2.17, 20]. The spatial processes are treated as being purely radial.

2.4 Thermal Recovery

After all the heat has been released from the vibrational modes and the pressure in the core has returned to equilibrium, a Gaussian radial temperature distribution remains. Since $\tau_D \sim \tau_\kappa$ and both are much greater than τ_P and τ_S, it is a good approximation to take this distribution as being the same as the initial laser pulse distribution. Then the problem is to solve the equation

$$\frac{\partial \delta T}{\partial t} = \kappa \nabla^2 \delta T \tag{2.4.1}$$

with the initial condition that

$$\delta T(r,0) = \delta T_0 e^{-(r/r_G)^2} \quad . \tag{2.4.2}$$

The solution can be obtained by the argument employed to solve (2.3.3), leading to

$$\delta T(r,t) = \delta T_0 \frac{\tau_\kappa}{t+\tau_\kappa} \exp\left[-r^2/4\kappa(t+\tau_\kappa)\right] \quad , \tag{2.4.3}$$

where

$$\tau_\kappa = r_G^2/4\kappa \quad . \tag{2.4.4}$$

The thermal lens signal depends on the radial curvature of the refractive index distribution on the axis, which is proportional to

$$-\left(\frac{\partial^2 \delta T}{\partial r^2}\right)_0 = 2\delta T_0 \frac{1}{r_G^2}\left(\frac{\tau_\kappa}{t+\tau_\kappa}\right)^2 \tag{2.4.5}$$

through (2.7.8).

Substituting (2.4.5) into (2.7.8) and rearranging gives

$$\left(1 - \frac{1}{\sqrt{1-S}}\right)^{-1/2} = \frac{ct}{\tau_\kappa} + C \quad , \tag{2.4.6}$$

where c is a constant. A plot of the left-hand side of (2.4.6) against t should be a straight line such that

$$\text{intercept/slope} = \tau_\kappa \quad . \tag{2.4.7}$$

The validity of this equation has been verified by measurements on inert gases doped with small quantities of absorbers and has been shown to be the basis of a potentially useful method for measuring thermal diffusivities [2.17, 18].

2.5 Short Time-Scale Measurements

The case where the time scale is much shorter than τ_κ but where τ_P and τ_S are relevant and of comparable magnitude is also amenable to a reasonably explicit mathematical analysis. In this case the thermal conduction and viscosity terms are neglected in (2.1.22, 25). The heat capacity of the gas has also been assumed to be nonrelaxing, so that the only time constants left in the problem are τ_P and τ_S. If δT is then eliminated between (2.1.22–24), the remaining relations are

$$\frac{\partial^2 \delta\varrho}{\partial t^2} = \nabla^2 \delta p \quad ,$$

$$\frac{\partial \delta p}{\partial t} - c^2 \frac{\partial \delta\varrho}{\partial t} = q \quad , \tag{2.5.1}$$

and elimination of $\delta\varrho$ gives the equation for δp:

$$\frac{\partial^2 \delta p}{\partial t^2} - c^2\nabla^2 \delta p = \frac{\partial q}{\partial t} = \omega_s Q_0 \exp\left[-(r/r_G)^2 - \omega_s t\right] \cdot \left[-\omega_s h(t) + \delta(t)\right]$$
$$= S(r,t) \quad , \tag{2.5.2}$$

where, again, $h(t)$ is the unit function and $S(r,t)$ is, essentially, the time derivative of the source function. The source is switched on by the laser pulse at $t = 0$, when the vibrationally excited molecules begin to decay. The time dependence of the source function is

$$q(t) = \omega_s Q_0 e^{-t/\tau_s} \cdot h(t) \quad , \tag{2.5.3}$$

and its derivative with the spatial distribution included is given by the right-hand side of (2.5.2). There Q_0 is the total heat released to translation and is equal to the energy absorbed from the laser per unit mass less any energy loss by radiative processes. If $\omega_s \to 0$, then q is constant, corresponding to a continuous heating effect, but $\omega_s Q_0$ remains finite and equal to the heat released per unit time. In this limit, only the δ-function term remains on the right-hand side of (2.5.2), which now describes the response of the system to a continuous laser switched on at time $t = 0$. Contact is therefore made with the treatments of thermal blooming for continuous lasers, which has been much discussed in the literature. Both terms on the right-hand side of (2.5.2) are essential for an accurate description of the type of system treated here.

Equation (2.5.2) can be solved by Fourier transformation. It is convenient to introduce dimensionless variables

$$r' = r/r_G \quad ;$$
$$s = t/\tau_p \quad ;$$
$$\varepsilon = \tau_p/\tau_s = \omega_s r_G/c \quad ;$$
$$l = k r_G \quad . \tag{2.5.4}$$

The unit of length is now the radius of the pulsed laser and the unit of time is the pressure wave transit time. Since ε is the ratio of the pressure wave to source relaxation times, $\varepsilon \to 0$ corresponds to a slow release of energy, and as will be shown, no pressure pulse, while $\varepsilon \to \infty$ implies instantaneous release of the source energy as heat and a maximum pressure disturbance. In these units, the Fourier transform solution for δp is

$$\delta p(r', s) = \frac{\varepsilon Q_0}{2} \int_0^\infty dl\, l(l^2 + \varepsilon)^{-1}\left[-\varepsilon e^{-ls} + l \sin(ls) + \varepsilon \cos(ls)\right]$$
$$\times \exp(-l^2/4) J_0(lr') \quad , \tag{2.5.5}$$

where J_0 is the zero-order Bessel function of the 1st kind. It is easy to obtain numerical evaluations of this integral for all values of r, s and ε. One important limiting case that can be treated analytically is that where the time dependence of the response on the axis is to be calculated for $\varepsilon \to \infty$, when the development of the signal is dependent only on the propagation of the pressure pulse. If terms $O(1/\varepsilon)$ and the $\exp(-\varepsilon s)$ term are neglected, we have

$$\delta p(0,s) = \frac{Q_0}{2}\int_0^\infty dl\, l \exp(-l^2/4)\cos(ls) \quad , \tag{2.5.6}$$

which, on integration by parts, gives

$$\delta p(0,s) = Q_0\left[1 - s\int_0^\infty \exp(-l^2/4)\sin(ls)ds\right] \quad . \tag{2.5.7}$$

The remaining integral is $2F(s)$, where $F(s)$ is Dawson's integral [2.28]

$$F(s) = e^{-s^2}\int_0^s e^{x^2}dx \qquad \text{so that} \tag{2.5.8}$$

$$\delta p(0,s) = Q_0[1 - 2sF(s)] \quad . \tag{2.5.9}$$

All the on-axis properties of the response of the system to an instantaneous pressure increment with Gaussian radial distribution can be written in terms of $F(s)$ and its derivatives. A plot of $F(s)$ is shown in Fig. 2.1. The function reaches a maximum value of 0.54 at $s = 0.91$. This maximum is the feature that is related to the change in sign of the pressure in the rarefaction part of the pulse and, through the associated density variation, to the "overshoot" of the thermal lens signal.

From (2.5.1) we find that

$$\delta\varrho(0,s) = -2sF(s)Q_0/c^2 \quad , \tag{2.5.10}$$

and by expanding the Bessel function around the axis and differentiating it can be shown that

$$\left(\frac{\partial^2\delta\varrho}{\partial r^2}\right)_0 = [(-4s^2 + 6s)F(s) + 2s^2]Q_0/c^2 \quad . \tag{2.5.11}$$

The thermal lens signal can then be calculated from (2.7.8).

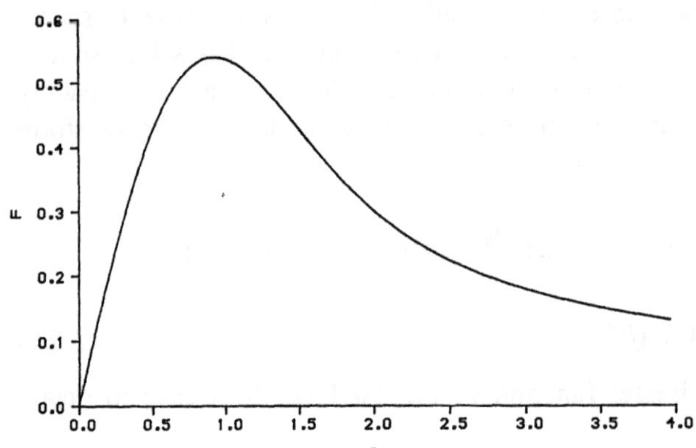

Fig. 2.1. Plot of $F(s)$ (Dawson's integral)

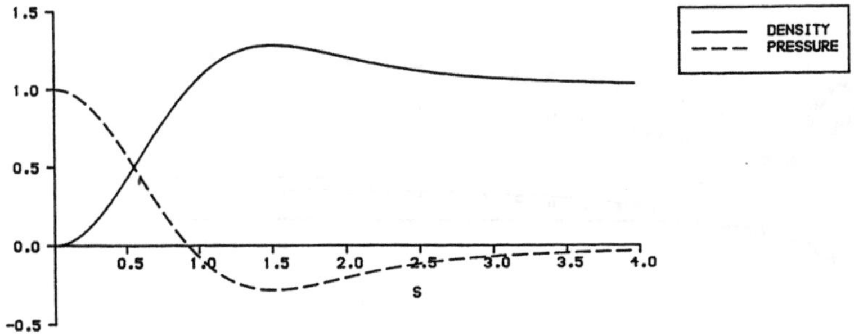

Fig. 2.2. Density and pressure vs s, the reduced time

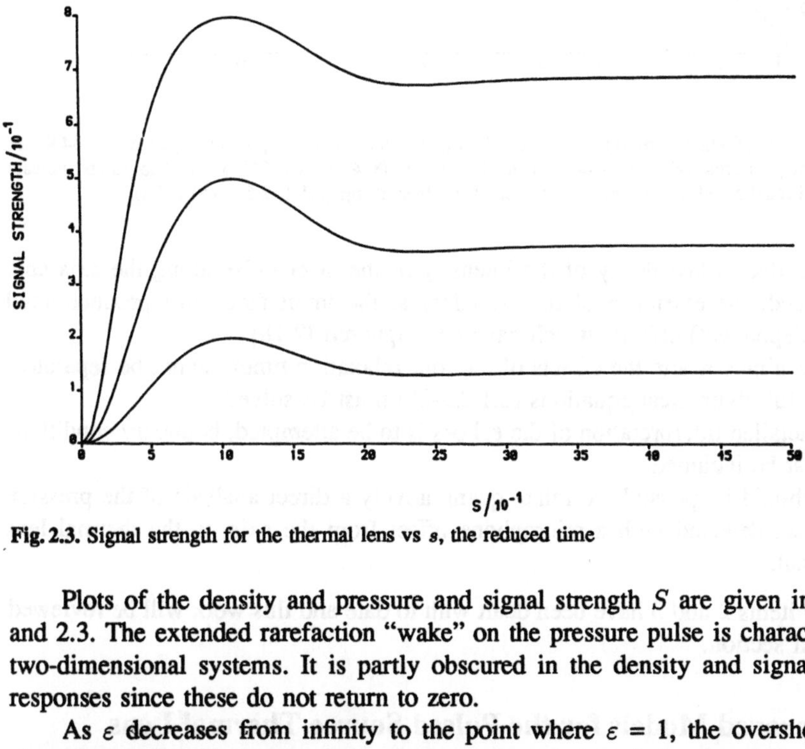

Fig. 2.3. Signal strength for the thermal lens vs s, the reduced time

Plots of the density and pressure and signal strength S are given in Figs. 2.2 and 2.3. The extended rarefaction "wake" on the pressure pulse is characteristic of two-dimensional systems. It is partly obscured in the density and signal strength responses since these do not return to zero.

As ε decreases from infinity to the point where $\varepsilon = 1$, the overshoot of the signal gradually disappears and, when $\varepsilon \ll 1$, the rise of the signal is essentially exponential, but with a characteristic flattening near the origin, which emerges from the solution of the equations and has been previously identified by *Siebert* [2.14] in a treatment concerned specifically with the onset of the signal at very short times. Figure 2.4 shows a plot of the signal strength against the reduced time s for various values of ε. The curves are normalized so that, in each case, the final signal strength (before the onset of thermal relaxation) is 0.5. By fitting, with the aid of a computer, the experimentally observed signals to plots of this kind, the ratio τ_S/τ_P can be determined.

The preceding analysis establishes the characteristic features of the pulsed source, parallel beam technique, but a more refined analysis is desirable in relation to a number of points:

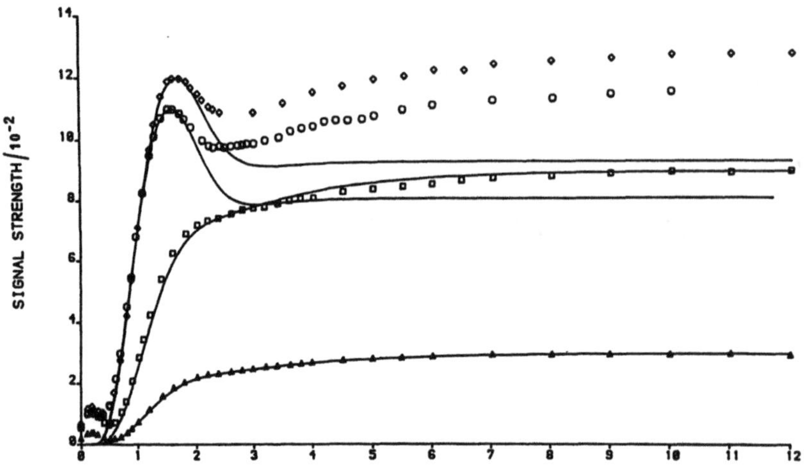

Fig. 2.4. Results of signal strength vs reduced time for the SF_6/cyclopropane system at 354 K for four different pressures, (△) 26.15 torr; (□) 42.1 torr; (◇) 206.4 torr, (○) 357.5 torr. The points indicate experimental results and the curves are attempts to fit these using (2.5.1, 2.5.5 and 2.7.8)

1. The Beer's law decay of the intensity of the laser pulse along the axis and, indeed, the existence of the boundary at the input face must produce axial (z-dependent) effects, which have been ignored [2.18].
2. In regimes where the effects of the four relaxation times cannot be separated, the full dynamical equations (2.1.22–25) must be solved.
3. If detailed interpretation of the echoes is to be attempted, boundary conditions must be included.
4. It should be possible to relate quantitatively a direct analysis of the pressure pulse, detected with a microphone offset from the axis, to the thermal lens signal.

Only items 2 and 3 have been dealt with to date and this work will be reviewed in the next section.

2.6 Improved Models for the Pulsed Source Thermal Lens

Barker and co-workers [2.21–23] have developed a model for a finite cylinder. The fourth-order equation (Sect. 2.1), including bulk viscous and thermal effects, is solved, but the expansion functions are the zero-order Bessel functions, which satisfy only the boundary condition on u_{rad}. The level of approximation therefore corresponds to the perturbation theory approach described at the end of Sect. 2.2. The results obtained confirm the analysis of the previous section at high pressure, except for features, localized in time, on the thermal recovery part of the curve, which arise from echoes off the cylindrical walls. At lower pressures, when, in particular, the thermal time scale becomes very much shorter and the regimes distinguished in the previous section are no longer separable, substantial quantitative differences from the above analytical theory are observed. It should be remarked that, for many systems, these differences are only really appreciable where the pressure is so low that the mean free path approaches the radial cell dimension. Surface effects are then likely to contribute substantially to the deexcitation process and the theory will require some major amendments.

A recent paper by *Jacobs* [2.24] provides a numerical treatment of the problem in the finite cylinder in which a complete set of boundary conditions are imposed. These conditions differ slightly from the set used in the accurate form of the resonator theory and the significance of the differences has still to be explored. The general numerical method is, nevertheless, applicable to all possible choices. Solutions of the unapproximated fourth-order equation are obtained, no expansion functions are used, and the method can equally well be applied to nonlinear disturbances. The computed results reproduce those of *Rothem* et al. [2.22] and the analytical treatment, but the potential value of the method for dealing with the nonlinear regime is considerable. But it should be noted that quite substantial amounts of computing time are required although the treatment is still restricted to the case of a purely radial distribution. Some work on nonlinear responses in the pulsed source thermal lens experiment has been reported by *Guckert* and *Carr* [2.29].

2.7 Optics of the Thermal Lens

The simplest approach is via geometric optics. This method will be presented first, before more general methods are described that will enable the range of applicability of the simple geometric formula to be assessed.

The equation governing the propagation of a paraxial ray in a medium with an inhomogeneous refractive index distribution is [2.30]

$$n_0 \frac{d^2 r}{dz^2} = \frac{\partial n}{\partial r} \quad , \tag{2.7.1}$$

where n_0 is the unperturbed index. The approximations made in deriving this equation from the general ray equation [2.30] are not important, although the application of ray optics is itself an approximation.

Since the probe beam is detected by a very small pinhole, only rays very close to the axis are detected, and consequently the refractive index can be expanded as

$$n(r) = n_0\left(1 + \tfrac{1}{2}\gamma r^2\right) \quad , \tag{2.7.2}$$

where

$$\gamma = \frac{1}{n_0}\left(\frac{\partial^2 n}{\partial r^2}\right)_0 \quad . \tag{2.7.3}$$

Substitution in (2.7.1) gives

$$\frac{d^2 r}{dz^2} = \gamma r \quad . \tag{2.7.4}$$

If it is assumed that the probe beam is parallel at the entry to the gas cell, then $r(0) = 0$, and if it is further assumed that n is independent of z the solution of

(2.7.4) giving the ray path is

$$r(z) = r(0)\cosh(\gamma^{1/2}z) \quad ,$$
$$r'(z) = \gamma^{1/2} r(0)\sinh(\gamma^{1/2}z) \quad . \tag{2.7.5}$$

The thermal lens signal is defined as

$$S = \frac{I_0 - I}{I_0} = 1 - \frac{I}{I_0} = 1 - \left(\frac{r(0)}{r(\text{pinhole})}\right)^2 \quad , \tag{2.7.6}$$

which gives explicitly

$$S = 1 - \left[\cosh(\gamma^{1/2}l) + \gamma^{1/2} L \sinh(\gamma^{1/2}l)\right]^{-2} \quad , \tag{2.7.7}$$

where l is the length of the cell and L the distance to the pinhole. It is usually the case that the quantity $\gamma^{1/2}l$ is small enough to allow this expression to be approximated as

$$S = 1 - (1 + \gamma l L)^{-2} \quad , \quad L \gg l \quad , \tag{2.7.8}$$

which is the equation used in most of the thermal lens work described in the previous sections. There is no great difficulty in modifying the expression within the framework of geometric optics to take account of beams that are diverging on entry to the gas cell, or of z-dependent refractive indices, which arise through Beer's law absorption [2.18]. A much more fundamental task is to establish the conditions under which these simple expressions can validly be used. The problem of predicting the thermal lens signals is not difficult if the parameters governing the shape, width and size of the probe and pulse beams are known very accurately, but once conditions deviate from those where geometric optics applies, the results become rather critically dependent on the values of these parameters. Since in practice they are not known with very great accuracy, it is important to establish the ranges over which limiting cases with simple functional dependences on the parameters apply. In particular, in the geometric optics approximation, the signal strength does not depend on the width of the probe beam.

An alternative approach to attempting to establish conditions under which ray optics applies is to attempt to treat the problem as one of propagation in a "lens-like" medium. The case often arises in modern optical systems that light is propagated entirely within regions where the refractive index is a quadratic radial dependence. Under these conditions the medium is referred to as a lens-like medium [2.31]. The propagation equations then have analytic solutions, which can be found in standard books [2.31].

In the following the case of the "thin" thermal lens is investigated with the object of establishing the range of validity of different treatments.

The equation governing the propagation of a paraxial ray in wave optics [2.30] is

$$\nabla^2 E + k^2(r)E = 0 \quad , \qquad \text{where} \tag{2.7.9}$$

$$k(r) = \frac{2\pi n(r)}{\lambda_{\text{vac}}} = \frac{2\pi}{\lambda_{\text{vac}}}\frac{n_0 n(r)}{n_0} = k_0\left(1 + \frac{\delta n(r)}{n_0}\right) \quad . \tag{2.7.10}$$

The constant k_0 and the associated rapid oscillatory variation in E can be removed by the substitution

$$E = \psi(r, z)e^{-ik_0 z} \quad , \tag{2.7.11}$$

giving

$$\nabla_r^2\psi - \frac{2k_0^2\delta n(r)}{n_0}\psi = 2ik_0\frac{\partial\psi}{\partial z} \quad , \tag{2.7.12}$$

where the term in $\partial^2\psi/\partial z^2$, which is very much less than $k_0\partial\psi/\partial z$, has been dropped.

The refractive index variation has the form

$$\delta n(r) = \delta n_0 e^{-(r/r_G)^2} \quad , \tag{2.7.13}$$

where the time dependence of n and r is suppressed for the purposes of the present argument. Transforming to dimensionless quantities by putting

$$\zeta = \frac{z}{2k_0 r_G^2} \quad ; \qquad \varrho = r/r_G \quad ; \qquad a = \frac{2(k_0 r_G)^2\delta n_0}{n_0} \quad , \tag{2.7.14}$$

leads to

$$\nabla_\varrho^2\psi - ae^{-\varrho^2}\psi = i\frac{\partial\psi}{\partial\zeta} \quad . \tag{2.7.15}$$

The thin lens theory is based on the observation that, at any given position along the z-axis, $a\exp(-\varrho^2)$ is much greater than $\nabla_\varrho^2\psi$. Hence, over a short interval of ζ (here, between $-l$ and 0), equation (2.7.15) can be written as

$$\frac{\partial\psi}{\partial\zeta} = iae^{-\varrho^2}\psi \quad , \tag{2.7.16}$$

with the solution

$$\psi(\zeta = 0) = \psi(\zeta = -l)\exp[ial\exp(-\varrho^2)] \quad . \tag{2.7.17}$$

If the wave entering the lens is described by the function

$$\psi_{-l} = Ae^{-\alpha\varrho^2} \tag{2.7.18}$$

then the emerging wave at $\zeta = 0$ is

$$\psi_0 = A\exp[-\alpha\varrho^2 - i\beta\exp(-\varrho^2)] \quad , \quad \beta = al \quad . \tag{2.7.19}$$

Propagation of this wave through the homogeneous medium from 0 to the pinhole can be dealt with by using the Green's function of the equation

$$\nabla_\varrho^2\psi = ik_0\partial\psi/\partial\zeta \quad ,$$

$$g(\varrho, \zeta/\varrho_0, 0) = \frac{i}{4\pi\zeta}\exp\left(-i\frac{|\varrho - \varrho_0|^2}{4\zeta}\right) \quad , \tag{2.7.20}$$

where ϱ and ϱ_0 are radial vectors. After the angular integrations have been performed, the value of ψ in the plane at ζ, which may be taken as the plane of the pinhole is

$$\psi(\varrho,\zeta)=\frac{\mathrm{i}A}{4\zeta}\exp\left(-\frac{\mathrm{i}\varrho^2}{4\zeta}\right)$$

$$\int_0^\infty \exp\left[-\left(\alpha+\frac{\mathrm{i}}{4\zeta}\right)\varrho_0^2+\mathrm{i}\beta\exp(-\varrho_0^2)\right]J_0\left(\frac{\varrho\varrho_0}{2\zeta}\right)\varrho_0 d\varrho_0 \quad . \tag{2.7.21}$$

In this equation, in dimensionless units, $1/\sqrt{\alpha}$ is the probe beam diameter at the cell, ζ the distance to the pinhole from the output face of the cell and $\beta = al$ is the strength of the thermal lens effect. The radial length unit is the radius of the pulse beam.

After removing unimportant phase factors, the on-axis value of ψ becomes

$$\psi(0,\zeta)=\frac{A}{4\zeta}\int_0^\infty \exp\left[-\left(\alpha+\frac{\mathrm{i}}{4\zeta}\right)\varrho_0^2+\mathrm{i}\beta\exp(-\varrho_0^2)\right]\varrho_0 d\varrho_0 \quad . \tag{2.7.22}$$

The integral as it stands is analytically intractable because of the exponential function in the exponent of the integrand. If the probe beam is narrower than the pulse beam and samples only the central part of the refractive index distribution, then the approximation

$$\mathrm{e}^{-\varrho_0^2}=1-\varrho_0^2 \tag{2.7.23}$$

will be valid and the integral can be evaluated to give

$$\psi(0,\zeta)=\frac{A}{4\zeta}\,\frac{1}{[\alpha+\mathrm{i}(1/4\zeta+\beta)]} \quad . \tag{2.7.24}$$

The intensity of the light at the pinhole is

$$I\propto|\psi(0,\zeta)|^2=\frac{1}{(4\zeta\alpha)^2+(1+4\beta\zeta)^2} \tag{2.7.25}$$

and the thermal lens signal is

$$S=1-[(4\zeta\alpha)^2+(1+4\beta\zeta)^2]^{-1} \quad . \tag{2.7.26}$$

This is the solution for a lens-like medium and could also have been written down using the standard rules for propagation in such media. If $\alpha\to 0$ the solution reduces to

$$S=1-(1+4\beta\zeta)^{-2} \quad , \tag{2.7.27}$$

which is the same result as that which would be obtained from (2.7.8) using the geometric optics approximation. The lens-like medium model therefore reduces to the geometric optics approximation when the probe beam becomes infinitely wide. But this result does not establish the validity of the geometric optics approximation in the limit of wide probe beams, because the lens-like medium approximation is invalid when the probe beam becomes wider than the pulse beam and is no longer confined to a region where the refractive index varies quadratically. To make progress it is therefore necessary to investigate the behavior of the integral in (2.7.22) more systematically.

Only the case of the infintely wide probe beam ($\alpha \to 0$) is considered here. The case of nonzero α can be treated in the same way. In the notation of (2.7.22), with $\alpha \to 0$, the geometric optics solution is

$$\psi(0,\zeta) = \frac{1}{4\zeta(1/4\zeta + \beta)} = \frac{1}{4\zeta\gamma} \quad , \tag{2.7.28}$$

where $\gamma = 1/4\zeta + \beta$.

The integral in (2.7.22) can be written, without approximation, as

$$\begin{aligned} \psi(0,\zeta) &= \frac{A\exp(\mathrm{i}\beta)}{4\zeta}\int_0^\infty \exp\{-\mathrm{i}\gamma\varrho_0^2 + \mathrm{i}\beta[\exp(-\varrho_0^2) - 1 + \varrho_0^2]\}\varrho_0 d\varrho_0 \\ &= \frac{A\exp(\mathrm{i}\beta)}{4\zeta}\int_0^\infty \varrho_0 \exp(-\mathrm{i}\gamma\varrho_0^2)\exp\{\mathrm{i}\beta[\exp(-\varrho_0^2) - 1 + \varrho_0^2]\}d\varrho_0 \quad , \end{aligned} \tag{2.7.29}$$

which can be integrated repeatedly by parts to give a series solution for ψ. The first few terms of the series are

$$\psi(0,\zeta) = \frac{Ae^{\mathrm{i}\beta}}{4\zeta}\left(\frac{1}{\mathrm{i}(\beta + 1/4\zeta)} - \frac{\beta}{(\beta + 1/4\zeta)^3} - \frac{\mathrm{i}\beta}{(\beta + 1/4\zeta)^4} + \dots\right) \quad . \tag{2.7.30}$$

The first term in this method of expansion gives the geometric optics solution. The absence of the squared term reflects the axial symmetry. The coefficients in the higher terms become rather complex, but an examination of some of them leads to the conclusion that the series is asymptotic rather than convergent. This is as would be expected, since geometric optics is an asymptotic approximation to wave optics. The essential condition for the applicability of the geometric optics solution is that $(\beta + 1/4\zeta)$ should be large compared with β. This is a condition that is most likely to be satisfied for a given thermal lens strength β at small ζ. The geometric optics approximation is essentially a near field approximation. Its precise range of validity for the thermal lens setup can be ascertained by computing the results from the integral (2.7.22), or by suitable numerical procedures for thick lens conditions, and comparing the results with those obtained from the geometric formula. In practice, the condition is not too restrictive, provided that the extremes of sensitivity associated with some analytical work are not required.

2.8 Conclusions

The fundamental differential equations of linear gas dynamics have been derived in a form applicable to media with a relaxing heat capacity. The origins of the approximations commonly introduced in the application of these equations to the optoacoustic resonator have been discussed. The treatment of the dynamic response of the pulsed source coaxial double beam thermal lens system, based on the separation of processes occurring on very different time scales, has been described and it has been shown that in certain important limiting cases analytical solutions can be obtained. Finally, an analysis of the range of applicability of the geometric optics (ray) approximation to the interpretation of the lens signal has been presented.

References

2.1 P.M. Morse, K.U. Ingard: *Theoretical Acoustics* (McGraw-Hill, New York 1968)
2.2 K.F. Herzfeld, T.A. Litovitz: *Absorption and Dispersion of Ultrasonic Waves* (Academic, New York 1959)
2.3 J.O. Hirchfelder, C.F.Curtiss, R.B. Bird: *Molecular Theory of Gases and Liquids* (Wiley, New York 1954)
2.4 R.M. Christensen: *Theory of Viscoelasticity* (Academic, New York 1971)
2.5 H. Jeffreys, B.L. Jeffreys: *Methods of Mathematical Physics* (Cambridge University Press, Cambridge 1950)
2.6 A.J. Matheson: *Molecular Acoustics* (Wiley-Interscience, New York 1971)
2.7 A. Karbach, J. Roper, P. Hess: Chem. Phys. **82**, 427 (1983)
2.8 A. Karbach, P. Hess: J. Chem. Phys. **83**, 1075 (1985)
2.9 A. Karbach, P. Hess: J. Chem. Phys. **84**, 2945 (1986)
2.10 C.M. Knobler: Pure Appl. Chem. **55**, 455 (1983)
2.11 M.R. Moldover, J.B. Mehl, M. Greenspan: J. Acoust. Soc. Am. **79**, 253 (1986)
2.12 D.R.Siebert, F.R. Grabiner, G.W. Flynn: J. Chem. Phys. **60**, 1564 (1974)
2.13 F.R. Grabiner, D.R.Siebert, G.W. Flynn: Chem. Phys. Lett. **17**, 189 (1972)
2.14 D.R. Siebert: Ph.D. Dissertation, Columbia University (1973)
2.15 A.J. Twarowski, D.S. Kliger: Chem. Phys., **20**, 153 (1977)
2.16 A.J. Twarowski, D.S. Kliger: Chem. Phys., **20**, 259 (1977)
2.17 R.T. Bailey, F.R. Cruickshank, D. Pugh, W. Johnstone: J. Chem. Soc., Faraday Trans. 2, **76**, 633 (1980)
2.18 R.T. Bailey, F.R. Cruickshank, D. Pugh, W. Johnstone: J. Chem. Soc., Faraday Trans. 2, **77**, 1387 (1981)
2.19 R.T. Bailey, F.R. Cruickshank, R. Guthrie, D. Pugh, I.J.M. Weir: J. Mol.Spectros. **80**, 433 (1982)
2.20 R.T.S. Bailey, F.R. Cruickshank, R. Guthrie, D. Pugh, I.J.M. Weir: Mol. Phys. **48**, 81 (1983)
2.21 J.R. Barker, T. Rothem: Chem. Phys. **68**, 331 (1982)
2.22 T. Rothem, P.L. Trevor, J.R. Barker: Chem. Phys. **68**, 341 (1982)
2.23 J.R.Barker, B.M. Toselli: In *Photothermal Investigations of Solids and Fluids*, ed. by J.A. Sell, (Academic, New York 1988)
2.24 S.J. Jacobs: Chem. Phys., in press
2.25 P.R. Longaker, M.M. Litvak: J. Appl. Phys. **40**, 4033 (1969)
2.26 S.R.J. Brueck, H. Kildal, J. Belanger: Opt. Commun. **34**, 199 (1980)
2.27 R.J.A. Tough, D.V. Willetts: J. Phys. D **15**, 2433 (1982)
2.28 M. Abramowitz, I.A. Stegun (eds.): *Handbook of Mathematical Functions* (National Bureau of Standards, Washington, DC 1964)
2.29 J.R. Guckert, R.W. Carr.: J. Chem. Phys. **90**, 4286 (1986)
2.30 M. Born, E. Wolf: *Principles of Optics*, 6th ed. (Pergamon, Oxford 1980)
2.31 A. Yariv: *Quantum Electronics* (Wiley, New York 1975)

3. Thermal Lensing

R.T. Bailey and F.R. Cruickshank

With 9 Figures

In this chapter, applications of the techniques of thermal lens measurements are described for all commonly used geometries. Particular attention is given to experimental design and important parameters are discussed with reference to each configuration. Quantitative results in the fields of energy transfer kinetics, transport phenomena, photochemistry, surface phenomena, spectroscopy, photoacoustics, and laser physics are described. Special attention is paid to the coaxial geometry, since other configurations are discussed at length in other chapters of this volume.

3.1 Introduction

3.1.1 Survey of Approaches

Since the first observation (in 1964) by *Whinnery* ([3.1] and references therein) of intracavity thermal lensing and the paper using the phenomenon extracavity, thermal lens studies have expanded in a very large number of fields where they make an essential contribution through their extreme sensitivity. The few papers per annum of the 1970s have become hundreds now. Just as in the mirage, a thermal lens consists of a graded refractive index lens caused by the density gradient induced in some medium by the heating effect consequent upon conversion of the energy of absorbed photons to translational energy. Careful control of the geometry of such effects leads to an accurately quantifiable technique invaluable as a diagnostic aid in many physical and chemical contexts, as described below [3.2, 3].

It is possible to use this effect in three distinct ways. In the direct approach, photons are absorbed from a pump laser beam by a medium transparent to a probe laser beam. The refraction of this probe beam is then detected. In the indirect photothermal deflection approach, pump beam photons are absorbed by one medium (generally comparatively opaque) and the resultant heating of this medium is transferred to another, transparent to the probe beam. The thermal lens created in this second medium is then responsible for the refraction of the probe laser beam. In all cases the pump laser may be replaced by an incoherent source and the refraction of the probe beam may be monitored as a "blooming" or as a deflection. In the former, the optical axes of pump and probe coincide; in the latter, these axes do not. Finally, inteferometry may probe the refractive index profile set up by the pump beam absorption.

There are several techniques which rely on *reflection* of the probe beam from the surface of the sample [3.3–6]. The surface is distorted by the thermal expansion induced by the heating effect of the absorbed energy. Consequent divergence or deflection of the probe beam is then detected. This technique is an area of photothermal spectroscopy adequately dealt with elsewhere in this volume. In this review signal detection by microphone will be omitted, being dealt with elsewhere in this series, and only detection by photorefractive techniques will be discussed.

3.1.2 Configurations

The *coaxial thermal lens* is an optothermal technique using the blooming effect. It is, of all variations, capable of being accurately modelled over a very wide dynamic range (0.05% to > 80% modulation of the probe beam). The equipment is shown in Fig. 3.1.

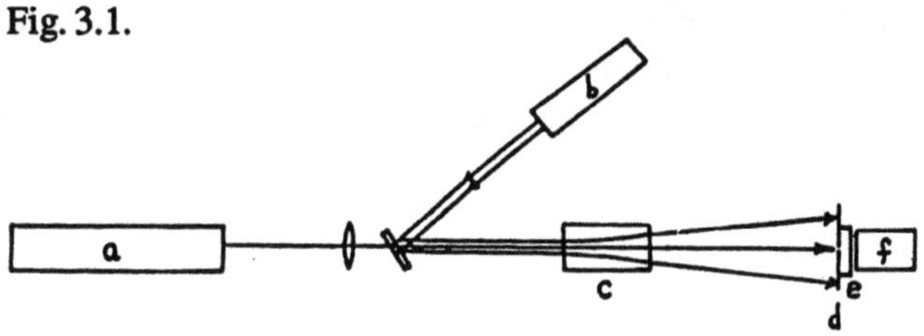

Fig. 3.1. The coaxial thermal lens apparatus. (*a*) pump laser, (*b*) probe laser, (*c*) sample, (*d*) pinhole, (*e*) narrow bandpass filter, tuned to pass only the probe laser beam, (*f*) photodiode or photomultiplier

Generally, two lasers are used. One, a pulsed (pump) laser (*a*), is tuned to the desired absorption frequency of the sample species. This laser has a pulse duration of length negligible in comparison with the lifetimes of those kinetic processes it is desired to follow. The transverse mode pattern is a high quality Gaussian variation of intensity with diameter (TEM_{00}). This laser is focused to a near-parallel beam waist throughout the sample (*c*). The other cw laser (*b*), coaxially aligned with the first and of greater diameter (> ~4 times), is used to probe the optics of the sample by being diverged or converged as the sample translational temperature rises or falls respectively. This laser should have the highest amplitude and pointing stability possible. There are several commercial helium/neon lasers that are suitable for this role.

The sample (*c*) may be any state of matter and, for the most accurate work, should be of high optical quality and have only modest absorption loss at the "pump" and much less at the probe laser frequencies.

The *crossed beam thermal lens*, in which focused pump and probe beam axes intersect exactly, refracts the probe beam as a cylindrical lens. This technique has been exploited to give highly sensitive absorbance detection. This configuration is particularly useful in probing microscopic volumes (e.g. ~0.2 picoliters!) since the lens occurs only where the beams intersect. This allows detailed mapping of a sample's absorbance.

Mirage detection, sometimes known as transverse photothermal deflection spectroscopy [3.7, 8], takes the form of a pump laser or monochromatised arc focused on the substrate at normal incidence with a probe laser propagating parallel with the substrate surface and intersecting the pump laser beam path at 90°, such that the center of the probe beam passes through the pump beam path off-axis in the medium

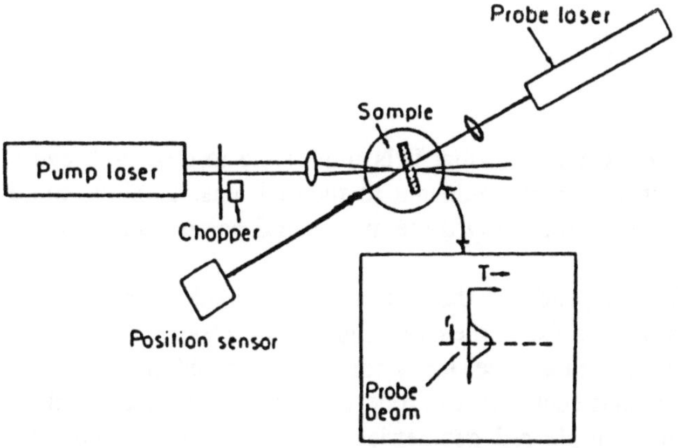

Fig. 3.2. Collinear photothermal deflection system

adjacent to the sample surface. The resulting temperature profile in this medium is detected as a deflection of the probe beam.

Typically, absorption of 1 mW cm^{-2} at a 2 mm diameter spot on a sample will deflect the probe beam 10^{-5} rad with a background noise of 10^{-9} rad, allowing measurement of 10^{-5} absorbance units if e.g. CCl_4 is used as the medium adjacent to the sample, since it enhances the signal considerably.

The *collinear photothermal deflection system* (Fig. 3.2) differs from a thermal lens system in that the counterpropagating pump and probe beams are not exactly collinear [3.7]. Additionally, the probe beam is focused to a smaller waist than the pump beam, contrary to the coaxial thermal lens requirement.

Interferometric detection of the refractive index changes has also been used. This is not strictly a refraction technique of course, but is included here briefly for completeness [3.9, 10].

In this approach both high-intensity and low-intensity laser beams are passed through the sample. These are then recombined at similar intensities to give the interference pattern caused by the alteration of refractive index due to the heating effect of the intense beam. For small absorbances,

$$P = P_0[1 - (a + a_s)l] \quad .$$

Here P is the emergent laser power and P_0 the incident power; a and a_s are the absorbance and scattering losses respectively and l is the sample path length. Clearly these are additive for small losses. The phase shift is most accurately measured as an intensity change near the maximum in d(intensity)/d(angle), i.e. when the phase shift is close to 90°. Typical shifts are around 0.05 rad, giving approximately 5% intensity change for an absorbance of 10^{-5}cm^{-1}. Since the laser intensity fluctuations are frequently 1% – 0.5%, the technique has a poor signal-to-noise ratio. Normalization to laser intensity is essential by observing this intensity on a separate detector simultaneously with the interference pattern. An incoherent source can be used to provide the heating effect (450 W xenon arc, monochromatized) with 2–3 mW focused on the sample, giving a detectability of 10^{-5}cm^{-1}. The technique is best for liquids with a temperature coefficient of up to 8×10^{-4}/°C.

This coefficient is an order of magnitude less for solids, so that the phase shift is correspondingly less [3.7, 11].

3.2 Experimental

3.2.1 Laser Sources

So great is the variety of lasers available that it is impossible to deal with every one. Instead, some of the characteristics relevant to thermal lensing of the more popular lasers will be discussed. Frequently these are general points which should be considered for any laser.

The *stability* of the laser depends on the resonator's mechanical and thermal properties as well as the discharge noise in the case of gas lasers. Cavity construction incorporating mixed elements, such as steel tubes for strength cored with graphite for thermal stability or aluminum-coated quartz rods will give the desired stability. Super Invar is also used in multiple rod geometries, which can give exceptional rigidity as well as negligible expansion. However, these rods should be chromium or nickel plated, since they rust!

The minimum discharge noise requires a linear passbank to reduce line ripple on the dc supply to the plasma tube. Typically, 0.5% rms noise over a 10 – 1MHz bandwidth is achieved in the current control mode, whereas 0.2% is possible if the dc supply is *power* controlled. It is not often realised that an "unpolarized" laser is in fact polarized, but the orientation of the polarization vector varies randomly with time. Such lasers should be avoided in thermal lens work because the reflectance of mirrors is a strong function of the polarization of the incident beam. Accordingly, randomly varying source polarization leads to excessive intensity fluctuations in the downstream optical train. Totally linearly polarized lasers should be used always.

Beam pointing stability is essential if the angular movement of the laser beams with temperature fluctuations is to be minimized so that the thermal lens geometry is temperature invariant. This is often determined by the frequency selection element in the cavity. This should be of temperature compensated design. Thermal gradients in the resonator structure can move cavity alignment enough to cause angular beam movement. This is reduced if all elements in the cavity are efficiently thermally coupled.

The *beam diameter* should generally be as large as possible consistent with the single transverse mode necessary to eliminate "hot spots". A large diameter allows the tightest focusing by lenses before the thermal lens cell. Beam divergence is generally quite low (a few milliradians full apex angle of cone) and can usually be neglected in calculations for a given experimental geometry.

Ion laser sources are ideal for pump and probe beam applications. However, these are generally cw, so that mechanical or electrooptic modulation must be used. Currently argon ion lasers provide a variety of wavelengths from the deep UV (~400 mW at 275–305 nm), UV (up to 5 W at 331–363.8 nm) and visible (over 10 W at 488 and 514 nm) with a number of weaker lines between 457.9 and 514.5 nm. Substitution of infrared mirrors allows operation at 1090 nm. Krypton lasers operate across a wider spectrum than argon lasers, but have less gain and therefore less power. The strongest line is at 647.1 nm, although there are significant lines in the blue, green, yellow, red and infrared.

Pulsed operation of ion lasers is possible if they are mode-locked and cavity dumped to yield very short pulses, for example, 150 ps pulses from single shot to megahertz repetition rates. A cw dye laser pumped by an argon or krypton laser can give tunable radiation from frequencies somewhat below the pump frequency to the near infrared.

Ion lasers have $1/e^2$ beam divergence of, for example, less than 0.7 mrad at 514 nm with a TEM_{00} transverse mode linearly polarized to better than 100:1.

Until recently *solid-state lasers* possessed poor mode quality. Nd:YAG lasers are now available with very stable TEM_{00} patterns. Noise and stability have been improved to $\pm 1\% - 1.5\%$ rms for the arc-pumped cw lasers. Q-switched Nd:YAG lasers still suffer pulse-to-pulse stability variation of up to $\pm 5\%$. Frequency multiplication of these lasers readily gives 532, 355 and 266 nm wavelenths. Pulse widths are typically adjustable from ~ 5 to 20 ns with energies of over 1 J. Again cw Nd:YAG lasers can be mode-locked to give shorter pulses of a few picoseconds.

Frequency multiplied, Q-switched Nd:YAG lasers are popular as dye laser pumps. Tunable UV to near IR radiation is then readily obtained in short pulses. The range can be extended into the UV and IR with Raman shifting in a high pressure gas (e.g. H_2, CH_4, etc.). This approach gives only discrete lines, but continuity of tuning is restored by mixing these with tunable dye radiation. Commercial units are now able to do this from 200 to 900 nm and from 1200 to 4500 nm. The recent introduction of diode-pumped Nd:YAG lasers, both Q-switched and cw, has greatly miniaturized such systems, giving 1064 nm and frequency-doubled to 532 nm light. These lasers are much more stable than arc-pumped systems and exhibit less noise. Typical maximum power is $\sim$1 W. These are also used for injection mode seeded, Q-switched YAG lasers of high power. This results in virtually single longitudinal mode operation with its attendant high temporal smoothness in the pulses. Pulse-to-pulse reproducibility is also greatly improved.

Dye lasers are most efficient in the visible, IR dyes being expensive and inefficient, while UV dyes photolyse and have a short lifetime. High-power argon lasers, specially modified for UV operation (up to 5 W) are popular cw dye laser pumps. Krypton lasers are preferred for pumping near-IR dyes. Efficiency is greatly improved if the pump lasers are mode-locked. Pulsed dye lasers can also be pumped by N_2 and excimer lasers.

The output stability characteristics of dye lasers usually depend on the pump laser and follow it nonlinearly. Moreover, their transverse mode pattern is best if they are longitudinally pumped. Again, the strategy of using a low Fresnel number oscillator followed by two or more amplifiers yields the best TEM_{00} patterns. Frequency selection is best performed by a high-quality grating in a high-quality monochromator mount to maintain narrow linewidth and power stability. Intracavity, temperature-controlled etalons can further narrow linewidths. Flashlamp-pumped dye lasers have poorer performance than the above in almost every respect, but are cheaper and frequently adequate.

Excimer lasers are not very satisfactory as precise thermal lens pumps because their aperture is large and therefore their mode pattern poor. They are also expensive to run with heavy maintenance requirements and venting problems.

The CO_2 *TEA laser* is a popular choice for energy transfer experiments, operating in the fingerprint region where most species absorb. Pulse widths are about 120 ns provided the "tail" on the pulse is removed by using less N_2 in the mixture to lower the gain slightly. It is spontaneously mode-locked with $\pm 2\%$ pulse-to-pulse stability. With a long cavity, pointing stability and transverse mode pattern are excellent. A low-pressure discharge tube inserted in the cavity can act as a longitudinal mode filter removing the mode-locking and giving a smooth temporal pulse.

3.2.2 Probe Lasers

The HeNe laser operating at 632.8 nm is the laser of choice from every point of view. It is cheap, has good polarization, transverse mode and energy stability, and good pointing stability, which is sometimes determined by the packaging between discharge tube and outer metal tube, which is inexcusable!

Generally, 5–7 mW power is adequate in a cw HeNe probe laser. Power stability of $\pm 2.5\%$ and $<0.1\%$ rms noise and 10μrad pointing stability are readily available. Selectable wavelengths are 543, 594, 612, 632, 1152, 1523, 2396 and 3392 nm, covering most requirements.

A low-power 1064 nm diode-pumped Nd:YAG laser provides an excellent probe laser.

3.2.3 Laser Beam Tailoring

If the pump is a laser, its beam should be a high-quality Gaussian (TEM_{00}) profile. This is usually produced by reducing the ratio of cavity cross section to its length such that the Fresnel number ($a^2/L\lambda$, where a is the mirror radius, L cavity length, and λ the wavelength) allows gain to be produced in the TEM_{00} mode only. Achieved by an intracavity aperture or by using a narrow laser rod etc., this is frequently accompanied by a loss of power and this is recovered by a laser amplifier. This adds to the size, expense and complexity of the system of course, but will retain the TEM_{00} mode quality. It is not advisable to use a spatial filter to purify the mode of a multimode laser in thermal lens work, since this leads to large fluctuations in the pump beam intensity.

In the case of an incoherent, monochromatized arc source, a true single mode is not possible, but satisfactory results can still be obtained if, as is usually the case, a lens is used to focus the pump beam into the sample, because the pattern at the focus serves the same purpose although of a more poorly defined shape and larger than that obtainable with a coherent source. Indeed, for laser sources, such a lens is also used and here the effect is similar. Often this results in useful mode "clean up" for systems where it is difficult to obtain satisfactory single-mode operation.

For thermal lenses, as compared to deflection methods, the pump beam 1/e diameter must be less than the probe beam diameter by at least a factor of four. This means that the quantitative description of the signal will be accurate, since the probe beam profile on-axis will be effectively flat, justifying the paraxial approximation. Additionally, it is experimentally much easier to align the laser beams within an acceptable eror in collinearity.

The diameter D of a laser beam at the focus of a lens of focal length f is given by

$$D = \frac{4\lambda f}{\pi d}, \tag{3.2.1}$$

Where λ is the laser beam wavelength and d its unfocused diameter. At a distance Z from the center of the focal waist the beam diameter has expanded to D' such that

$$\left(\frac{D'}{2}\right)^2 = \left(\frac{D}{2}\right)^2 \left(1 + \frac{\lambda Z}{\pi (D/2)^2}\right) . \tag{3.2.2}$$

These equations allow calculation of the lens and beam diameters necessary for setting up thermal lens and photothermal experiments. In practice, focal waist diameters obtained, particularly at long wavelengths, are far larger than those calculated if simple lenses are used. In fact one of the most reliable ways to measure the diameter of a pulsed laser beam is by thermal lens measurement of the thermal diffusivity of a standard, e.g. water (see below), followed by calculation. For cw lasers the beam may be scanned across a pinhole masking a phototube and the profile displayed on an oscilloscope. Alternatively, a detector array can be used, but these tend to have rather coarse resolution and it is not advisable to use lenses, telescopes, etc., to alter the beam to be measured to improve this resolution because of the unpredictable effects of these. It should be remembered that the diameter of lenses used should be far larger than the laser beams passing through them, since when a beam fills a lens the Airy disc (with rings) results, rather than a Gaussian image.

Great care must be exercised if it is decided to use an iris to determine a laser beam diameter. At long wavelengths it is very easy to forget the effects of Fresnel diffraction [3.12]. Here, an axial peak in intensity is obtained when $\delta = (n + 1/2)\lambda$, and an axial mimimum when $\delta = (n + 1)\lambda$ where δ is defined as the path difference from aperture edge to beam axis at a given distance D and aperture center to the same point at distance D. Table 3.1 shows that comparatively large apertures can significantly distort the beam profile, e.g. 6.4 mm aperture at 19.8 cm distance from the aperture with a wavelength of 10431.6 nm. The factor $2\,\delta/\lambda$ gives the number of maxima expected. Small apertures can safely be used [3.13] at shorter wavelengths. For example, 355 nm and a 2 mm diameter aperture predicts 14 maxima at 19.8 cm, i.e. essentially a smooth profile.

Table 3.1. Data at 10431.6 nm demonstrating distortion of the beam profile

Iris diameter [mm]	D [cm]	δ/λ
6.4	19.8	2.7–2.5
3.8	19.8	0.95–1.0
2.5	19.8	0.41–0.5

3.2.4 Laser Beam Alignment Techniques

Accurate alignment of the laser beams is essential if the appropriate theoretical model is to be accurately applicable and the results quantifiable. The position of a focused laser beam may be detected by a variety of ways depending on its wavelength. Luminescent phosphors are available for wavelengths from the ultraviolet down to the near infrared (1300 nm). At a focus there is often enough power to give a visible flash on a carbon plate or on a card shaded over with a graphite pencil (useful at 10 000 nm). In coaxial alignment, it is best to fix the path of the pump laser beam by locating it at two points in space with an appropriate visible method and marking these with irises closed down to their minimum apertures. The second probe laser, usually of totally different wavelength, is then centered on these irises using the appropriate visible method and opening the irises to a suitable diameter to ensure accurate concentricity. For typical probe laser diameters (e.g. HeNe, 2 mm) and pump diameters (e.g. 0.5 mm) this procedure is perfectly satisfactory. In the case of crossed beam geometries, particularly with visible lasers focused to microscopic spot sizes, a needle tip can be used to define the region of intersection for preliminary alignment. Thereafter, some experimental optimization may be necessary. However, some probe lasers do not have truly uniform TEM_{00} profiles and such experimental optimization can lead to alignment on a profile "hot spot". The true test of correct final alignment is to observe an equal decrease in probe beam intensity when it is deliberately offset by equal angles in all four $\pm X$ and $\pm Y$ directions.

3.2.5 Beam Splitting and Combining

Beams need be combined and split in the case of coaxial geometry only. There are four methods of doing this. The simplest is with dichroic filters which efficiently reflect one of the wavelengths and transmit the other. These should be antireflection coated for the transmitted wavelength. For experiments where polarization of the beams is important, the filters and all mirrors in the optical train should be aligned with the incidence planes orthogonal to the polarization electric vectors. Even so, it is advisable to purify the polarization with a polarizer close to the sample. Diversity of wavelengths sometimes makes this impossible, however. Prisms and gratings also allow beam mixing and separation and again the polarization effects of these, especially gratings, must be considered. The reflectivity of a grating for a given polarization depends on the angle of incidence *and* the blaze angle. Manufacturers' specifications should be consulted! Finally, polarization differences in pump and probe beams can be used to mix and split beams. This is particularly useful if the wavelengths are close, as in the case of stimulated Raman experiments with both one pulsed pump and the probe in the red.

3.2.6 Sample Cells

Sample cells should be of far greater cross-sectional area than either pump or probe beam. A factor of 20 is typical. This ensures that the usual approximation of an infinite medium is justifiable. Thermal lens signals may be generated by the windows (or indeed other elements in the optical train). This is usually checked by testing for the absence of signal when the sample cell is empty. This is not adequate in the case of windows if very small absorbances are to be observed. If a window absorbs some energy, the consequent heating effect may not be observable with the cell empty because the window itself and the vacuum or air in the empty cell have too low a temperature coefficient of refractive index. If a liquid is in the cell, however, it will be heated by contact with the window and its enhanced refractive index temperature coefficient may give a signal which could be erroneously attributed to the liquid! This effect is readily assessed by placing an additional piece of window material in the liquid-filled cell and testing for an increased signal.

3.2.7 Signal Detection

Divergence or convergence of probe beams can be detected at pinholes or slits [3.14], whereas deflection is detected at a sharp edge or V-shaped aperture [3.15]. Position-sensitive detector arrays can be used for either application. Previously their spatial resolution was poor, but this is improving. Nevertheless, these arrays are expensive and processing their signals can be complicated and expensive also.

When pinholes and slits are used they are usually placed in front of a narrow bandpass filter to reject the pump beam. This is the best order for these elements, since narrow bandpass interference filters produce fine fringe patterns which can be resolved by the pinhole or slit, if it is behind the filter, giving rise to gross signal distortion.

Such masking elements are then followed by a photodiode or photomultiplier tube. The latter must be fast coupled via a low load resistor. Although this gives the requisite speed, it results in low signal-to-noise ratios. Since lens experiments allow large light intensities onto the detector, care is required to prevent phototube overload. Neutral density filters may be necessary to reduce the intensity and allow the tube to operate in its linear regime. A further consequence of the large light intensity is that the principal source of noise is shot noise.

The fraction of the signal which is shot noise is

$$\frac{\Delta V}{V} = \sqrt{2e\Delta f(G/I)} \quad , \tag{3.2.3}$$

where e is the electron charge, Δf the electronics bandwidth, G the phototube gain and I the signal current, which is the signal voltage/load resistor. Substitution of typical figures, e.g. Δf = 1MHz, $G = 10^7$, signal = 20 mV and load resistor= 50 Ω, gives a shot noise figure of 0.09, i.e. about 10%. Higher bandwidths are not uncommon and increase this figure accordingly.

The performance of photodiodes is now such that they can be used as thermal lens detectors. PIN photodiodes can be obtained with sensitivities of the order 3×10^7 VW^{-1} with an area of 7 mm^2. The noise of these devices is characterized by their noise-equivalent power (NEP), where figures around 17×10^{-12} $WHz^{-1/2}$ are common, but coupled with low bandwidth, in this case, 10 kHz. The linearity of current response against photon flux is typically 2%. The bandwidth of diodes depends on their capacitance, so that higher bandwidths are available from smaller area devices only; e.g. 2 ns risetime for a 1 mm^2 diode and 2 ps risetime for 0.01 mm^2 with a sensitivity of 10 VW^{-1} and NEP of 10^{-1} $WHz^{-1/2}$. A photodiode has unity gain, so that for weak signal detection, avalanche diodes should be used, these having gain of up to 200. Both PIN and avalanche photodiodes have low dark currents, typically $(1-5) \times 10^{-9}$ A. For operation into the near infrared, InGaAs diodes seem best (800–1650 nm) with slightly higher dark currents.

Such photodiodes can be arranged as quadrant position-sensitive detectors and arrays with spatial resolution of 26 μm. Two- and three-dimensional arrays are available with up to 1048 elements and a dynamic linear range of up to 10 000 to 1. Silicon arrays (200–1100 nm) are frequently supplied on Peltier cooled substrates operating from $-5°$ to $-35°$ C. For low light levels a proximity-focused microchannel plate intensifier can be added. Arrays are useful in checking laser transverse mode profiles, although the resolution is often disappointing, and can be used as position-sensitive detectors, although readout can be expensive.

The choice of photomultiplier or photodiode detector depends on the minimum incident power level to be detected and the wavelength of the probe laser. For incident powers above 10^{-7} W and a wavelength of 500 nm a photodiode is best. For powers down to 10^{-11} W, avalanche diodes and photomultipliers are similar and are preferable to the PIN diode. At longer wavelengths the photodiodes have the best performances, and silicon diodes are best in the red with germanium diodes for the near infrared. The small areas of these devices can make the experimental arrangement less easy and optical fiber coupled systems can facilitate this.

3.2.8 Signal Retrieval

In the *pulsed source experiment* the signal waveform yields kinetic information while the peak signal amplitude can be used as a spectoscopic signal. For such systems where the mark-to-space ratio is far from unity, phase-sensitive (lock-in) detection is highly inefficient and undesirable. The alternatives are boxcar and digital waveform averagers. The boxcar is less expensive, but records only one point per experiment, so that for equivalent signal-to-noise improvement it is far slower. More seriously, variations in laser pulse amplitude, often appearing nonlinearly in the lens signal, produce serious noise on a boxcar-averaged recording. With a digital waveform averager such irreproducibilities are equivalent to summing more or fewer signals, so that they have no other effect on the signal-to-noise ratio and do not alter the shape of the waveform. It is also very easy to use dual timebases with prerecording of baseline signals, both very desirable features in kinetic analysis of the signals. In signals normalized for number of shots recorded the amplitude will still vary and appear as noise in spectroscopic applications.

In spectroscopic applications the waveform averager has, however, an important advantage in that signal saturation effects which occur with lock-in systems are not so readily encountered and a wide, linear, dynamic range (typically 3 decades) obtains.

If the signal from *modulated source experiments* is detected by a phase-sensitive (lock-in) amplifier, the phase shift between incident radiation waveform and thermal energy waveform emergent from the irradiated sample will have a marked effect on the detected signal strength. Lock-in systems have, of course, the merit of very narrow bandwidth, thereby greatly reducing noise, but in a system to be used for quantitative spectrophotometry the signal must be optimized for every sample by phase-shift adjustment. This phase shift depends on the rate of heat energy release in the radiationless relaxation processes. In gases these rates can be very slow and comparable with the period of the excitation waveform. This is particularly so in infrared excitation.

In the UV-visible range of the spectrum, internal conversion processes are generally "immediate" on this time scale, introducing negligible phase shift. However, intersystem crossings involving triplet states may be far slower and require phase correction. If the thermal energy has to travel some way to the probe beam, a further phase shift is induced and must be compensated for. The transit time for this process will depend upon the thermal diffusivity of the medium above the sample (this is not always air, for example CCl_4 increases the signal 1000-fold compared with air) and of the sample itself. These are governed by expressions similar to $D(f)$ above, (3.2.1) [3.16].

3.3 Theory

3.3.1 The Pulsed Source Experiment

The theory of pulsed source coaxial thermal lens experiments is well developed and accurately quantifies the experimental data, as shall be seen below in discussing applications of this thermal lens. This theory has already been described in Chap. 2 and will not be repeated here. It should be remembered that the simplest equations result from probe beams larger than pump beam diameters coupled with detection in the near field, where there are no diffraction effects.

3.3.2 The Modulated Source Experiment

In modulated source experiments, the source may be a laser or a powerful incoherent arc (xenon emission or similar light passed through a monochromator), modulated by a chopper to an approximately sinusoidal or trapezoidal waveform and focused onto the sample. In many applications the probe laser beam is made smaller than the pump beam. In this geometry it can be shown [3.17] that the largest response is obtained when the sample cell (assumed to be thin) is placed one confocal length beyond the focus of the lens focusing the pump beam. A confocal length is where the pump beam has expanded to twice the *area* it has at the focus. This result is based on matrix transfer optics, so that the interpretation of the optics is quite different from the coaxial lens theory.

The absorbance of the sample can be characterized by an absorption length $L(\lambda)$ given by

$$L(\lambda) = 1/A(\lambda) \quad , \tag{3.3.1}$$

where $A(\lambda)$ is the optical absorption coefficient of the sample in reciprocal length units. Such units are more familiar to the physicist than the chemist for whom Beer's law uses the notation

$$I = I_0 \times 10^{-act} \quad , \tag{3.3.2}$$

where a is the decadic absorption coefficient, c the sample concentration and t the sample thickness (i.e. the optical path length). Clearly,

$$A = ac \quad , \tag{3.3.3}$$

so that the absorption coefficient $A(\lambda)$ depends on concentration for solutions (pressure for gases at the same temperature) as well as the intrinsic absorption properties represented by the Beer's law coefficient a.

When the energy absorbed relaxes by radiationless processes, the resultant heat energy evolved diffuses into the sample with a characteristic diffusion length

$$D(f) = \sqrt{K/\pi dCF} \quad , \tag{3.3.4}$$

where K is the sample thermal conductivity coefficient, d its density and C its heat capacity. The amplitude of the thermal wave resulting from the modulated source absorption is attenuated by $\exp(2\pi)$ in one diffusion length $D(f)$, i.e. by 500. Accordingly, it is usually assumed that only heat generated within one diffusion length of the surface will reach the surface to be detected by an appropriate technique. From this assumption comes the concept of thermal depth profiling. Clearly, the depth accessed by the technique is accurate only to the order of a sizeable fraction of this diffusion length. Frequently this is quite adequate for the analysis in hand. This diffusion length will not generally be equal to the characteristic absorption length $L(\lambda)$. If $D(f) > L(\lambda)$, and the sample thickness is greater than both, the incident energy is entirely absorbed and small changes in the absorption coefficient have no effect on detected signal strength, i.e. it is saturated and depends only on the incident energy and the reflectivity of the sample. This condition is obviously useless for spectroscopy. However, $D(f)$ can be made less than $L(\lambda)$ by increasing f, whereupon a satisfactory spectrum can be obtained by varying λ with all other parameters constant. Detected signal strength will depend upon the energy absorbed in the length $D(f)$. This is a function of absorption coefficient a, concentration c where apporopriate and, through $D(f)$, of K, d, C and f. Since $D(f)$ is a function of K, samples with the same optical spectra but different thermal properties will give different signal strengths, as will samples with identical thermal properties and different absorption spectra. For example, the technique is frequently used to characterize optically thick materials [3.7], e.g. semiconductors, giving a depth profile of their spectra.

Thermal diffusion or conductivity processes are relatively slow, so that in the attempt to reduce $D(f)$ by increasing f it usually follows that temperature integration occurs in the sample. This arises because the heat energy from one incident wave does not have time to dissipate completely before another wave arrives. Accordingly, the system settles down into a quasi steady state with superimposed modulation. This has two very important consequences. Firstly, the signal is not modulated to its full depth and so is weaker than it need be. Secondly, the sample temperature changes, sometimes markedly if incident energy is increased to compensate for signal reduction due to the first effect, and this can alter spectra, absorption coefficient, K, energy relaxation rates and so phase, and may lead to optical damage of the sample or its degradation. The phase relationships can be used to discriminate between signals from surface and bulk sources, as when a film deposited on some substrate is to be examined.

3.3.3 The Transverse and Collinear Photothermal Lens

Detailed quantitative analysis of signal waveforms in these geometries is given in [3.2] and perhaps the key difference between deflection and blooming experiments is that the former measures the refractive index gradients produced, while the latter measures the second derivative or curvature of this gradient [3.7, 8]. These equations are very complex, but can be solved numerically, and extensive experiments have by now verified the accuracy of these solutions. The dependence of signal on chopper frequency and probe and pump beam axis offsets is accurately predicted. In the transverse experiment, the signal strength decreases exponentially with chopper frequency, so that for high frequencies, including pulsed work and solids immersed in liquids, the probe beam must be placed close to the sample surface. Since this requires tight focusing, the sample must be of small lateral dimensions. Maximum beam deflection is obtained when the probe is offset from the pump beam by $1/2w$.

Typically, absorption of $1\,\mathrm{mW\,cm^{-2}}$ at a 2 mm diameter spot on a sample will deflect the probe beam 10^{-5} rad with a background noise of 10^{-9} rad, allowing measurement of 10^{-5} absorbance units if, e.g. CCl_4 enhancement is used.

3.4 Applications

3.4.1 Determination of Energy Transfer Rate Constants

The process of chemical activation is poorly understood and represents a major barrier in the a priori calculation of Arrhenius activation energies and, therefore, reaction rate constants. It is not possible to follow experimentally the detailed route for accumulation of translational energy in a particular reaction coordinate leading to bond rupture or molecular rearrangement, for example. However, this chemical activation process can be followed in reverse as the vibrational energies in various molecular vibrations are exchanged and relax into translational energy, following excitation of one vibrational mode only. Such experiments are termed energy transfer studies. Boltzmann ratios give the equilibrium constants between such energy relaxation rate constants and the activation rate processes, so that the latter can then be readily calculated. Accordingly, energy transfer studies are essential to the chemist trying to understand the origins of the Arrhenius activation energy.

A more recent use of energy transfer data is in the improvement of laser efficiency by means of additives with the best energy transfer characteristics to facilitate "pumping" of the upper lasing state and quenching of the lower lasing state.

In the coaxial thermal lens approach, the analytical procedure requires the solution of the equations for the rate of release of energy into translation by V-V and V-T exchanges. The resulting expression for E_{trans} as a function of time is fitted directly to the thermal lens signal curve. Figure 3.3 shows the energy levels of CO_2, N_2 and CO. Figure 3.4a shows the signal obtained from CO_2 alone, Fig. 3.4b in the presence of N_2 and Fig. 3.4c in the presence of CO. In every case the positive-going modulation signal is indicative of sample heating, i.e. divergence of the probe beam, i.e. *decrease* in the photomultiplier tube signal. From Fig. 3.4a it is clear that

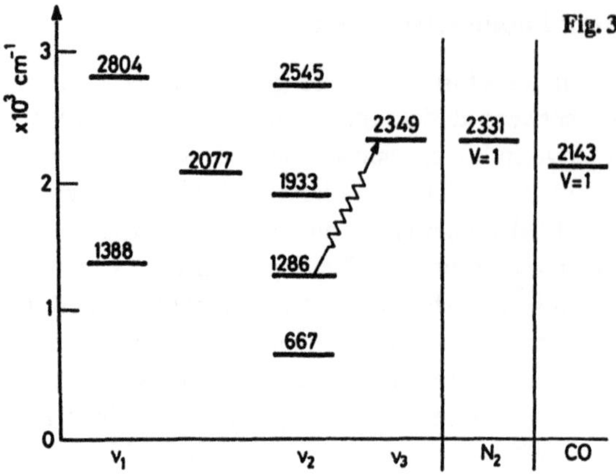

Fig. 3.3. The energy levels of CO_2, N_2 and CO

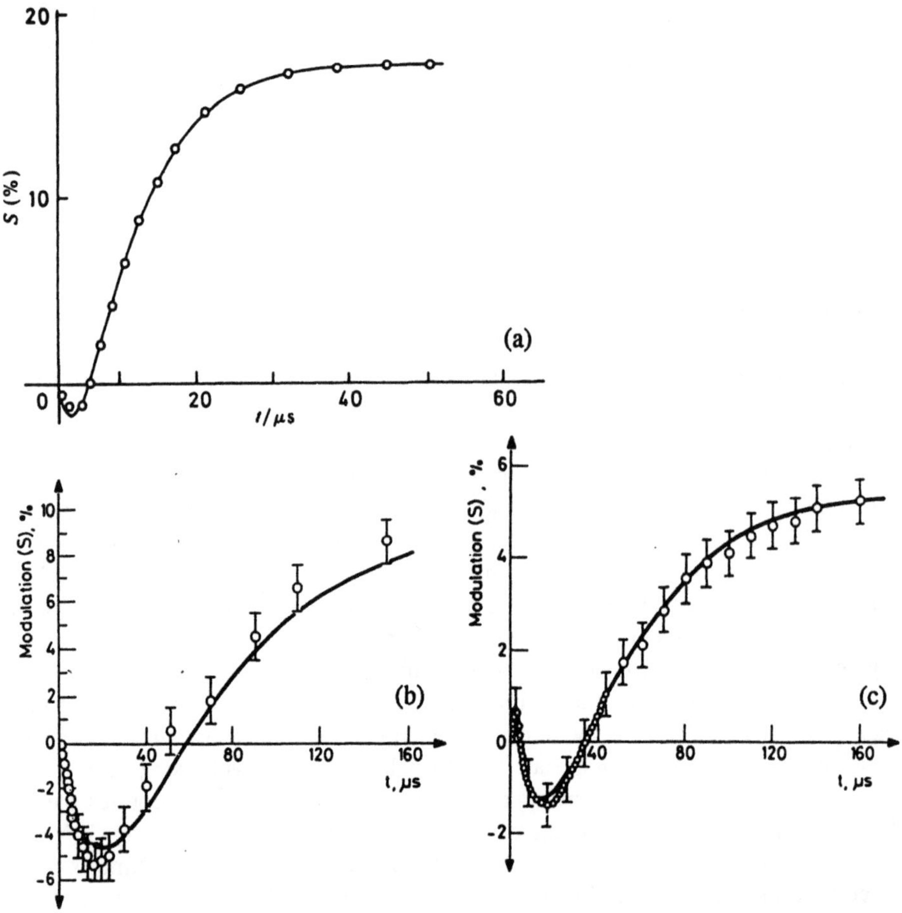

Fig. 3.4. Thermal lens signals from (a) 644 torr pure CO_2, (b) 144 torr CO_2 + 445 torr N_2, and (c) 744 torr CO_2 + 154 torr CO

on receipt of the absorbed energy CO_2 initially *cools* translationally. This unlikely event results from the finely balanced kinetics of relaxation of the 001 level in competition with those of the 010 level [3.18]. Addition of the N_2 further depletes the 001 population with which the N_2 $V = 1$ level is within $18\,cm^{-1}$ of exact match (resonance). The consequently much larger translational cooling, evidenced again by the minimum in the signal curve, is obvious from Fig. 3.4b. With CO added instead of N_2, an initial transient heating of the sample occurs (Fig. 3.4c). This is due to the initial equilibration of the CO $V = 1$ level with the CO_2 001 level. This process is $206\,cm^{-1}$ exothermic as opposed to $18\,cm^{-1}$ in the case of N_2, so that the transient is now more obvious. From a detailed analysis of such signals [3.18, 19] the data of Table 3.2 were obtained.

Table 3.2. Energy transfer rate constants for CO_2 in the presence of various quench gases (see, e.g., Fig. 3.4)

Quench gas	k_{x3} [s^{-1} $torr^{-1}$]	k_{10} [s^{-1} $torr^{-1}$]	k_{equil} [s^{-1} $torr^{-1}$]
He	45	1500	-
Ar	60	27	-
N_2	60	45	> 15000
CO	450	160	10000

3.4.2 Transport Phenomena

The pressure dependence of thermal diffusivity is the reason for the existence of one of the explosion limits in the hydrogen/oxygen system. In order to avoid such conditions and in a variety of chemical engineering design considerations, thermal diffusivities must be known with some accuracy. Nevertheless, it is surprisingly difficult to determine this parameter with accuracy in all states of matter at high and low temperatures. Most traditional techniques are invasive and often grossly interfere with the nature of the sample. Optothermal approaches remove these difficulties and permit very rapid determinations to be made noninvasively.

If the decay of the thermal lens signal to ambient conditions is expressed as a percentage modulation S, it has been shown that

$$\frac{1}{\sqrt{(1-S)^{-1/2}-1}} = \frac{t}{\tau C} + \frac{1}{C} \quad . \tag{3.4.1}$$

Plots of the left-hand side of this relation against time t are linear and the ratios of intercept to slope (in which the arbitrary constant C disappears) yield τ values which, when plotted against pressure P result in thermal conductivity coefficients K, since [3.1]

$$\tau = \frac{C_p R_G^2 P}{4RTK} \quad . \tag{3.4.2}$$

Here C_p is the heat capacity at constant pressure, R_G is the radius of the Gaussian

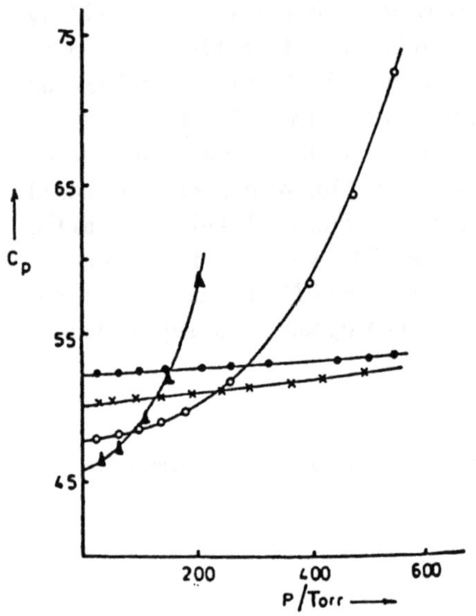

Fig. 3.5. The variation of heat capacity C_p of methanol vapor with pressure P at (▲) 321 K, (o) 341 K, (×) 377 K and (•) 405 K

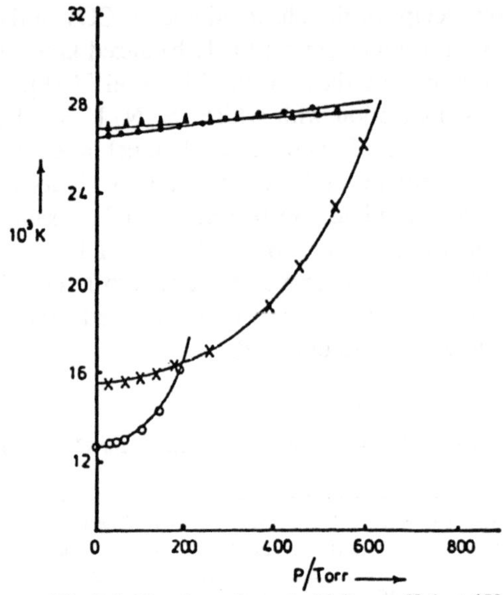

Fig. 3.6. The thermal conducitivity coefficient (K in $\mathrm{J\,K^{-1}\,s^{-1}\,m^{-1}}$ of methanol vapor as a function of pressure (P) at (o) 321 K, (×) 341 K, (•) 377 K and (▲) 405 K

laser beam profile at half height, P is the gas pressure, R is the gas constant, T the temperature in Kelvin, and K is the thermal conductivity coefficient.

A particularly interesting example investigated recently [3.20] is that of methanol, which exists as dimers and tetramers in an equilibrium which, at higher temperatures, lies far over on the dissociative side. Predictably, the heat capacity C_p of methanol in such an equilibrium mixture, as a function of pressure, follows a curve at low temperature, but a straight line of nearly zero slope at higher temperatures. This is illustrated in Fig. 3.5.

The resultant plots of τ against pressure are strictly linear and give the thermal conductivity coefficients shown in Fig. 3.6. This linearity implies that the thermal diffusivities of *all* equilibrium mixtures of methanol species are *constant*, so that the thermal conducitivity coefficients simply reflect the heat capacity dependence on temperature. This is obvious from Fig. 3.6.

Liquid- and solid- (polymer as well as crystal) state thermal conductivities can be similarly obtained by this technique. The thermal diffusivities of rabbit and baboon corneas were measured to determine the maximum rate at which the eye should be subjected to laser pulses [3.21]. Clearly, if a second laser pulse impinges on the cornea before the thermally induced optical distortion caused by the first pulse (the thermal lens) has had time to dissipate, this second pulse will not focus at the desired point. A succession of such pulses will also lead to a steadily rising temperature in the cornea and this would eventually cause irreparable damage. The thermal lens experiments determined the maximum rate for laser pulses such that there was none of this integration of temperature. It is about 1/12 Hz for a 2 mm diameter laser beam.

The thermal conductivities of oriented, stretched polyethylene terephthalate sheet have been determined as a function of direction. The greater conductivity is observed in the direction of greatest alignment [3.14]. The values obtained can be used to noninvasively monitor the degree of orientation induced in such films on a production line, or to select high thermal conductivity polymers for packaging semiconductor integrated circuits, for example.

Thermal diffusivities of well-characterized materials: standard Poco-graphite and S.S. 1.4970 stabilized austenitic steel, may be accurately measured [3.8] (Table 3.3).

Table 3.3. Photothermal deflection measurements of thermal diffusivities [$cm^2\ s^{-1}$]

	Standard value	Experimental value
Poco-graphite	0.67	0.66
S.S. 1.4970 steel	0.041	0.040

Although thermal transport has been investigated in many studies, flow velocity has also been recently monitored quantiatively [3.22] using a crossed beam geometry with 4 mW of HeCd pump power (422 nm) modulated at 25 Hz and focused via a 25 mm lens onto a 1 mm square capillary. The similarly focused 5 mW HeNe probe laser was offset to the $1/2w$ point to give a deflection system. Flow rates of $0.016\,mm\,s^{-1}$ to $> 100\,mm\,s^{-1}$ could be measured with a resolution of ~ 0.1 mm. The expected parabolic profile was obtained. Here again the high spatial resolution of this crossed beam, focused system could be used to great effect.

3.4.3 Photochemistry

In laser-induced photochemistry it is essential to identify the thermal contribution to the reaction. This problem is approached in several ways, e.g. by addition of a species capable of unimolecular decomposition with rate constants well-characterized in terms of their Arrhenius parameters. The observed rate of decomposition of this "chemical thermometer" is then used to deduce a temperature. Alternatively, a large pressure of "buffer" gas (e.g. argon) is used to provide chemically inert heat capacity. The reasoning is that, if the thermal contribution to the reaction is significant, the extra heat capacity will reduce the temperature and, with it, the rate. Both these approaches are too simplistic. In the first, it is assumed that the "thermometer" kinetics are unperturbed by the presence of laser radiation and/or intermediates of the laser photochemical decomposition. In both methods it is assumed that the added gases do not, in themselves, affect the temperature. This last assumption is always false! The temperature reached at the hottest spot in the usual, coaxial laser photochemical geometry is determined by the *kinetic* balance between release of translational energy from the energy transfer pathways and the thermal diffusivity of the sample. Since the energy transfer kinetics are first order in added inert gas and the thermal diffusivity (i.e. *rate* of dissipation of on-axis heat) is *inversely* proportional to pressure of added inert gas, adding argon will clearly *increase* the temperature. When the compensating effect of the increased heat capacity is included in the calculation, a pressure is found where the temperature, as a function of added inert gas pressure, reaches a maximum. The question of temperature as a function of space *and* time can be answered by the thermal lens signal amplitude [3.2, 3]. Observations on the CO_2-laser-induced isomerization of cyclopropane [3.22] indicate that the *largest*, *on-axis* temperature rise was far too low to account for the substantial extent of reaction! Clearly the reaction was essentially photochemical. This is of special interest since it has been pointed out [3.23] that previous experiments were completely ambiguous in their interpretation.

Since the optothermal techniques allow observation of translational energy, they are complementary to many more traditional approaches. For example, in energy transfer kinetics studies of the type mentioned above, they complement IR fluorescence. Likewise, in electronic spectroscopy, fluorescence studies are complemented by thermal lens measurement of the translational energy lost to internal conversion. This allows sensitive monitoring of excited states with low quantum yields of fluorescence and measurement of quantum yields of highly reactive intermediates. For example, $O_2\,{}^1\Delta_g$ production in haem-porphyrin derivative photochemical treatment of tumors. Alternative photoacoustic or IR emission techniques cannot be used in the biologically relevant media, since there is no possibility of finding a reference material with known quantum yield of $O_2\,{}^1\Delta_g$ under the same conditions. Thermal lens techniques avoid the need for such reference systems, since the total translational energy is measured in the same experiment as the various time-resolved quantities of heat. Thus the quantum yield for the production of $O_2\,{}^1\Delta_g$ in haem-porphyrin derivative photochemistry can be measured. The component of this highly complex mixture with the greatest quantum yield for singlet oxygen production will be most active in destroying tumors, if singlet oxygen is indeed the important intermediate [3.24]. Thus this quantum yield becomes a monitor of drug activity. The technique can be checked with $O_2\,{}^1\Delta_g$ quantum yield measurement in various solvents where IR emission values are available. The ageement is very satisfactory.

Such quantum yields can be similarly measured in the solid and liquid states also. Solid pyridazine has been studied [3.25]. At 60 K the quantum yield for $T_1 - S_2$ intersystem crossing is found to be 0.66 and the lifetime of the triplet state 3.9 μs. These parameters are of prime importance to the chemist in understanding the kinetics of reaction of excited molecules.

3.4.4 Surface Phenomena

This topic has recently been reviewed [3.26]. α-Si:H on the surface of silicon was investigated [3.27] spectroscopically from 1550 to 620 nm using phase discrimination to distinguish surface from substrate absorbances, the latter lagging by 180° behind the surface signal. CCl_4 was used to enhance the mirage deflection technique, it having been shown that this did not influence the surface involved. The system had a sensitivity of 10^{-8} W absorbed power, which with 1 mW at 1550 nm converts to an absorbance of 10^{-5}. A signal strength of 10 V cm^{-2} °C^{-1} was obtained compared with 0.1 in the same units from a photoacoustic system.

In this study it was possible to check various density of states models for the α-Si:H, determine the energy and number of defects, doping concentrations and film inhomogeneity. The probing of nonradiative processes this way, together with luminescence studies, produces a complete picture of the branching ratios between radiative and nonradiative deexcitations.

A microwave source (up to 7 W, 9.35 GHz) has also been used [3.28] in the case of the ferromagnetic resonance spectrum determination on a plate of sintered nickel ferrite with a HeNe probe beam to investigate spectra of both front and rear faces. Depths of 200-900 μm were examined (f = 5kHz-10Hz) with 200μm lateral resolution. Computer graphics allow a three-dimensional plot of the spectrum (scanning applied magnetic field 0-0.5 T) as a function of depth and lateral position to be displayed. The presence of inhomogeneities in the power absorbance throughout the three-dimensional scan is of importance in the construction of microwave components from such material.

The crossed beam thermal lens has been used in a photothermal microscope [3.29]. This has a resolution of 1000 nm at present, but has unique advantages over the usual optical microscope. The sensitivity is proportional to the pump laser power, allowing the study of highly transparent samples down to 10^{-3} of the optical density observable in traditional microscopy.

The dynamic range of signal is greater, being about 10^8 compared with the usual 10^3 contrast ratio of optical microscopy. The sample can be depth profiled with high resolution, eliminating the need for thin histological, microtomed samples in many cases. Finally, thermal diffusivity mapping is possible. In Fig. 3.7, 100 × 100 pixel images of *Tilia americana* are shown as (a) absorbance contrasted, and (b) thermal diffusivity contrasted. The sample was stained with safranin and fast green, which stain both cell nuclei and lignified cell walls red. Clearly different thermal properties allow distinction between like-colored areas.

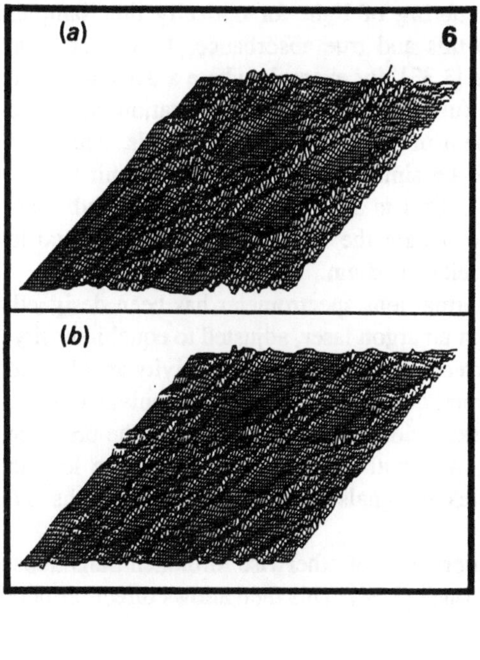

Fig. 3.7. (a) Absorbance-contrasted and (b) thermal diffusivity contrasted images obtained with the photothermal microscope. The sample is stained *Tilia americana*

3.4.5 Spectroscopies

As a detection system for the absorption of light the thermal lens offers extremely high sensitivity. The fractional modulation of the probe laser, S, is given, for weakly absorbing samples by

$$S = 2.303\,E\,A + (2.303\,E\,A)^2/2, \tag{3.4.3}$$

where A is the absorbance and E is the approximate improvement factor of the technique over traditional spectrophotometry; E is given [3.17] by

$$E = \frac{(P\,dn/dT)}{\lambda K}\ , \tag{3.4.4}$$

where P is the excitation laser power, dn/dT is the temperature coefficient of the refractive index of the absorbing medium, λ is this excitation laser wavelength and K is the thermal conductivity coefficient of this absorbing medium. Clearly this improvement factor can be increased by increasing P.

For example, with a 0.25 mm diameter reverse phase chromatography column fitted with thermal lens detection, 0.75×10^{-15} mol quantities of 18 different (dimethylamino)azo-benzenesulphonyl derivatized amino acids can be detected and separated. Only 50 analyte molecules were present in the 0.2 pl detection volume of this crossed beam geometry thermal lens! This detection limit equals that of laser-induced fluorescence. Moreover, the technique is complementary to fluorescence, being at its best when the quantum yield for fluorescence is very low [3.30]. Detection limits of 10^{-5} cm^{-1} absorbance corresponding to 5×10^{-8} M U(VI) in 0.1M HCO_3 have been reported [3.31]. An excimer laser (XeCl, 308 nm) was used as a pump and HeNe as probe laser. This is orders of magnitude lower than conventional spectrophotometry for U(VI) and the method has the added advantage of very high speed so that no movement of chemical equilibria occurs during measurement.

The method is not sensitive to scattering of light for optically thin samples [3.32] and easily distinguishes between this and true absorbance. This is of great advantage in the analyses of humic acids [3.33], for example where a detection limit of $20\,\mu g\,l^{-1}$ is attainable for Armadale fulvic acid. No preconcentration is needed and this system easily allows measurement of naturally occurring levels, which are typically $1 - 25\,mg\,l^{-1}$. Also, Tb^{3+} can be similarly detected with a limit around 5×10^{-7} M in reversed micelles [3.34]. This is 100 times lower than traditional methods for Nd or Pr in perchloric acid. Again the insensitivity of the method to scattering losses is invaluable in the micellar medium.

A double beam dual wavelength thermal lens spectrometer has been designed, using two lines (514.5 and 457.9 nm) from an argon laser, adjusted to equal intensity, and HeNe probe. Solvent background correction and improved selectivity are claimed [3.35]. Praseodymium ions were determined down to 4.7×10^{-5} M, equivalent to an absorbance of 2.8×10^{-5}, with 25 mW excitation modulated at 1 Hz in the presence of 10^{-2} M nickel glycinate complexes. A signal averager was used since lock-in detection is not possible with the multiplexed signals from the two wavelengths and baseline.

Two-photon spectroscopy allows observation of otherwise forbidden transitions, since this is subject to different selection rules [3.36]. This then allows direct probing of the nature of the states involved. Thermal lens detection has the required very high sensitivity to monitor such two-photon transitions. In pyridine, for example, which is noted for its lack of fluorescence or phosphorescence, absorbance at $38130\,cm^{-1}$ was assigned to the $1B_2$ state and that at $41000\,cm^{-1}$ to the $1A_2$ state of the second $n-\pi^*$ transition. Polarization ratios were also measured by the lens technique and this was essential for the assignment. Higher resolution spectra of pyridine determined this way will allow detailed examination of the nature of these otherwise inaccessible low-lying $n-\pi^*$ states.

If two pulsed lasers (e.g. Nd:YAG and a dye laser) are tuned so that they differ in frequency by the Raman transition frequency of a molecule (e.g. CO_2), the upper Raman level can be "pumped", so that high population levels obtain. This population relaxes to the ground state with the release of translational energy, and so the system lends itself to thermal lens monitoring as a function of time. Energy transfer kinetics of the CO_2 Raman level, 100, have been directly observed by this method, but the amplitude of the signal can be used as a direct sprectroscopic signal [3.37].

When the dye laser frequency is tuned so that the frequency difference is made to scan across the Raman band, the amplitude yields the CO_2 Raman band shape. This technique is not plagued by fluorescence interference and can be made to yield very high signal-to-noise ratios when laser pulse width, energy and detection amplifier bandwidth are optimized. Likewise, "hot band" spectra can be observed with high sensitivity. Figure 3.8 shows the apparatus and Fig. 3.9 typical spectra of benzene and toluene.

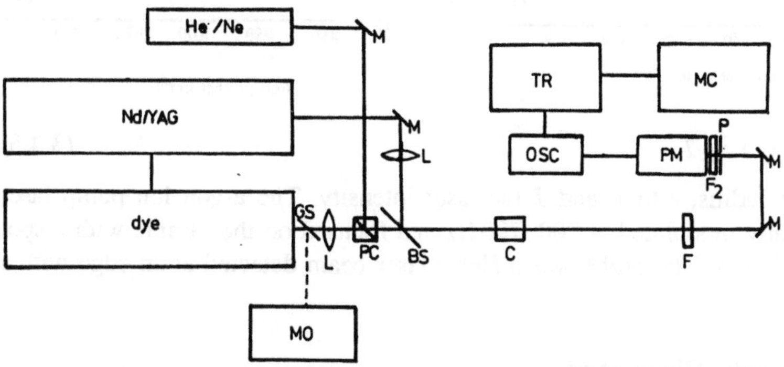

Fig. 3.8. Coaxial thermal lens system for Raman gain measurements. M: dielectric mirrors, L: lens, TR: transient recorder, MC: microcomputer, PM: photomultiplier, OSC: oscilloscope, P: pinhole, F: 632.8 nm narrow bandpass filter, C: Sample cell, bs: beamsplitter, PC: polarizing cube, MO: power monitor

Photodeflection spectroscopy has recently been reviewed [3.38]. It has been used to measure propagation loss in waveguides [3.39] and to obtain spectra of shallow donors in n-GaAs in a magnetic field $\leq 4\,T$ at 1.4–12 K. The final states of the observed transitions can be unambiguously assigned even under these difficult experimental conditions [3.40].

Photothermal deflection has been used [3.41] to measure the thermal nonlinear coefficients of the refractive indices in glasses containing $CdS_x\,Se_{1-x}$. These are of importance in high-speed optical bistable devices. The refractive index is written

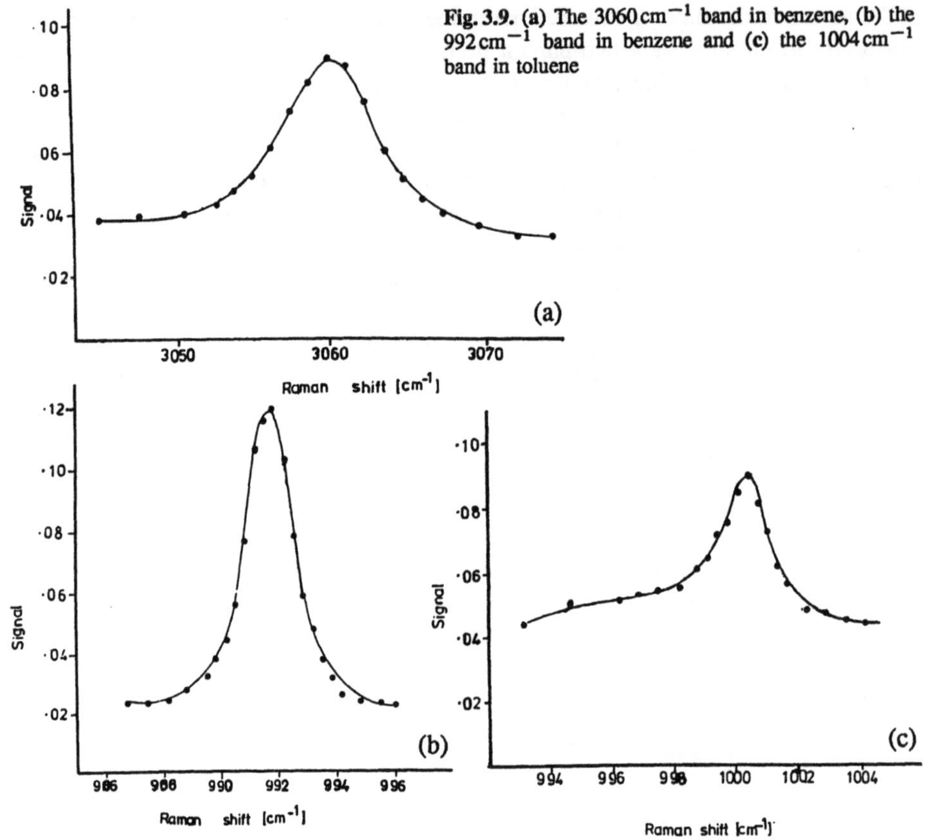

Fig. 3.9. (a) The $3060\,\mathrm{cm}^{-1}$ band in benzene, (b) the $992\,\mathrm{cm}^{-1}$ band in benzene and (c) the $1004\,\mathrm{cm}^{-1}$ band in toluene

$$n(r,t) = n_0 + n_2 I, \tag{3.4.5}$$

where r is the radius, t time and I the laser intensity. The argon ion pump laser was mechanically modulated at 200–700 Hz and focused on the sample with a spot size ($1/e^2$) of 60 μm. The probe was a HeNe laser beam detected at an edge with a photodiode.

3.4.6 Photoacoustic Diagnostics

Although no thermal lens ensues until the acoustic time has elapsed, it is possible to observe the acoustic wave traverse the probe laser beam after reflection at the wall as an "echo" on the thermal diffusivity controlled decay of the thermal lens signal curve. Several such "echoes" may be seen. If they are digitized, their shape can be compared with those predicted by the theory above [3.42]. This analysis has been done for CO_2 and SO_2. In neither case does the shape of the observed profiles agree precisely with those calculated from the thermal lens theory. Undoubtedly this is because several factors intervene. In the case of CO_2 as seen above, there is an initial contraction, although this is absent in SO_2. Additionally, there is the bandwidth of the microphone (dc to >1 MHz) and the wave distortion at the wall due to energy loss.

The arrival time of the acoustic wave is exactly as predicted [3.43] from the nondispersive sound velocity v given by

$$v = \sqrt{\gamma RT/M} \tag{3.4.6}$$

and, in the case of SO_2, the bandwidth of the highly damped sound wave is in agreement with the acoustic time at the higher pressures. This use of thermal lens signals to verify models of the acoustic wave should prove invaluable, especially at high frequencies where microphone bandwidths are very limiting and other transducers suffer from severe resonances.

3.4.7 Laser Physics

Thermal lens techniques have been used to provide the feedback mechanism for stabilization of lasers. A waveguide CO_2 laser has been frequency stabilized to one part in 10^8 at a variety of frequencies by this means.

In a pulsed Nd:YAG laser, the laser rod is subject to coaxial thermal lensing because the pumping flashlamps heat the exterior of the rod with that part of the spectrum not used to create the lasing population. This produces a convergent lens and the focal length of this is a function of laser flashlamp energy (i.e. laser power) and repetition rate. The effect can be compensated by refocusing an intracavity telescope to optimize the output power for each set of conditions. Laser dyes are similarly affected, particularly in cw dye lasers, because of both flow and thermal effects.

3.5 Summary

Space does not permit a full survey of the increasing utility of this technique. However, the applications described show the ability of the optothermal approach to solve a very wide range of problems of chemical interest. This technique has several unique advantages over more traditional methods:

1) It has a wide dynamic range and can be modelled to yield accurate physical transport parameters, usually noninvasively.
2) It is complementary to laser-induced fluorescence and is equally sensitive in many cases, allowing nonradiative processes to be monitored directly.
3) It does not suffer from the limited bandwidth, resonances and ambient noise problems of photoacoustics and has greater sensitivity, especially if compared to the greatest bandwidth microphone systems.
4) It can distinguish between absorbance and light lost by scattering.
5) It can distinguish between surface and bulk spectra, allowing depth profiling without sectioning samples.
6) It rapidly leads to a result in most applications, so that system stability is not a problem.
7) It can be used on all states of matter.
8) It is simple to construct.

References

3.1 C. Hu, J.R. Whinnery: Appl. Opt. **12**, 72 (1973)
3.2 R.T. Bailey, D. Pugh, F.R. Cruickshank: J. Chem. Soc., Faraday Trans. 2, **76**, 633 (1980)
3.3 R.T. Bailey, D. Pugh, F.R. Cruickshank: J. Chem. Soc., Faraday Trans. 2, **77**, 1387 (1981)
3.4 R.T. Bailey, F.R. Cruickshank, D. Pugh, A. McLeod: Proc. of the Max Born Centenary Conference, Edinburgh, ed. by M.J. Colles, D.W. Swift, Proc. SPIE **369**, 88 (1982)
3.5 H. Coufal, J.F. McClelland: J. Mol. Structure **173**, 129 (1988)
3.6 H. Vargas, L.C.M. Miranda: Phys. Rep. **161**, 43 (1988)
3.7 W.B. Jackson, N.M. Amer, A.C. Boccara, D. Fournier: Appl. Opt. **20**, 1333 (1981)
3.8 G. Suber, M. Bertolotti, C. Sibilia, A. Ferrari: Appl. Opt. **27**, 1807 (1988)
3.9 J. Stone: J. Opt. Soc. Am. **62**, 327 (1972)
3.10 J. Stone: Appl. Opt. **12**, 1828 (1973)
3.11 M. Bertolotti, A. Ferrari, C. Sibilia, G. Suber, D. Apostol, P. Jani: Appl. Opt. **27**, 1811 (1988)
3.12 R.T. Bailey, F.R. Cruickshank, D. Pugh, W. Johnstone: Chem. Phys. Lett. **59**, 324 (1978)
3.13 K. Heihoff, S.E. Braslavsky: Chem. Phys. Lett. **131**, 183 (1986)
3.14 R.T. Bailey, F.R. Cruickshank, A. McLeod, D. Pugh, A.G. Faraday: Polym. Commun. **26**, 23 (1985)
3.15 C.S. Vikram, D.K. Agrawal, R. Roy, H.A. McKinstry: Opt. Laser Technol. **20**, 45 (1988)
3.16 A. Asano, T. Ichimura, Y. Uchida, H. Sakai: J. Appl. Phys. **63**, 2346 (1988)
3.17 J. Georges, J.M. Mermet: Analysis **16**, 203 (1988)
3.18 R.T. Bailey, F.R. Cruickshank, D. Pugh, K.M. Middleton: J. Radioanal. Nucl. Chem. **101**, 383 (1986)
3.19 R.T. Bailey, F.R. Cruickshank, D. Pugh, K.M. Middleton: J. Chem. Soc., Faraday Trans. 2, **81**, 255 (1985)
3.20 R.T. Bailey, F.R. Cruickshank, D. Pugh, I.J.M. Weir: Chem. Phys. **134**, 311 (1987)
3.21 S. Venkatesh, S.M.G. Guthrie, F.R. Cruickshank, R.T. Bailey, W. Foulds, W. Lee: Br. J. Opthal. **69**, 92 (1985)
3.22 W.A. Weimer, N.J. Dovici: Appl. Opt. **24**, 2981 (1985)
3.23 R.T. Bailey, F.R. Cruickshank, D. Pugh, R. Guthrie, W. Johnstone, J. Mayer, K.M. Middleton: J. Chem. Phys. **77**, 3453 (1982)
3.24 R.W. Redmond, K. Heihoff, S.E. Braslavsky, T.G. Truscott: Photochem. Photobiol. **45**, 209 (1987)
3.25 M. Terazima, T. Azumi: Chem. Phys. Lett. **145**, 286 (1988)
3.26 J.C. Murphy, L.C. Aamodt, J.W. Maclachlan: Proc. SPIE **897**, 135 (1988)
3.27 N.M. Amer, W.B. Jackson: Semicond. Semimetals **21 B**, 83 (1984)
3.28 O. von Geisau, U. Netzlemann, J. Pelzl: J. Appl. Phys. **63**, 3347 (1988)
3.29 D.J. Burgi, N.J. Dovici: Appl. Opt. **26**, 4665 (1987)
3.30 T.G. Nelson, N.J. Dovici: Anal. Chem. **59**, 2803 (1987)
3.31 G. Bidoglio, G. Tanet, P. Cavalli, N. Omenetto: Inorg. Chim. Acta **140**, 293 (1987)
3.32 Z.A. Yasa, W.B. Jackson, N.M. Amer: Appl. Opt. **21**, 21 (1982)
3.33 J.F. Power, C.H. Longford: **60**, 842 (1988)
3.34 C.D. Tran: Anal. Chem. **60**, 182 (1988)
3.35 F. Mladen, C.D. Tran: Anal. Chem. **60**, 1925–1928 (1988)
3.36 P.R. Salvi, P. Foggi, R. Bini, E. Castellucci: Chem. Phys. Lett. **141**, 417 (1987)
3.37 R.T. Bailey, F.R. Cruickshank, S.M.G. Guthrie, C.H. Nelson, D. Pugh: J. Raman Spectrosc. **18**, 13 (1987)
3.38 S. Nonomura, H. Hayashi, R. Takagi, S. Nitta: Kotai Butsuri **23**, 249 (1988)
3.39 R.K. Hickernell, D.R. Larson, R.J. Phelan, L.E. Larson: Appl. Opt. **27**, 2626 (1988)
3.40 H.P. Wagner, W. Prestl: Solid State Commun. **66**, 367 (1988)
3.41 M. Bertolotti, A. Ferrari, C. Sibilia, G. Suber, D. Apostol, P. Jani: Appl. Opt. **27**, 1811–1813 (1983)
3.42 R.T. Bailey, F.R. Cruickshank, R. Guthrie, D. Pugh, I.J.M. Weir: Mol. Phys. **48**, 81 (1983)
3.43 R.T. Bailey, F.R. Cruickshank, D. Pugh: In *Photoacoustic and Photothermal Phenomena*, ed. by P. Hess, J. Pelzl, Springer Ser. Opt. Sci., Vol. 58 (Springer, Berlin, Heidelberg 1988) p.99
3.44 E.V. Anufriev, V.P. Biryulin, S.T. Kornilov, I.V. Ostreikovskii, N.M. Prokopova, E.D. Protsenko: Prib. Tekh. Eksp. **2**, 164 (1988)

4. Spherical Acoustic Resonators [1]

J. B. Mehl and M. R. Moldover

With 6 Figures

Gas-filled spherical resonators are excellent tools for measurements of the speed of sound. The radially symmetric gas resonances are nondegenerate and have high quality factors (typically 2,000–10,000). These resonances can be used with very simple instrumentation and unsophisticated analysis to measure the speed of sound in a gas with an accuracy on the order of 0.01%. With data analysis based on a complete theoretical model of the acoustical system, the accuracy can be increased to better than one part per million. The model includes the effects of the coupling between acoustic and thermal waves, thermal and viscous effects at the shell boundary, shell motion, and imperfect shell geometry. Other boundary effects, including the effects of holes in the resonator wall and precondensation effects, have also been considered. The results of the theoretical model are described in detail and compared with experimental results in this chapter. There is also a brief review of the thermophysical importance of acoustic measurements, including a discussion of the determination of ideal-gas specific heats and information about intermolecular interactions.

4.1 Introduction

Gas-filled spherical resonators are mechanical systems whose symmetry enables them to be modeled with very high accuracy. The state of the art is best represented by the recent acoustic measurement of the gas constant R, in which the speed of sound in argon at a temperature of 273.16 K and pressures in the range 100–500 kPa was measured with an uncertainty of less than one part in one million [4.1,2]. The theory of spherical acoustic resonances was first discussed in detail by *Rayleigh* [4.3]. Important work before 1980 includes the studies of dissipation in liquids begun by *Leonard* [4.4], the measurement of the speed of sound in moist air by *Harris* [4.5], *Campbell*'s discussion of the advantages of the spherical geometry [4.6], and *Bancroft*'s pedagogical article [4.7]. The rate of development increased substantially during the past decade, probably associated with the growth in the use of laboratory computers to control experiments and analyze data. These developments include applications of the spherical resonator technique, with generally increasing accuracy, in the determination of thermophysical properties of gases [4.8–15], to thermometry [4.16,17], to determination of isotopic composition of gases [4.18–20], and to photoacoustic spectroscopy [4.21–23]. Recent development and testing of the theory of spherical resonators is reported in [4.24–29].

Practical spherical resonators are generally fabricated from machined metal hemispheres. To first approximation, when the spherical cavity contains a gas at low to moderate pressures, the boundary is very rigid. Under these conditions the effects of shell motion on the acoustic eigenfrequencies is small (except for accidental coincidences of gas resonance with shell resonances), and the eigenfrequencies depend most strongly on the speed of sound in the gas. The theory of these resonances is described in detail in this chapter. The discussion includes some comparisons with experiments; however, few experimental details are presented. The resonator used by ***Moldover*** et al. [4.1,2] is shown in Fig. 4.1. This resonator had an unusually clean geometry. The transducers were mounted so that the front surfaces were flush with the inner surface of the resonator. Similarly, the front surface of the fill-valve (when closed) was flush with the inner surface. The resonator was constructed with stainless steel. It had an approximate inner radius of 8.89 cm. For brevity, it will be referred to as "the 8.89 cm resonator" throughout this chapter. Aluminum and brass have also been used to construct resonators. The arrangements generally are similar to that of Fig. 4.1. The resonators are normally used in pressure vessels in well-regulated thermal environments, and are normally supported by relatively firm mechanical parts. Depending on the design, the sphere may be sealed off from the chamber or the filling duct can be left open. Transducers can be embedded in the shell wall or mounted outside the shell, with ducts communicating with the gas inside. One transducer is driven from a source of stable frequency f; the other is used to detect the acoustic pressure in the resonator. It is advantageous to use a two-phase lock-in voltmeter to measure the acoustic signal, both for improvement of the signal-to-noise ratio and because measuring both phases of the acoustic signal gives more information.

4.2 Basic Theory

The various "small" quantities of linear acoustics will be assumed to depend upon time through an implicit factor $\exp(\mathrm{i}\omega t)$ where ω is 2π times the source frequency f. Let the acoustic particle velocity in the gas be described by a velocity potential $\Psi(\mathbf{r})$, which is related to the particle velocity $\mathbf{u}$ through

$$\mathbf{u}(\mathbf{r}) = \boldsymbol{\nabla}\Psi(\mathbf{r}). \tag{4.2.1}$$

In the approximation where the dynamics of the gas are governed by the Euler equation, and an adiabatic equation of state is used for the gas, the acoustic pressure p is related to the velocity potential through

$$p(\mathbf{r}) = -\mathrm{i}\omega\rho\Psi(\mathbf{r}), \tag{4.2.2}$$

where ρ is the mass density of the gas. The velocity potential satisfies the steady-state wave equation

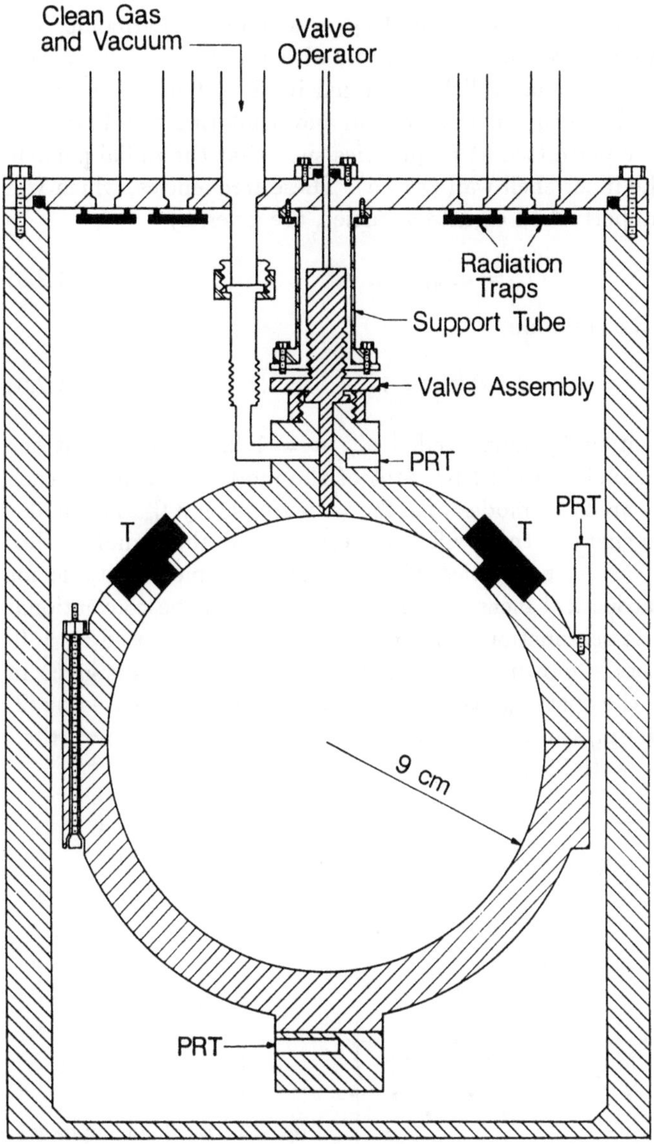

Fig. 4.1. Resonator and pressure vessel used in [4.1,2]. The transducer assemblies are indicated by T, and the locations of thermometers by PRT

$$(\boldsymbol{\nabla}^2 + k^2)\Psi(\mathbf{r}) = 0. \tag{4.2.3}$$

Here $k = \omega/c$, with c equal to the speed of sound in the gas. The regular solutions of (4.2.3) in spherical coordinates have the form

$$\Psi_N(\mathbf{r}) \propto j_l(kr)Y_{lm}(\theta, \phi), \tag{4.2.4}$$

where $j_l(z)$ is a spherical Bessel function and Y_{lm} is a spherical harmonic. The modes are labeled with a triplet index $N = lnm$, with n defined below. A real form for the spherical harmonics will be assumed in this chapter, except as noted. In a spherical shell which is perfectly rigid and insulating, free harmonic vibrations of the gas are permitted at frequencies such that the radial particle velocity vanishes at the inner shell wall ($r = a$). These frequencies, which will be referred to as the unperturbed eigenfrequencies, are given by

$$f_{ln}^0 = cz_{ln}/2\pi a, \tag{4.2.5}$$

where z_{ln} is a root of the equation

$$dj_l(z)/dz = 0. \tag{4.2.6}$$

These roots are conventionally numbereded z_{ln}, $n = 1, 2, \cdots$. Note that the unperturbed eigenfrequencies are independent of the mode index m; for each pair of indices l, n there are $2l+1$ modes with the same frequency f_{ln}^0. The modes with $l = 0$ are nondegenerate; they will be referred to as the radial modes. They have numerous special properties which make them most suitable for use in high-accuracy acoustical measurements. According to the conventional numbering of roots of Bessel functions, the first $l = 0$ root is designated $z_{01} = 0$. The distribution of roots is shown in Fig. 4.2; a sorted list of roots is given in Table 4.1. For large n the roots z_{ln} asymptotically approach $(2n + l - 1)\pi/2$. For a gas with $c \approx 300$ m/s in a resonator with $a = 5$ cm, the unperturbed eigenfrequencies are $f_{ln}^0 \approx (0.95 \text{ kHz})z_{ln}$.

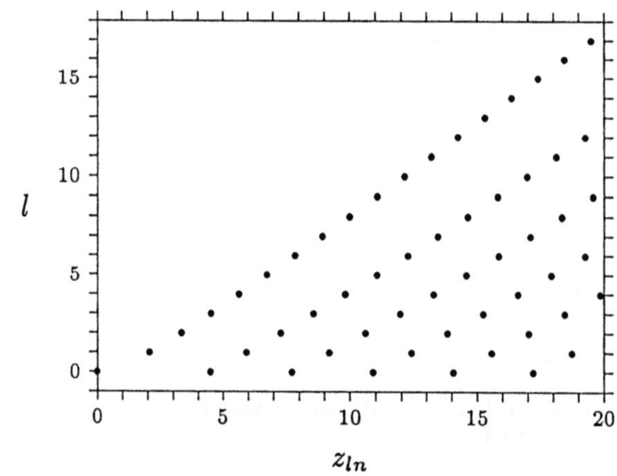

Fig. 4.2. Distribution of roots z_{ln} of (4.2.6)

A more complete acoustical model predicts complex eigenfrequencies $F_N = f_N + ig_N$ which differ from the f_{ln}^0 owing to numerous effects. Imperfect spherical geometry, for example, normally removes the degeneracy of the nonradial modes, and the thermal and viscous boundary layer effects both lead to the imaginary contributions and cause shifts in the real parts. The thermal boundary layer effects shift the eigenfrequencies by fractional amounts which are typically on the order of 10^{-4}. These are usually the dominant eigenfrequency

Table 4.1. Roots z_{ln} of (4.2.6)

l	n	z_{ln}	l	n	z_{ln}	l	n	z_{ln}
0	1	0.000000	2	5	17.042902	13	3	24.382038
1	1	2.081576	7	3	17.117506	5	6	24.474825
2	1	3.342094	0	6	17.220755	22	1	24.689872
0	2	4.493409	15	1	17.408034	3	7	24.850085
3	1	4.514100	5	4	17.947180	17	2	24.899636
4	1	5.646704	11	2	18.127564	1	8	25.052825
1	2	5.940370	8	3	18.356318	8	5	25.218652
5	1	6.756456	16	1	18.453241	14	3	25.561873
2	2	7.289932	3	5	18.468148	11	4	25.604057
0	3	7.725252	1	6	18.742646	23	1	25.724794
6	1	7.851078	6	4	19.262710	6	6	25.846084
3	2	8.583755	12	2	19.270294	18	2	26.012188
7	1	8.934839	17	1	19.496524	4	7	26.283265
1	3	9.205840	9	3	19.581889	9	5	26.516603
4	2	9.840446	4	5	19.862424	2	8	26.552589
8	1	10.010371	2	6	20.221857	0	9	26.666054
2	3	10.613855	0	7	20.371303	15	3	26.735177
0	4	10.904122	13	2	20.406581	24	1	26.758685
5	2	11.070207	18	1	20.538074	12	4	26.837518
9	1	11.079418	7	4	20.559428	19	2	27.121164
3	3	11.972730	10	3	20.795967	7	6	27.199245
10	1	12.143204	5	5	21.231068	5	7	27.693716
6	2	12.279334	14	2	21.537120	25	1	27.791618
1	4	12.404445	19	1	21.578052	10	5	27.801885
11	1	13.202620	3	6	21.666607	16	3	27.902527
4	3	13.295564	8	4	21.840012	3	8	28.023874
7	2	13.472030	1	7	21.899696	13	4	28.062137
2	4	13.846112	11	3	21.999955	1	9	28.203361
0	5	14.066194	6	5	22.578058	20	2	28.226838
12	1	14.258341	20	1	22.616601	8	6	28.536461
5	3	14.590552	15	2	22.662493	26	1	28.823658
8	2	14.651263	4	6	23.082796	17	3	29.064420
3	4	15.244514	9	4	23.106568	11	5	29.075817
13	1	15.310887	12	3	23.194996	6	7	29.084346
1	5	15.579236	2	7	23.390490	14	4	29.278731
9	2	15.819215	0	8	23.519452	21	2	29.329454
6	3	15.863222	21	1	23.653839	4	8	29.470637
14	1	16.360674	16	2	23.783192	2	9	29.710280
4	4	16.609346	7	5	23.906450	0	10	29.811599
10	2	16.977550	10	4	24.360789	27	1	29.854862

perturbations. Thus even with no further analysis, measurements of the resonance frequencies with very simple instrumentation would permit the speed of sound to be determined to an accuracy on the order of 0.01%.

In steady-state experiments where the system is excited at a single frequency f the response function measured at a detector microphone can be expressed as a superposition of contributions from the normal modes. The strongest contributions come from normal modes with eigenfrequencies near f. In most cases, particularly at pressures above 100 kPa, the radial modes are sufficiently well isolated to permit excitation with negligible excitation of nearby modes. There are, however, some troublesome near-degeneracies involving the radial modes: $z_{0,2}$ is 0.4,5% below $z_{3,1}$ and $z_{0,7}$ is 0.17% below $z_{13,2}$.

4.3 Steady-State Response

Attempts to excite two or more closely spaced modes result in a complex signal which must be described by a superposition of theoretical resonance functions. Consider a source transducer at position $\mathbf{r}_s$ on the inner surface of the shell, with an inward velocity u_s which can be considered independent of frequency over the range of resonances to be considered. The acoustic pressure $p(\mathbf{r}')$ at any point $\mathbf{r}'$ in the cavity is

$$p(\mathbf{r}') = \sum_N \frac{\mathrm{i} f \rho c^2 \Psi_N(\mathbf{r}')}{2\pi V \Lambda_N (f^2 - F_N^2)} \int_{S'} \Psi_N(\mathbf{r}) u_s(\mathbf{r}) dS, \qquad (4.3.1)$$

where Λ_N is the average of Ψ_N^2 over the cavity volume V. The complex voltage at the output of a detector transducer, assumed proportional to the acoustic pressure $p(\mathbf{r}_d)$ at the detector, thus has the form [4.29,30]

$$u + \mathrm{i}v = \sum_N \frac{\mathrm{i} f A_N}{f^2 - F_N^2} + B + C(f - \tilde{f}), \qquad (4.3.2)$$

where, in the limit of small transducers, A_N is proportional to the product of $\Psi_N(\mathbf{r}_s)$ and $\Psi_N(\mathbf{r}_d)$. The sum in (4.3.2) is taken over one or a small number of modes of interest; the contributions of the "tails" of the other modes is represented as a two-term Taylor's series $B + C(f - \tilde{f})$, where B and C are complex constants, and $\tilde{f}$ is an arbitrary fixed frequency near the modes of interest. Data are normally taken by measuring u and v as a function of frequency at a set of discrete values of f. The theoretical resonance formula is then fit to the data using a nonlinear least-squares procedure [4.31]. This technique is extremely powerful. When signal strengths are adequate, the resonance frequencies f_N and halfwidths g_N can generally be determined with a precision on the order of $10^{-4} g_N$, typically 10^{-7} of the frequency, for nondegenerate modes and triplets formed by the splittings of $l = 1$ modes [4.2,29]. *Moldover* et al. [4.2] found, for measurements with argon in the 8.89 cm resonator, that the signal strengths were not limited by noise when the pressure was above 100 kPa and the frequency was above 6 kHz. They explained the dependence on the ambient pressure P as a combination of two effects: a signal strength proportional to $P^{3/2}$ and a resonance width proportional to $P^{-1/2}$.

4.4 Wave Modes

The most complete description of the theory of the acoustic resonator is given in [4.29]. The basis of the theoretical model is a set of equations first derived by *Kirchhoff* in 1868 [4.3,32]. The dynamics of the gas are described by the

Navier-Stokes equation, which consists of Newton's second law and a constitutive relation giving the stresses in terms of the spatial derivatives of the velocity of the gas. The relevant kinetic coefficient is the shear viscosity η. Heat flow in the gas is assumed to be governed by Fourier's law; the relevant kinetic coefficient is the coefficient of thermal conductivity λ. Two additional equations are statements of the equations of continuity for mass flow and for heat flow. An equation of state for the gas is used to relate changes in pressure, density, and temperature. The second law of thermodynamics is also used to relate changes in the entropy of the gas to temperature and pressure variations. These principles are used to construct a linear acoustic theory. That is, the squares and products of certain "small" quantities are neglected in the equations. The pressure, temperature, and velocity fields are all coupled in this theory. The temperature is governed by a biquadratic partial differential equation whose solutions divide into two classes: acoustic solutions and thermal solutions. The acoustic solutions are closely approximated by the zero-order model when k is replaced by k_p, given by

$$k_p^2 = (\omega/c)^2\{1 - (\mathrm{i}/2)(\omega/c)^2[(\gamma-1)\delta_t^2 + \delta_v'^2] + O(\omega\delta/c)^4\}, \tag{4.4.1}$$

where $\gamma = C_p/C_v$ is the ratio of the specific heat capacities. The thermal solutions are evanescent waves with propagation parameters k_t given by

$$k_t^2 = (-2\mathrm{i}/\delta_t^2)\{1 + (\mathrm{i}/2)(\omega/c)^2(\gamma-1)(\delta_t^2 - \delta_v'^2) + O(\omega\delta/c)^4\}. \tag{4.4.2}$$

The notation here indicates that the solutions are correct through third order in the ratio of any of the characteristic lengths δ to the wavelength of an acoustic wave. The characteristic lengths are defined by

$$\delta_t = \sqrt{2D_t/\omega} \tag{4.4.3}$$

$$\delta_v = \sqrt{2D_v/\omega} \tag{4.4.4}$$

$$\delta_v' = \sqrt{(4/3)\delta_v^2 + \eta_b/\rho}, \tag{4.4.5}$$

where $D_t = \lambda/(\rho C_P)$, the thermal diffusivity, is the ratio of the thermal conductivity to the isobaric specific heat per unit volume, $D_v = \eta/\rho$ is the viscous diffusivity, and η_b is the bulk viscosity. A complete description of the fields also requires a shear wave with a propagation parameter k_v given by

$$k_v^2 = -2\mathrm{i}/\delta_v^2. \tag{4.4.6}$$

For argon at 273 K and 100 kPa, with a frequency of 10 kHz, approximate values for the characteristic lengths are $\delta_t = 25$ μm, $\delta_v = 20$ μm, and $\delta_v' = 23$ μm. (For argon and other monatomic gases, the bulk viscosity term is negligible.) The thermal and shear waves decay rapidly over a length scale of the order of the appropriate characteristic length; they are only important near the boundaries.

The fields must satisfy five boundary conditions at the gas-shell interface. First, the radial components of the gas and shell velocities must match. Second, the radial component of the shell velocity is equal to the radial force per unit area exerted on it by the gas times an effective acoustic admittance β_{sh}.

Third, the transverse component of the gas velocity must match the transverse component of the shell velocity. The fourth boundary condition relates the temperatures of the gas and the shell boundary. In [4.29] and earlier work it was assumed that the temperatures of the gas and wall were equal at the interface. *Herzfeld* [4.33] had shown that the effects of penetration of the thermal wave into the solid boundary and the effects of the temperature jump at the interface were small for air under ambient conditions. More recently *Ewing* et al. [4.28] provided quantitative evidence that at low densities a temperature discontinuity should be included in the boundary condition. According to kinetic theory the temperature of the gas (extrapolated to the wall) should exceed that of the shell by $J_t l_a/\lambda_g$, where J_t is the normal heat flux across the interface, l_a is the accommodation length in the gas, and λ_g is the thermal conductivity of the gas. The accommodation length is given by

$$l_a = \frac{\lambda_g}{P}\sqrt{\frac{\pi M T_g}{2R}}\,\frac{(2-h)/h}{C_V/R+1/2}, \tag{4.4.7}$$

where M is the molar mass of the gas, T_g is the temperature of the gas, R is the gas constant, C_V is the molar specific heat at constant volume of the gas, and h is the thermal accommodation coefficient [4.28]. *Ewing* et al. found that the thermal accommodation coefficient between argon and the machined aluminum wall of their resonator was 0.84 ± 0.05; later *Moldover* et al. [4.2] determined a value 0.93 ± 0.07 for argon on stainless steel. With $h = 1$, the accommodation length of argon at 273 K and 100 kPa is 118 nm.

In a recent monograph on applications of the Boltzmann equation, *Cercignani* [4.34] discusses the temperature jump more broadly. His formulation includes additional parameters, including some which describe a discontinuity in the velocity field near a solid boundary. The effects of including the additional terms in the acoustics boundary condition have not yet been investigated.

The fifth boundary condition is a statement of continuity of heat flow at the gas-shell interface. The five boundary conditions constitute a set of equations which can be solved exactly for the spherical geometry. For details see [4.2,29]. A series expansion of the exact solution shows that the complex eigenfrequencies are given to first order in the fractional frequency perturbation $\Delta F/f$ by $F_N = f^0_{ln} + \Delta f_N + \mathrm{i}g_N$, where

$$\Delta f_N = \Delta f_t + \Delta f_v + \Delta f_{\mathrm{sh}} \tag{4.4.8}$$

$$g_N = g_t + g_v + g_{\mathrm{bulk}} \tag{4.4.9}$$

include the contributions from shell motion (Δf_{sh}), thermal waves ($\Delta f_t + \mathrm{i}g_t$), shear (viscous) waves ($\Delta f_v + \mathrm{i}g_v$), and bulk dissipation (g_{bulk}). Each of these contributions arises from either bulk phenomena or a uniform boundary effect. There are additional eigenfrequency perturbations whose effects cannot be calculated within an exact formalism. Such effects, which include imperfect spherical geometry, openings in the shell wall, and transducer coupling, can be treated using boundary perturbation theory. Let $\beta(\omega, \mathbf{r}) = \rho c u/p$ be the specific acoustic admittance of the surface at boundary point $\mathbf{r}$, and let $\Delta\beta(\omega, \mathbf{r})$

be the fraction of the specific acoustic admittance which must be treated using boundary perturbation theory. The fractional eigenfrequency shift is [4.30]

$$\frac{\Delta F_N}{f_{ln}^0} = \frac{i\omega a^2}{2cz_{ln}^2} \frac{\int_S \Psi_N^2 \Delta\beta(\omega, \mathbf{r})\, dS}{\int_V \Psi_N^2\, dV}. \tag{4.4.10}$$

The integral in the numerator is over the region of the boundary perturbation, and the integral in the denominator is over the volume of the resonator. In this and other equations for frequency perturbations, any explicit frequency terms on the right side should be evaluated at the drive frequency for steady-state measurements and at the eigenfrequency for free-decay experiments. The next series of sections describes in detail the important frequency perturbations.

4.5 Thermal and Viscous Boundary Layers

The thermal wave near the boundary is the source of the terms Δf_t and g_t, which are most conveniently expressed as fractions of the unperturbed frequency:

$$\begin{aligned}(\Delta f_t + ig_t)/f_{ln}^0 = {} & \\ M_{ln}\left[(-1+i)\frac{\gamma-1}{2}\frac{\delta_t}{a} + (\gamma-1)\frac{l_a}{a} + (1+i)\frac{\gamma-1}{2}\frac{\delta_{sh}}{a}\frac{\lambda_g}{\lambda_{sh}}\right]. &\end{aligned} \tag{4.5.1}$$

The factor

$$M_{ln} \equiv z_{ln}^2/[z_{ln}^2 - l(l+1)] \tag{4.5.2}$$

is unity for radial modes and approaches unity as n becomes large for nonradial modes. For modes with $l < 10$ it is significantly greater than unity only for $n = 1$. This means that the thermal boundary effects on the first member of any nonradial series are significantly larger than that for radial modes only for the first member of the series. For the higher-frequency modes the thermal boundary layer term approaches the expression for a radial mode at the same frequency. Within the square brackets on the right side of (4.5.1), the first term is the one obtained if the temperature of the gas and shell match; the second term accounts for imperfect thermal accommodation at the boundary; and the third term accounts for penetration of the thermal wave into the shell. In (4.5.1) λ_{sh}, δ_{sh}, and $k_{t,sh} = (1-i)/\delta_{sh}$ are the thermal conductivity, thermal penetration length, and propagation parameter for thermal waves in the shell, respectively. For argon at 273 K, 100 kPa, with the 8.89 cm radius resonator used in [4.2], the magnitudes of the three terms at the frequency of the 0, 2 radial mode (2476 Hz) are

$$\frac{\gamma-1}{2}\frac{\delta_t}{a} \approx 1.8\times10^{-4} \tag{4.5.3}$$

$$(\gamma-1)\frac{l_a}{a} \approx 8.8\times10^{-7} \tag{4.5.4}$$

$$\frac{\gamma-1}{2}\frac{\delta_{sh}}{a} \approx 1.0\times10^{-7}. \tag{4.5.5}$$

The first of these is proportional to $1/\sqrt{fP}$, the second to $1/P$, and the third to $1/\sqrt{f}$. The relative importance of the second term increases at pressures below 100 kPa.

The dominant contribution to the thermal boundary layer perturbation (4.5.1) is proportional to $3\delta_t/a$, the fraction of the resonator volume where the thermal wave is significant. In this region the phase relation between the particle velocity and the pressure differs from that of a pure adiabatic sound wave. This phase shift is directly responsible for the real part of the frequency perturbation and, less obviously, to the imaginary part. *Herzfeld* [4.33] pointed out that the dissipation described by the imaginary term occurs through heating of the gas through compression. In normal sound propagation the temperature and density oscillations are in phase and the net work per cycle is zero. The altered phase relations in the boundary layer lead to a finite loss.

The major contributions to g_t/f and $\Delta f_t/f$ are equal to $(\gamma-1)\delta_t/(2a)$. In [4.2] this quantity, for the 0,2 mode, varied from 340 ppm at 25 kPa to 76 ppm at 500 kPa. For the 0,6 mode it was lower by a factor of 0.47. These are typical magnitudes for the thermal boundary layer effects. Most experiments are done with somewhat smaller resonators. For a given mode, the effect is inversely proportional to the square root of the cavity radius. For polyatomic gases it is generally smaller because of the smaller factor $(\gamma-1)$.

The viscous boundary layer perturbation is similar in functional form and magnitude to the dominant term in (4.5.1)

$$\frac{\Delta f_v + \mathrm{i} g_v}{f^0_{ln}} = \frac{-1+\mathrm{i}}{2}\frac{\delta_v}{a}(1-M_{ln}). \tag{4.5.6}$$

It vanishes, as expected, for radial modes. The factor $(1-M_{ln})$ approaches zero for large n, so that viscous boundary layer phenomena are only important for

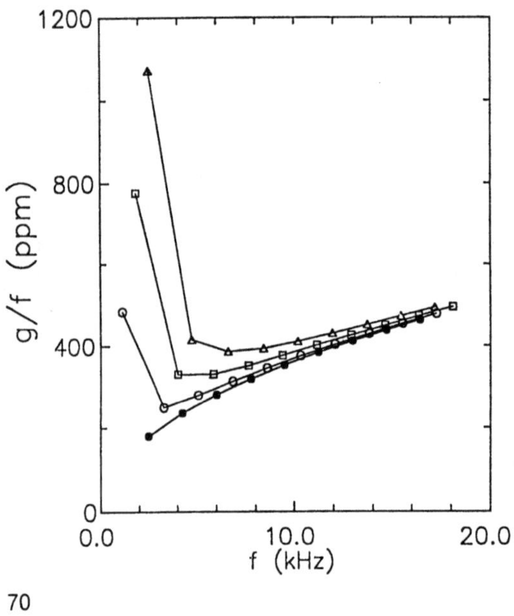

Fig. 4.3. Major boundary layer contribution to g/f as a function of frequency, for an 8.89 cm sphere filled with argon at 273 K and 100 kPa. •, $l=0$; ∘, $l=1$; □, $l=2$; ▽, $l=3$

nonradial modes with small n. The major boundary layer contribution to g/f, the sum of the dominant imaginary term in (4.5.1) and the imaginary part of (4.5.6), is shown in Fig. 4.3 as a function of frequency. Note that there are significantly greater losses only for $n = 1$, and that the losses approach those of radial modes in the high-frequency region.

4.6 Precondensation Effects

When the experimental gas is a vapor at a pressure near saturation a thin liquid film coats the cavity walls. Reversible condensation and evaporation occur during an acoustic cycle. This dramatically alters the boundary condition which couples the acoustic and thermal waves. *Mehl* and *Moldover* [4.10] developed a model for these effects which they showed was in qualitative agreement with measurements made in propane at 288 K. They observed excessive surface contributions to $\Delta f/f$ and g/f. With a 6.35 cm radius aluminum resonator they observed excess values of g/f which increased with pressure, reaching a value of 0.2% at a pressure within 0.5% of saturation. The excess value of $\Delta f/f$ reached a values of 0.1%. Both figures apply to the 0,2 mode; the effects vary with frequency. Similar results were found with a 3.17 cm brass spherical resonator; considerably larger effects were observed in a still smaller cylindrical resonator.

4.7 Bulk Dissipation and Relaxation

Classical bulk dissipation is described by the imaginary terms in (4.4.1). The contribution to the fractional halfwidth is

$$g_{\text{bulk}}/f = (\pi f/c)^2 \left[(\gamma - 1)\delta_t^2 + \delta_v'^2\right]. \tag{4.7.1}$$

For argon at 273 K this equals approximately

$$g_{\text{bulk}}/f = 18.2 \times 10^{-6}(f/10\ \text{kHz})(100\ \text{kPa}/P). \tag{4.7.2}$$

For polyatomic gases there is a contribution to η_b in the δ_v' term due to vibrational relaxation. In the approximation of a single relaxation time τ associated with a relaxation specific heat ΔC the relaxation contribution to the fractional halfwidth is, when $\omega\tau \ll 1$,

$$g_{\text{rel}}/f = \pi f \tau (\Delta C/R)(\gamma - 1)^2/\gamma. \tag{4.7.3}$$

In the low-frequency region ($\omega\tau \ll 1$) the relaxation correction to the eigenfrequencies is proportional to $(\omega\tau)^2$ and thus often negligible compared with the loss term (4.7.3). In ethylene in the range 273–373 K, 0.1–100 MPa, relaxation losses greatly exceeded the classical bulk losses and the losses due to the thermal boundary layer [4.9]. Acoustical data taken with a 6.35 cm spherical resonator at frequencies below 13 kHz were used to determine a relaxation time τ which was in close agreement with independent ultrasonic measurements taken at far higher frequencies. The measurements, while far less accurate than those of more recent work with spherical resonators, were also capable of resolving the pressure dependence of the quantity $(\gamma - 1)^2/\gamma$.

4.8 Shell Motion

The effects of shell motion on the eigenfrequencies have been discussed by *Greenspan* [4.29], *Mehl* [4.26], and *Moldover* et al. [4.29]. The effects are small except when there is an accidental near-coincidence of acoustic and shell eigenfrequencies. The theories treat isotropic elastic shells which are isolated from the outside environment. The effects of radiation from the outside surface of the shell have however been included and shown to be negligible for cases where the shell is surrounded by a low-density gas, providing that there is no near-coincidence of acoustic and shell eigenfrequencies. Mechanical radiation through the support structures is probably more important; this has not been formally treated.

The frequency perturbation is given by

$$\frac{\Delta f_{\text{sh}}}{f_{ln}^0} = \frac{-\mathrm{i}\beta_{sh}/z_{ns}}{1 - l(l+1)/z_{ln}^2} = \frac{\rho_g c_g^2}{\rho_{\text{sh}}(c_{\text{sh}}^l)^2} \frac{S_l(k_{\text{sh}}^l a)}{1 - l(l+1)/z_{ln}^2}. \tag{4.8.1}$$

Here c_{sh}^l is the longitudinal speed of sound in the shell material, ρ_{sh} is the shell density, and $k_{\text{sh}}^l = \omega/c_{\text{sh}}^l$. The dimensionless admittance function $S_l(k_{\text{sh}}^l a)$ is of order unity except near shell resonances. It depends on the frequency, the elastic constants, and the dimensions of the shell. Let b be the outer radius of the shell, c_{sh}^t the speed of transverse waves in the shell, and k_{sh}^t the propagation parameter for transverse waves in the shell, $\mathcal{A} = k_{\text{sh}}^l a$, and $\mathcal{B} = k_{\text{sh}}^l b$. The transverse and longitudinal quantities are related through

$$\chi \equiv \frac{k_{\text{sh}}^t}{k_{\text{sh}}^l} = \frac{c_{\text{sh}}^l}{c_{\text{sh}}^t} = \sqrt{\frac{2(1-\sigma)}{1-2\sigma}}, \tag{4.8.2}$$

where σ is Poisson's ratio. The admittance function $S_l(k_{\text{sh}}^l a)$ can be calculated as follows [4.26]. The first step is solution of the linear equations

$$\mathbf{AB} = \mathbf{C} \tag{4.8.3}$$

for the vector $\mathbf{B}$. The elements of the first and third rows of the matrix $\mathbf{A}$ are

$$\begin{aligned} A_{11} &= [2l(l+1)/\mathcal{A} - \chi^2\mathcal{A}]j_l(\mathcal{A}) + 4j_{l+1}(\mathcal{A}) \\ A_{12} &= [2l(l+1)/\mathcal{A} - \chi^2\mathcal{A}]y_l(\mathcal{A}) + 4y_{l+1}(\mathcal{A}) \\ A_{13} &= 2l(l+1)[(l-1)j_l(\chi\mathcal{A})/(\chi\mathcal{A}) - j_{l+1}(\chi\mathcal{A})] \\ A_{14} &= 2l(l+1)[(l-1)y_l(\chi\mathcal{A})/(\chi\mathcal{A}) - y_{l+1}(\chi\mathcal{A})] \\ A_{31} &= (l-1)j_l(\mathcal{A})/\mathcal{A} - j_{l+1}(\mathcal{A}) \end{aligned} \tag{4.8.4}$$

$$A_{32} = (l-1)y_l(\mathcal{A})/\mathcal{A} - y_{l+1}(\mathcal{A})$$
$$A_{33} = [(l^2-1)/(\chi\mathcal{A}) - \chi\mathcal{A}/2]j_l(\chi\mathcal{A}) + j_{l+1}(\chi\mathcal{A})$$
$$A_{34} = [(l^2-1)/(\chi\mathcal{A}) - \chi\mathcal{A}/2]y_l(\chi\mathcal{A}) + y_{l+1}(\chi\mathcal{A}),$$

where the functions y_l are spherical Neumann functions. The elements of the second and fourth rows of $\mathbf{A}$ are obtained from the first and third rows by substitution of $\mathcal{B}$ for $\mathcal{A}$. The first element of the vector $\mathbf{C}$ is unity and the others are zero. The admittance function is given in terms of the solution of (4.8.3) by

$$S_l(k^l_{\rm sh}a) = \chi^2 \sum_{i=1}^{4} B_i D_i, \tag{4.8.5}$$

where

$$\begin{aligned} D_1 &= lj_l(\mathcal{A})/\mathcal{A} - 4j_{l+1}(\mathcal{A}) \\ D_2 &= ly_l(\mathcal{A})/\mathcal{A} - 4y_{l+1}(\mathcal{A}) \\ D_3 &= l(l+1)j_l(\chi\mathcal{A})/(\chi\mathcal{A}) \\ D_4 &= l(l+1)y_l(\chi\mathcal{A})/(\chi\mathcal{A}). \end{aligned} \tag{4.8.6}$$

Numerical calculations of the shell admittance for the 8.89 cm resonator with $b/a = 1.215$ are shown in Fig. 4.4. The shell admittance diverges at the shell resonances, where the determinant of the matrix $\mathbf{A}$ vanishes. Figure 4.5 shows the dependence of the latter on the ratio b/a for stainless-steel shells.

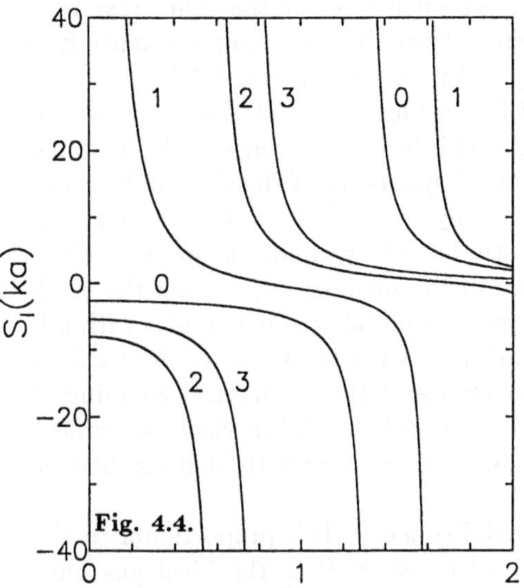

Fig. 4.4. Shell admittance factor $S_l(k^l_{\rm sh}a)$ for a stainless-steel ($\sigma = 0.297$) shell with $b/a = 1.215$. The curves are labeled with the mode index l. The admittance factor diverges at the shell resonances

Fig. 4.5. Shell resonances as a function of b/a for stainless-steel shells. Conventional frequency units can be obtained by multiplying the dimensionless quantities $k^l_{\rm sh}a$ by (10.4 kHz)(a/8.89 cm). The curves are labeled with the mode index l. The $l = 0$ and $l = 1$ modes are extensional; the other modes are flexural. Extensional modes for indices l = 2–5 occur at higher frequencies

The results in both figures are only weakly dependent on Poisson's ratio (0.297 for stainless steel), and thus apply qualitatively to other shell materials. The resonance frequencies parameters $k_{sh}^l a$ depend only on the quantities b/a and σ and thus can be applied to shells of arbitrary radius.

Closed form expressions for the admittance function have not been found except for the radial case, for which [4.24,26,29]

$$S_0 = -\chi^2 \frac{(4 + 4\mathcal{A}\mathcal{B} - \chi^2\mathcal{B}^2)\tan(\mathcal{H}) - 4\mathcal{H} - \mathcal{A}\chi^2\mathcal{B}^2}{[(\chi^2\mathcal{A}^2 - 4)(\chi^2\mathcal{B}^2 - 4) + 16\mathcal{A}\mathcal{B}]\tan\mathcal{H} - 4\mathcal{H}(4 + \chi^2\mathcal{A}\mathcal{B})}, \tag{4.8.7}$$

where a thickness parameter $\mathcal{H} = \mathcal{B} - \mathcal{A}$ has been introduced. The radial ($l = 0$) shell resonances occur at frequencies for which the denominator of (4.8.7) vanishes. Normally only the lowest, or *breathing* resonance is of interest; its frequency will be designated f_{br}. The 8.89 cm resonator has a breathing resonance of approximately 13.6 kHz. In the low-frequency region an excellent approximation to (4.8.7) is obtained by dividing the zero-frequency limit by a suitable resonance denominator

$$S_0 = -(\chi^2/4)\frac{3\chi^2 ab^2 - 3bh(4a - \chi^2 b) - 4h^3}{3\chi^2 h(a^2 + ab + b^2) - 12abh - 4h^3}\,\frac{1}{1 - (f/f_{br})^2}. \tag{4.8.8}$$

For the 8.89 cm resonator this approximation lies below the exact result by an amount which increases with frequency. Near the breathing resonance it is about 3% low; at the highest frequency in Fig. 4.4 it is about 7% low.

Note that the elements A_{ij}, B_i, and D_i, and hence the admittance function, are all independent of the mode index m. This is a consequence of the assumed isolation of the spherical shell. The assumed symmetry of the shell and its surroundings implies that an acoustic mode only drives shell oscillations with the same angular factor Y_{lm}. A more realistic boundary condition would include the effects of the mechanical impedance of the member supporting the shell. This would introduce a dependence on m. It would also couple modes with different indices through the boundary condition. A full treatment of these effects has not yet been worked out. The importance of the mechanical coupling of the shell to the external environment should probably be checked experimentally in applications. This can be done by experimenting with high compliance suspensions, for example.

Equation (4.8.1) predicts a fractional frequency shift proportional to the shell admittance factor. The quantity ρc^2 equals $\gamma_0 P$ in the ideal gas limit (with γ_0 equal to the ideal-gas specific heat ratio). The quantity $\rho_{sh}(c_{sh}^l)^2$ is approximately 1.11×10^{11} Pa for aluminum and 2.63×10^{11} Pa for stainless steel. Thus, for a monatomic gas ($\gamma_0 = 5/3$) at a pressure of 100 kPa, the prefactor in (4.8.1) is 1.5×10^{-6} for aluminum and 6.34×10^{-7} for stainless steel. In the low-frequency limit the admittance factor is negative for all modes except those with $l = 1$. For an aluminum shell with $b/a = 4/3$ and a 5 cm radius, the fractional correction for the radial modes is approximately -4 ppm at low frequencies. For the 8.89 cm stainless-steel sphere the fractional correction at low frequencies is −1.7 ppm. Thus the frequency perturbations of the radial modes due to shell motion are small for gases at moderate pressures, except for frequencies near the breathing resonance of the shell, where large perturbations of the real and imaginary parts of the eigenfrequencies are observed [4.2,29]. The $l = 1$ nonradial modes exert a net (oscillatory) pressure on the shell which drives translational motion at low frequencies. This effect is responsible for the zero-frequency divergence in Fig. 4.4. Measurements on the $l = 1$ triplets [4.29] over a wide frequency range are in good agreement with the predictions of (4.8.1).

4.9 Imperfect Spherical Geometry

The shape of experimental resonators differs somewhat from the ideal spherical geometry. When a spherical cavity is fabricated from two separately machined parts of nearly hemispherical shape two types of errors are likely. The individual parts are generally highly axisymmetric due to the turning process. They must, however, be cut to the proper length and assembled. This is likely to lead to errors in the alignment of the axes of the two parts, and to some differences between the "polar" and "equatorial" diameters of the cavity.

Let the cavity shape be described by

$$r(\theta, \phi) = a[1 - \epsilon\mathcal{F}(\theta, \phi)], \tag{4.9.1}$$

where ϵ is a small parameter and the shape function $\mathcal{F}$ is of order unity. *Greenspan* [4.24] made a general argument, based on the Ehrenfest adiabatic principle, that the frequencies of the radial modes of resonators of equal volume should not differ by more than terms of order ϵ^2. He also calculated the frequencies of a spheroidal resonator by developing a series approximation for the exact solution (expressed in spheroidal wave functions). He found

$$(\Delta f_{\text{geom}}/f)_{0n} = (4/135)z_{0n}^2\epsilon^2 + O(\epsilon^3). \tag{4.9.2}$$

Mehl [4.25,27] applied boundary shape perturbation theory [4.35] to the problem. Let the shape function be expressed as a series of (complex) spherical harmonics

$$\mathcal{F}(\theta, \phi) = \sum_{l=0}^{\infty}\sum_{m=-l}^{l} c_{lm}Y_{lm}(\theta, \phi). \tag{4.9.3}$$

Then, for the radial modes, the geometric eigenfrequency perturbation is given by

$$\left(\frac{\Delta f_{\text{geom}}}{f}\right)_{0n} = -\frac{\epsilon^2}{2\pi}\sum_{l'=2}^{\infty} T_{nl'} \sum_{m'=-l'}^{l'} |c_{l'm'}|^2 + O(\epsilon^3). \tag{4.9.4}$$

The coefficients T_{nl} are given by

$$T_{nl} = (z_{0n}/2)j_l(z_{0n})/j_l'(z_{0n}) = (z_{0n}^2/2)/[\mathcal{R}_l(z_{0n}) - l - 1] \tag{4.9.5}$$

where

$$\mathcal{R}_l(z_{0n}) = z_{0n} j_{l-1}(z_{0n})/j_l(z_{0n}) = z_{0n}^2/[2l - 1 - \mathcal{R}_{l-1}(z_{0n})]. \tag{4.9.6}$$

The second form for $\mathcal{R}_l$ comes from recurrence formulas for the spherical Bessel functions. Starting with $j_1(z_{0n}) = 0$ it is possible to determine simple forms for coefficients with low l. For example

$$T_{n2} = -z_{0n}^2/6 \tag{4.9.7}$$
$$T_{n3} = (5/2)z_{0n}^2/(z_{0n}^2 - 20) \tag{4.9.8}$$
$$T_{n4} = -(z_{0n}^2/10)(z_{0n}^2 - 35)/(2z_{0n}^2 - 35). \tag{4.9.9}$$

The first of these, with the choice $\mathcal{F} = \cos^2\theta$ for a spheroidal shape, reproduces *Greenspan*'s result (4.9.2). Numerical calculations of the T_{nl} can be carried out for other cases; some results are tabulated in [4.25]. The coefficients of ϵ^2 in (4.9.4) for some simple shape deformations are also listed. They are typically on the order of unity. With excellent machine work cavities have been constructed with $\epsilon \sim 3 \times 10^{-4}$; the radial eigenfrequencies of such cavities should not differ from those of a perfect spherical cavity of the same volume by more than an amount of order 10^{-6}.

The nonradial modes are sensitive to shape to order ϵ. This can be understood qualitatively by considering the $1, 1, m$ triplet. The angular dependence of the acoustic pressure for the $1, 1, 0$ case is proportional to $\cos\theta$. The gas motion for this mode is mainly along the "polar" axis of symmetry; the frequency is mainly dependent upon the cavity length in that direction. The other two components have orthogonal symmetry axes perpendicular to that of the first; their eigenfrequencies are mainly dependent on the "equatorial" diameter of the cavity. A volume-preserving spheroidal deformation which lengthens the polar diameter must decrease the equatorial diameter by half as much. The polar-symmetric mode would thus have its frequency decreased; the other two modes would have their frequency increased by half as much. The mean frequency of the triplet would remain constant to order ϵ.

Boundary shape perturbation theory was applied to the nonradial modes in [4.27]. A calculation to order ϵ for the lnm modes requires diagonalization of a matrix of dimension $2l + 1$. The elements of the matrices depend on the $c_{l'm'}$; only the coefficients with even l' in the range $0 \le l' \le 2l$ contribute. The diagonalization of the matrix determines new (and real) combinations of the angular factors Y_{lm} so that m may not be a good mode index in the presence of a shape perturbation. The fractional eigenfrequency perturbations are equal to the eigenvalues of the matrix whose components are

$$\Lambda_{mm'}^{ln} = \epsilon \sum_{l''=0}^{2l} \sum_{m''=-l''}^{l''} c_{l''m''} \left[1 + \frac{l''(l''+1)/2}{z_{ln}^2 - l(l+1)}\right] \langle lm|Y_{l''m''}|lm'\rangle, \tag{4.9.10}$$

where the matrix elements of the spherical harmonics are

$$\langle lm|Y_{l''m''}|lm'\rangle = \int Y_{lm}^* Y_{l''m''} Y_{lm'}\, d\Omega. \tag{4.9.11}$$

For $l = 1$ modes this matrix has contributions from c_{00} and the $c_{2m''}$. The former describe a change of resonator volume which affect all modes equally. The other terms contain the relevant information about shape. Their contribution to the matrix is given by

$$A^{1n} = \frac{\epsilon}{4\sqrt{5\pi}}(1 + M_{1n})\mathbf{C}_1^2, \tag{4.9.12}$$

where

$$\mathbf{C}_1^2 = \begin{pmatrix} -c_{20} & \sqrt{3}c_{21} & -\sqrt{6}c_{22} \\ \sqrt{3}c_{21} & 2c_{20} & -\sqrt{3}c_{21} \\ -\sqrt{6}c_{22} & -\sqrt{3}c_{21} & -c_{20} \end{pmatrix}. \tag{4.9.13}$$

The symmetry of $\mathbf{C}_1^2$ comes from the requirement that $\mathcal{F}$ be real; its eigenvalues are

$$\begin{aligned} \lambda_0 &= -(c_{20} - \sqrt{6}c_{22}) \\ \lambda_\pm &= (c_{20} - \sqrt{6}c_{22})/2 \\ &\quad \pm\sqrt{(c_{20} - \sqrt{6}c_{22})^2/4 + 2c_{20}^2 + 6c_{22}^2 - 2\sqrt{6}c_{20}c_{22}}. \end{aligned} \tag{4.9.14}$$

The eigenvalues, and hence the perturbed frequencies, depend only on the combinations $c_{20} - \sqrt{6}c_{22}$ and $c_{20}^2 + 3c_{22}^2 - \sqrt{6}c_{20}c_{22}$. The fractional eigenfrequency perturbations are

$$(\Delta f_{\text{geom}}/f)_{1n} = (\epsilon/4\sqrt{5\pi})(1 + M_{1n})\lambda. \tag{4.9.15}$$

To gain some insight into these terms consider a cavity constructed of two parts which, when their axes are aligned, have an inner surface described by a term in $\epsilon\mathcal{F}$ equal to the Legendre polynomial $\epsilon_0 P_2(\cos\theta)$; i.e. the polar radius is decreased by a fractional amount ϵ_0 while the equatorial radius is increased by half this amount. The effect of transverse displacement of the axes by $2\epsilon_1 a$ of the two parts can be described by a term $(5/8)\epsilon_1 P_2^1(\cos\theta)\cos\phi$. This deformation corresponds to a particular choice of the coefficients c_{20} and c_{21}. With this choice the fractional eigenvalue perturbations in (4.9.15) are

$$\left(\frac{\Delta f_{\text{geom}}}{f}\right)_{1n} = \frac{\epsilon_0}{10}(1 + M_{1n})\left\{\begin{array}{l} -1 \\ (1/2) \pm (3/2)\sqrt{1 + (25/16)(\epsilon_1/\epsilon_0)^2}. \end{array}\right. \tag{4.9.16}$$

When the axes are aligned each degenerate triplet is split into a doublet and singlet with a fractional frequency separation of $3\epsilon_0/5$ for the higher-n modes for which $M_{1n} \approx 1$. The doublet is split by the ϵ_1 term. With the transducer locations used in [4.29] all components of the $l = 1$ triplets were weak but resolvable. The fractional frequency shifts were observed to be approximately +190, −70 and −120 ppm. With a different choice of transducer locations in [4.2] only a doublet could be observed; the splitting was 215 ppm. Microwave modes are also perturbed by imperfect geometry. The frequency perturbations of the microwave triplets are proportional to the perturbations of the acoustic triplets [4.36]. Thus, microwave measurements can be used as a check on the geometric interpretation of the triplet splitting. Measurements [4.37] on the microwave TM-11 and TM-12 triplets yield values of the parameter ϵc_{20} of

−550 and −546 ppm; measurements on the 1,3 and 1,8 acoustic triplets yield the values −548 and −563 ppm. The agreement of these numbers strongly supports a geometric interpretation of the splittings. However, an unresolved difficulty of this interpretation is that a term ϵc_{20} will, according to (4.9.4), contribute a term $(\epsilon c_{20} z_{0n})^2/6\pi$ to the fractional frequency perturbations of the radial modes. With $\epsilon c_{20} \approx 5.5 \times 10^{-4}$ there should be a relative shift of the 0,6 and 0,2 modes of about 2 ppm. An effect of this size should have been observable in the data of [4.2]. A possible explanation is the contributions of other terms in (4.9.4). More work with resonators with larger, known deformations will perhaps clarify this.

The mean eigenfrequency perturbation of any acoustic multiplet has been shown theoretically [4.27] and experimentally [4.38] to be insensitive to shape perturbations. A similar theoretical argument applies to the microwave multiplets. Formally, the mean frequencies of these multiplets, like the frequencies of the radial acoustic modes, are sensitive to geometry only in order ϵ^2 and higher. Thus, except for the experimental inconvenience in dealing with nearly degenerate modes, the mean frequencies of the multiplets are as useful as the radial acoustic modes in determining propagation speeds.

4.10 Ducts and Slits in the Shell Wall

Some resonator designs include ducts through the shell wall for filling and/or coupling to acoustic transducers. Coupling to remote transducers will probably be required in applications of spherical resonators at high temperatures. The effects of ducts through the shell wall can be calculated using boundary perturbation theory. Equation (4.4.10) gives the fractional eigenfrequency shifts in terms of the input admittances of the ducts. For the radial modes and a single duct of cross-sectional area A_{duct} the eigenfrequency perturbation is

$$[(\Delta f + \mathrm{i}g)_{\text{duct}}/f]_{0n} = (\mathrm{i}\beta_{\text{duct}}/z_{0n})A_{\text{duct}}/4\pi a^2, \tag{4.10.1}$$

which will be on the order of the ratio of the duct area to the sphere area for duct admittances of order unity.

Sound propagation in a circular duct has a propagation parameter given by the Kirchhoff-Helmholtz theory [4.32] as

$$k_{\text{KH}} = (\omega/c)\{1 + (1-\mathrm{i})[(\gamma-1)\delta_t + \delta_v]/2r_h\}, \tag{4.10.2}$$

where r_h is the duct radius. Let β_0 be the terminal specific admittance of a duct of length L. Then the input specific admittance of the duct β_{duct} is given by

$$\mathrm{i}\beta_{\text{duct}} = (\mathrm{i}\beta_0 - \tan k_{\text{KH}}L)/(1 + \mathrm{i}\beta_0 \tan k_{\text{KH}}L). \tag{4.10.3}$$

For ducts used for filling the cavity which couple to the larger space of a sample chamber, $\mathrm{i}\beta_0$ can be estimated by assuming a radiation boundary condition appropriate for an opening with an infinite flange:

$$i\beta_{rad} = [8(\omega r_h/c)/3\pi - i(\omega r_h/c)^2/2]^{-1}. \tag{4.10.4}$$

Under typical conditions where $|i\beta_{rad}| \gg |\tan k_{KH}L|$, the right-hand side of (4.10.3) simplifies to $\cot k_{KH}L$, which diverges at the open-pipe resonances of the duct. Ducts which couple the cavity to transducers have a termination which can be modeled by an effective volume parameter V_{eff}. The corresponding terminal admittance is

$$i\beta_V = -\omega V_{eff}/\pi r_h^2 c. \tag{4.10.5}$$

Under typical experimental conditions the duct dimensions can be chosen so that the magnitudes of both the real and imaginary parts of $i\beta_{duct}$ are of order unity except for frequencies near the resonance frequencies of the ducts. *Goodwin* [4.14] has pointed out the advantages of using a duct of length $L \approx a$ to minimize accidental coincidences of the frequencies of duct and sphere resonances. For ducts terminated by an effective volume the lowest such frequency is that of the Helmholtz resonance of the volume coupled to the duct. Small volumes enhance transducer sensitivity but also tend to raise this Helmholtz resonance into the range of the spherical-cavity resonances.

The acoustic admittance of a slit has been calculated by *Trusler* [4.39]. For a slit of depth D, rigidly terminated at the end, the result is

$$\beta_{slit} = \frac{(1+i)\sqrt{3\gamma}}{6\delta_v/d_{slit}} \tanh[(1+i)(\delta_v/d_{slit})\omega D\sqrt{3\gamma}/c]. \tag{4.10.6}$$

This expression can be used to estimate the effects of the annular slots surrounding plugs and/or transducers fitted closely in holes through the cavity wall. It can also be used to estimate the effect of a small opening at the junction of the two hemispheres.

4.11 Measurement of the Speed of Sound

Measurements of a set of acoustic eigenfrequencies as a function of pressure at a fixed temperature provide data from which the speed of sound $c(T, P)$ can be determined to high accuracy. The first step in the data reduction is the correction of the eigenfrequencies for the various frequency perturbations. This requires information about kinetic coefficients like the thermal conductivity and (for nonradial modes) the viscosity, and about the mechanical properties of the shell. The accommodation length term in the thermal boundary layer correction must be determined from the data, however. The corrected experimental eigenfrequencies should then equal $cz_{ln}/2\pi a$. The frequencies of the radial modes and the mean frequencies of nonradial multiplets are insensitive to the shell geometry. Accurate values of the ratio c/a can be obtained from the corrected radial eigenfrequencies and/or the mean frequencies of multiplets. To obtain the speed of sound from this ratio the mean cavity radius a must be determined.

In [4.1,2] the cavity volume was determined by filling it with mercury whose mass was separately determined. This tedious process becomes increasingly difficult at temperatures far from ambient. Microwave techniques are currently being developed as a more practical way of determining the mean cavity radius as a function of temperature [4.37,38]. For many measurements it is practical to use a calibration gas, such as argon, for which the speed of sound is well known, to determine the mean radius.

A set of $c(T,P)$ determinations at a fixed T can be analyzed by representing the speed of sound in a series

$$c^2 = A_{-1}/P + A_0 + A_1 P + A_2 P^2 + \cdots, \tag{4.11.1}$$

in which $A_0 = \gamma_0 RT/M$ is the ideal gas result. The term A_{-1} is introduced because the accommodation parameter in the thermal-boundary-layer correction must be determined experimentally. This term is not intended as part of an analytic representation of the speed of sound. It is only important for measurements of very high accuracy at lower pressures. In [4.1,2] the data set consisted of five determinations of c^2 in argon at fourteen pressures. In fitting (4.11.1) to the data the low pressure measurements were given lower weight because of poorer signal-to-noise ratio. Figure 4.6 shows the fractional deviations of the data from the best fit to the polynomial (4.11.1). Note that the five measurements at each pressure above 100 kPa typically agree to within 1 ppm in c^2. In [4.1,2] the experimental value of A_0 was used to determine the gas constant R in an experiment near the triple point. The random uncertainty in $(c/a)^2$, expressed as a standard deviation, contributed only 0.7 ppm to the uncertainty of 1.7 ppm assigned to R [4.1,2]. The main systematic errors in $(c/a)^2$ in that work resulted from imperfect knowledge of the transducer positions (0.55 ppm) and from the uncertainty in the thermal-boundary-layer perturbation (0.3 ppm).

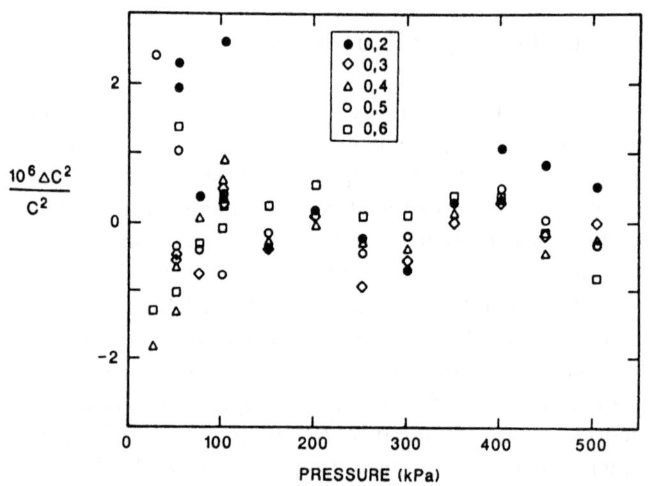

Fig. 4.6. Fractional deviations of 70 observations of c^2 from the best fit to the polynomial (4.11.1). The symbols used to identify the modes are indicated in the inset

(The latter uncertainty is due to the uncertaintly in the thermal conductivity argon.) Similar speed-of-sound measurements at other temperature are being used in the development of more accurate temperature scales [4.17].

4.12 Thermophysical Information from the Speed of Sound

From measurements in polyatomic gases the ideal gas specific heat ratio $\gamma_0(T)$ can be obtained from the coefficient A_0. From this, values of the specific heat with a typical accuracy on the order of 0.1% can be obtained [4.11–15]. The parameter A_1 is related to the coefficient B of the virial equation of state through

$$MA_1 = 2\gamma_0 B + 2\gamma_0(\gamma_0 - 1)T\frac{dB}{dT} + (\gamma_0 - 1)^2 T^2 \frac{d^2B}{dT^2}. \tag{4.12.1}$$

This is a differential equation in T with variable coefficients. It can be integrated numerically to determine $B(T)$ if two constants of integration can be determined. This process can yield values of $B(T)$ accurate to better than 1 cm^3/mole. An alternative approach, explored by *Ewing* et al. [4.40] is to attempt to obtain an intermolecular potential directly from $A_1(T)$.

Acknowledgements: The authors are grateful to their late colleague Martin Greenspan for many inspiring discussions, and to Mike Ewing, Tony Goodwin, and Martin Trusler for sharing their unpublished results with us, and for useful discussions.

References

4.1 M. R. Moldover, J. P. M. Trusler, T. J. Edwards, J. B. Mehl, R. S. Davis: Measurement of the universal gas constant R using a spherical acoustic resonator. Phys. Rev. Lett. **60**, 249–252 (1988)

4.2 M. R. Moldover, J. P. M. Trusler, T. J. Edwards, J. B. Mehl, R. S. Davis: Measurement of the universal gas constant R using a spherical acoustic resonator. J. Res. N. B. S. **93**, 85–144 (1988)

4.3 J. W. S. Rayleigh: *Theory of Sound* (1896), 2nd ed. (Dover, New York, 1945) Sect. 331

4.4 R. W. Leonard: The attenuation of sound waves in water by a reverberation method. J. Acoust. Soc. Am. **18**, 252 (1946)

4.5 C. M. Harris: Effects of humidity on the velocity of sound in air. J. Acoust. Soc. Am. **49**, 890–893 (1971)

4.6 I. D. Campbell: Gaseous oscillations within a perturbed spherical resonator. Acustica **5**, 145–148 (1955)

4.7 D. Bancroft: Measurement of velocity of sound in gases. Am. J. Phys. **24**, 355–358 (1956)

4.8 M. R. Moldover, M. Waxman, M. Greenspan: Spherical acoustic resonators for temperature and thermophysical property measurements. High Temp. High Pressure **11**, 75–86 (1979)

4.9 J. B. Mehl M. R. Moldover: Precision acoustic measurements with a spherical resonator: Ar and C_2H_4. J. Chem. Phys. **74**, 4062–4077 (1981)

4.10 J. B. Mehl, M. R. Moldover: Precondensation phenomena in acoustic measurements. J. Chem. Phys. **77**, 455–465 (1982)
4.11 J. B. Mehl, M. R. Moldover: "Specific heat and virial coefficient measurements with a spherical acoustic resonator," *Proc. Eighth Symp. Thermophysical Properties*, ed. by J. V. Sengers, (Am. Soc. Mech. Eng., New York 1982) pp. 134–141
4.12 M. B. Ewing, A. R. H. Goodwin, M. L. McGlashan, J. P. M. Trusler: Thermophysical properties of alkanes from speeds of sound determined using a spherical resonator. 1. Apparatus, acoustic model, and results for dimethylpropane. J. Chem. Thermodyn. **19**, 721–739 (1987)
4.13 M. B. Ewing, A. R. H. Goodwin, M. L. McGlashan, J. P. M. Trusler: Thermophysical properties of alkanes from speeds of sound determined using a spherical resonator. 2. n-butane. J. Chem. Thermodynamics **20**, 243–256 (1988)
4.14 A. R. H. Goodwin: Thermophysical properties from the speed of sound. PhD Thesis, University of London (1988)
4.15 M. B. Ewing, J. P. M. Trusler: Speeds of sound in CF_4 between 175 and 300 K measured with a spherical resonator. J. Chem. Phys. **90**, 1106–1115 (1989)
4.16 M. R. Moldover, J. B. Mehl: "Spherical acoustic resonators: promising tools for thermometry and measurement of the gas constant," *Precision Measurements and Fundamental Constants II*, ed. by B. N. Taylor, W. D. Phillips (National Bureau of Standards, Washington 1984) pp. 281–286
4.17 M. R. Moldover, J. P. M. Trusler: Accurate acoustic thermometry I: The triple point of gallium. Metrologia **25**, 165–187 (1988)
4.18 R. Keolian, S. Garrett, J. Maynard, I. Rudnick: Acoustic measurement of isotopic concentration of gases. J. Acoust. Soc. Am. **64** (S1), 561 (1973)
4.19 R. Keolian, S. Garrett, J. Maynard, I. Rudnick: Acoustic measurement of isotopic concentration of gases. Bull. Am. Phys. Soc. **24**, 673 (1979)
4.20 J. S. Brooks, R. B. Hallock: Simple apparatus for concentration determinations in binary-gas mixtures. Rev. Sci. Instrum. **54**, 1199–1201 (1983)
4.21 P. Hess: *Resonant Photoacoustic Spectroscopy*, Vol. III, (Springer, Berlin, Heidelberg 1983) pp. 1–32
4.22 A. Karbach, J. Roeper, P. Hess: Computer controlled perfomance of photoacoustic experiments. Rev. Sci. Instrum. **44**, 892–895 (1984)
4.23 A. Karbach, P. Hess: Laser excitation of acoustic resonances in a spherical resonator. J. Appl. Phys. **58**, 3851–3855 (1985)
4.24 M. Greenspan: Notes on the spherical acoustic resonator (1978), unpublished
4.25 J. B. Mehl: Acoustic resonance frequencies of deformed spherical resonators. J. Acoust. Soc. Am. **71**, 1109–1113 (1982)
4.26 J. B. Mehl: Spherical acoustic resonator: effects of shell motion. J. Acoust. Soc. Am. **78**, 782–788 (1985)
4.27 J. B. Mehl: Acoustic resonance frequencies of deformed spherical resonators II. J. Acoust. Soc. Am. **79**, 278–285 (1986)
4.28 M. B. Ewing, M. L. McGlashan, J. P. M. Trusler: The temperature-jump effect and the theory of the thermal boundary layer for a spherical resonator. Speeds of sound in argon at 273.16 K. Metrologia **22**, 93–102 (1986)
4.29 M. R. Moldover, J. B. Mehl, M. Greenspan: Gas–filled spherical resonators: theory and experiment. J. Acoust. Soc. Am. **79**, 253–272 (1986)
4.30 P. M. Morse, K. U. Ingard: *Theoretical Acoustics* (McGraw-Hill, New York 1968) pp. 554–576
4.31 J. B. Mehl: Analysis of resonance standing-wave measurements. J. Acoust. Soc. Am. **64**, 1523–1525 (1978)
4.32 G. Kirchhoff: Über den Einfluss der Wärmeleitung in einem Gase auf die Schallbewegung. *Ann. Phys. Chem. (Fifth Ser.)* **134**, 177–193 (1868) [English translation "On the Influence of Heat Conduction in a Gas on Sound Propagation," by R. B. Lindsay in *Benchmark Papers in Acoustics: Physical Acoustics*, ed. by R. B. Lindsay (Dowden, Hutchinson, & Ross, Stroudsburg, PA 1974), pp. 7–19]

4.33 K. F. Herzfeld: Reflection of sound. Phys. Rev. **53**, 899–906 (1938)
4.34 C. Cercignani: *The Boltzmann Equation and its Applications* (Springer, New York 1988) pp. 248–260
4.35 P. M. Morse, H. Feshbach: *Methods of Theoretical Physics* (McGraw-Hill, New York 1953) pp. 1038–1064
4.36 J. B. Mehl, M. R. Moldover: Notes on spherical microwave resonators (1988), unpublished
4.37 M. B. Ewing, J. B. Mehl, M. R. Moldover, J. P. M. Trusler: Microwave measurements of the thermal expansion of a spherical cavity. Metrologia **25**, 211-219 (1989)
4.38 J. B. Mehl, M. R. Moldover: Measurement of the ratio of the speed of sound to the speed of light. Phys. Rev. A **34**, 3341–3344 (1986)
4.39 J. P. M. Trusler: Notes on the acoustic admittance of slits (1987), unpublished
4.40 M. B. Ewing, M. L. McGlashan, J. P. M. Trusler: Intermolecular forces from the speed of sound I. Theory. Mol. Phys. **60**, 681–690 (1987)

5. Laser Excitation of Acoustic Modes in Cylindrical and Spherical Resonators: Theory and Applications

M. Fiedler and P. Hess

With 25 Figures

The principles and applications of photoacoustic resonance spectroscopy in the frequency domain are reviewed. An experimental setup is described for the accurate computer-controlled detection of the resonance profiles of laser-excited acoustic modes in spheres and cylinders. This yields the resonance frequencies, halfwidths of the resonance profiles, and signal amplitudes as a function of pressure. A theoretical model is developed which allows a quantitative analysis of the experimental results. The application of the resonance method to different problems is discussed in detail. This includes the determination of accurate relaxation times for energy exchange between different degrees of freedom and of rate constants in chemical kinetics. The investigation of thermophysical properties is also considered. Finally, results are reported on the intracavity operation of acoustic resonators far below but also near the laser threshold.

5.1 Introduction

5.1.1 History

In 1880 *Bell* discovered the photoacoustic effect in solids [5.1]. He noticed that a rapidly interrupted beam of sunlight focused onto a solid substance produced an audible sound. The effect immediately attracted the interest of other researchers. *Röntgen* performed experiments with gases and showed that the photoacoustic effect is not confined to solids [5.2]. He also recognized the nature of the effect and gave a simple explanation: "We have not to do with any new property of the rays; the heating and expansion produced by absorption, and the subsequent cooling and contraction of the absorbing body, are the causes of the phenomena" [5.2]. *Tyndall* investigated not only gases, but also liquids and discovered important properties of the audible tones produced by an intermittent beam of light, for example, "The amount of the absorption and the intensity of the sound go hand in hand" [5.3]. The first clear description of the resonant photoacoustic effect was given by *Bell* in his second paper on this subject: "When the beam was thrown into a resonator, the interior of which had been smoked over a lamp, most curious alternations of sound and silence were observed. The interrupting disk was set rotating at a high rate of speed, and was then allowed to come gradually to rest. An extremely feeble musical tone was at first heard, which gradually fell in pitch as the rate of interruption grew less. The loudness of the sound produced varied in the most interesting manner. Minor reinforcements were constantly occurring, which became more and more marked as the true pitch of the resonator was neared. When at last the frequency of interruption corresponded to the frequency of the fundamental of the resonator, the sound produced was so loud that it might have been heard by an audience of hundreds of people" [5.4]. After these early activities in 1880 and 1881, no further photoacoustic experiments were reported for more than 50 years. This was mainly due to the limitations set by the light sources and sound detectors available during that time. The resonance enhancement of the photoacoustic effect, so clearly described by *Bell*, was not investigated in more detail for nearly a century. Nevertheless, *Bell* anticipated a wide and independent field of usefulness in the investigation of absorption spectra for his so-called "spectrophone" [5.4].

5.1.2 Recent Developments

The renaissance in photoacoustic research was connected with the progress made in the field of electronics that provided sensitive microphones and lock-in detection and the availability of lasers as radiation sources. One of the first important applications was trace gas analysis, starting around 1970. Either accidental absorption coincidences with laser lines or line tunable lasers were employed for trace detection. A variety of photoacoustic setups were developed for the sensitive detection of minute quantities of molecules in the ppb-ppt range and the reliable analysis of atmospheric and stratospheric constituents. To improve the sensitivity, resonant photoacoustic cells were used in many of these investigations [5.5–18]. The photoacoustic signal could easily be amplified by a factor of 10–100 by exciting an acoustic resonance, without careful optimization of the geometry of the acoustical resonator. The acoustic resonances themselves were not studied in more detail in these experiments. It was assumed and verified in many cases that the photoacoustic signal was proportional to the trace gas concentration over many orders of magnitude. As long as the signal is proportional to the beam power, it can be increased by a multipass arrangement and intracavity operation of the photoacoustic cell. The early work on trace gas analysis was reviewed in [5.19] and recent reviews can be found in the present book, Chaps. 7 and 8. Due to its high sensitivity, photoacoustic spectroscopy is ideally suited for studying processes with very low absorption coefficients, such as highly forbidden vibrational overtone transitions. Such weak transitions cannot be saturated with present cw lasers even by intracavity operation of the photoacoustic cell. To improve the signal strength, acoustical resonators of cylindrical type were used in most experiments, again without investigating the acoustical resonances in any detail [5.20–22]. The photoacoustic cells were equipped with wedged Brewster angle windows, which destroy the ideal geometry. Nevertheless, amplification factors above 500 have been achieved with this design. Intracavity dye laser photoacoustic spectroscopy allows the detection of $0 \rightarrow 9$ transitions with absorption coefficients nearly 10 orders of magnitude smaller than the $0 \rightarrow 1$ vibrational transition. The main goal of these overtone experiments was the investigation of vibrational bands of highly vibrationally excited states. These bands could be recorded by tuning the dye laser in the spectral region of the corresponding overtone transitions. The characteristic shift of the band positions with vibrational quantum number yields new insight into the excitation process, e.g. local mode behavior, and the band width gives information on the intramolecular dynamics in polyatomic molecules.

5.1.3 Scope of Review

Within the last decade the investigation of acoustical resonances in simple resonators such as spheres or cylinders has been the subject of increasing interest. Acoustical resonators provide a powerful tool, if combined with the photoacoustic technique, for two main reasons: they are sensitive detectors for photoacoustic signals, as they act as acoustical amplifiers, and in addition they allow the very precise measurement of the velocity of sound and the sound absorption. The velocity can be obtained from the acoustical resonance frequency, and the absorption from the width of the resonance profile. Therefore, acoustical spectroscopy provides similar information to optical spectroscopy. However, resonant photoacoustics possesses not only a high sensitivity and precision on the experimental side, but also provides a quantitative understanding of the experimental results on the theoretical side. This possibility of a detailed theoretical description is a consequence of the high symmetry involved.

The apparatus consists of an acoustical resonator, for example of cylindrical symmetry, filled with the gas under investigation. An amplitude-modulated laser beam, which is absorbed by the gas, excites standing pressure waves if the modulation frequency agrees with one of the resonance frequencies of the acoustical cavity. The amplitude of the sound wave can be detected with an electret microphone and a lock-in amplifier. Sweeping the modulation frequency near an acoustic resonance allows the corresponding Lorentzian profiles to be recorded and the evaluation of the three quantities obtained by such an experiment, namely resonance frequency, width of resonance and signal amplitude. These measured quantities are related to the sound velocity, the sound absorption and the excitation strength, respectively. To extract the full information provided by the method several acoustical modes must be studied in a large pressure range. This is only possible if data acquisition and analysis is controlled by a computer [5.23].

The method has been used to study energy exchange processes such as vibrational and rotational energy transfer by recording the frequency dispersion, resonance broadening and signal change caused by these kinetic effects [5.24–30]. It was also shown that chemical reactions can be studied with this technique, very accurately [5.31]. The delayed excitation of internal degrees of freedom or reactions between different chemical species by collisions leads to a change in the specific heat, and thus to changes in the velocity and absorption of sound waves.

Acoustical resonators have also been used to compare the photoacoustic signal for an extracavity configuration with that of an intracavity setup [5.32–34]. In the latter case the laser behavior can be studied as a function of the losses induced by the absorbing gas in the photoacoustic cell. These measurements yielded new information on the laser behavior near threshold [5.34].

Acoustical resonances can be excited not only optically but also acoustically, employing two microphones, one for excitation and one for detection. Such experiments have been performed using carefully designed spherical acoustic resonators. For this type of resonator a complete theoretical model has been developed as described in Chap. 4. Based on this theoretical understanding of the acoustical system the accuracy was increased to better than one part per million. The state of the art is best represented by the recent acoustic measurement of the gas constant R with an uncertainty of less than one part in a million [5.35].

5.2 Optical Excitation of Acoustic Modes

5.2.1 General Considerations

If a single short laser pulse is employed, for example for vibrational excitation of a gas in a photoacoustic cell, the pressure wave caused by vibrational relaxation is reflected back and forth until it is completely damped. A steady-state acoustic signal can be obtained only if a cw laser is modulated to achieve a fast repetition of excitation processes. However, the maximum signal observed in this case is smaller than that obtainable in a single-pulse experiment. The reason for this is destructive interference of acoustic waves from previous cycles with waves of the following cycle. Thus, the energy available in the acoustic chamber from previous cycles is not used to amplify successive acoustic signals but to produce noise. It is not possible to give a detailed theoretical description of the complicated processes occurring in the chamber in this case.

In resonant photoacoustic experiments an acoustic resonator with an optimized geometry such as a cylinder or sphere is used as an absorption cell. In addition, the spatial arrangement of the laser beam in the resonator is selected to achieve a minimal disturbance of the acoustic standing wave built up in the resonator by successive laser pulses. Finally, the cw laser beam is modulated with one of the resonance frequencies of the acoustic resonator. This energy input with correct timing leads to the formation of a standing wave in the resonator. Energy from many cycles is accumulated in the standing wave and the system works as an acoustic amplifier. The final signal amplification obtainable in this resonance system is determined by the resonator losses. After an initial transient state during which energy is accumulated in the standing wave, a steady state is reached in which the energy lost by various dissipation processes is equal to the energy gained by absorption of laser photons.

The quality of the acoustical resonator and of the excited resonance can be characterized by the Q-factor, which is defined as the ratio of the resonance frequency to the frequency bandwidth between half-power points:

$$Q = \frac{f_0}{\Delta f} \quad . \tag{5.2.1a}$$

The amplitude of the microphone signal is $1/\sqrt{2}$ of the maximum amplitude at these half-power points, because the power stored in the resonator varies as the square of the pressure. The acoustic Q-factor should be independent of the level of excitation. In acoustic resonance experiments with cylinders or spheres Q-factors up to 10^3–10^4 have been achieved.

Physical insight into the resonance process is provided by the definition

$$Q = \frac{\omega E_{st}}{P} \quad , \tag{5.2.1b}$$

where the Q-factor is given by the product of the angular frequency ω and the maximum energy stored E_{st} in the resonator, divided by the power dissipated $P = -dE_{st}/dt$.

Thus, a small halfwidth of the resonance indicates a large amount of energy stored in the resonator and inefficient loss processes. Energy dissipation is caused, for example, by transport processes, e.g. heat conduction, or viscous losses at the resonator walls or relaxation processes, e.g. vibrational energy exchange in the bulk.

5.2.2 Acoustical Resonances in a Cylinder

As discussed in [5.19] the simplest approach to calculating the acoustic eigenfrequencies of a cylinder is to consider a lossless gas in a rigid cavity with perfectly insulating walls. This ideal model yields a simple expression for the resonance frequencies [5.19]:

$$f_{nmn_z} = \frac{c}{2}\left[\left(\frac{\alpha_{mn}}{R_0}\right)^2 + \left(\frac{n_z}{L}\right)^2\right]^{1/2} , \tag{5.2.2}$$

where R_0 and L are the radius and length of the cylinder, c is the sound velocity, n, m, n_z refer to the eigenvalues of the radial, azimuthal and longitudinal modes, respectively, and α_{nm} is the nth zero of the derivative of the mth Bessel function divided by π. Figure 5.1 presents a graphic representation of the different oscillations connected with these normal modes. In a real resonator dissipation processes occur, which are responsible for a broadening of the resonance described by the halfwidth Δf. Equation (5.2.2) gives a total of 12 acoustic modes within the frequency range from 4 kHz to 8 kHz for a cylinder with the dimensions $L \approx 2R_0 \approx 10$ cm. However, laser excitation with the beam propagating along the cylinder axis and detection with a microphone inserted flush with the inside wall surface midway between the ends of the cylindrical cell leads to only three detected resonances in this frequency interval, as shown in Fig. 5.2. Thus, symmetry constraints imposed by laser excitation and microphone detection drastically reduce the number of modes actually observed. Azimuthal modes ($m \neq 0$) are not excited by the coaxial laser beam possessing cylindrical symmetry. This reduces the number of cavity modes in the selected frequency region to six resonances. In addition, longitudinal modes with a node at the microphone position, e.g. the first longitudinal mode (001), are not detected. This reduces

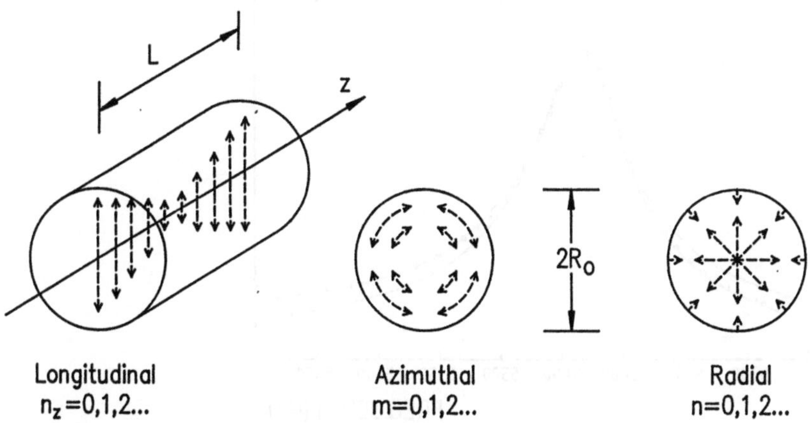

Fig. 5.1. Schematic representation of longitudinal, azimuthal and radial acoustic modes in a cylindrical resonator

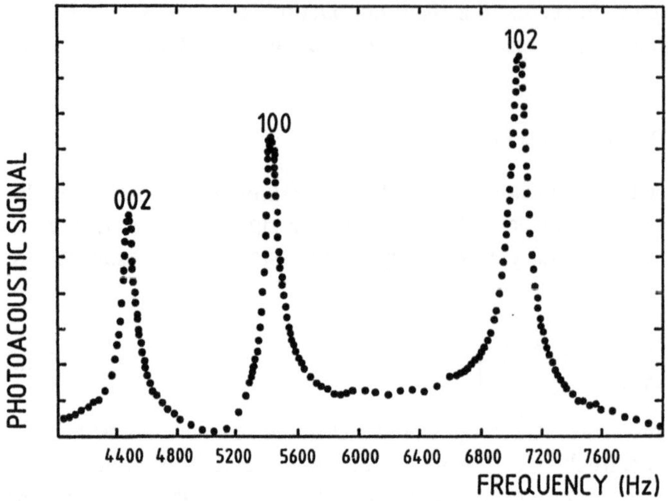

Fig. 5.2. Acoustic modes excited with a modulated laser beam in a cylinder filled with 36 torr of methane; the modes are the second longitudinal mode (002) the first radial mode (100) and the combination mode (102)

the number of modes observed experimentally to three resonances, namely the second longitudinal mode (002), the first radial mode (100) and the combination mode (102), as can be seen in Fig. 5.2.

Despite the drastic reduction of observed modes by symmetry constraints, overlap of the resonance profiles may occur at low pressures, where increasing dissipation leads to a strong broadening effect. The coherent superposition of the standing acoustic waves in the detection region of the microphone is responsible for the asymmetric resonance profiles and the constructive and destructive interference between modes which can be seen in Fig. 5.2. A detailed description of the procedure that allows correction of this distortion effect is given in [5.28]. Figure 5.3 shows for illustration

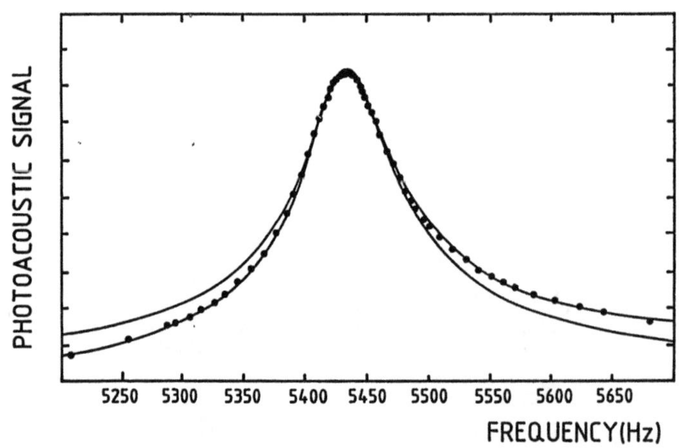

Fig. 5.3. Comparison between the measured disturbed profile of the first radial mode shown in Fig. 5.2 and the corrected line profile

a comparison between the originally measured and the corrected line profile of the first radial mode of a cylinder filled with CH_4 at 36 torr.

5.2.3 Acoustical Resonances in a Sphere

The eigenfrequencies of the acoustic modes in an ideal sphere without dissipation processes and with rigid resonator walls are given by [5.36]

$$f_{ln} = \frac{c\,a_{ln}}{2\pi R_0} \quad , \tag{5.2.3a}$$

where a_{ln} is the nth root of

$$\frac{dj_l(kr)}{dr} = 0 \quad , \tag{5.2.3b}$$

where $j_l(kr)$ is the spherical Bessel function of lth order and $k = \omega/c$.

Laser excitation of the first radial mode ($l = 0, n = 1$) and several nonradial modes ($l \neq 0$) was studied using a laser beam travelling along the symmetry axis of the two halves clamped together to obtain the spherical resonator [5.36]. Analysis of the microphone signal shows that the first radial mode changes its phase with decreasing pressure. This indicates that this mode is not excited by the laser radiation at some intermediate pressure. This behavior of the disappearance and reappearance of a resonator mode can be explained in terms of the overlap integral, which takes into consideration the spatial distribution of the translational energy resulting from the relaxation of the optical excitation energy and the symmetry of the acoustic modes. In this case the photoacoustic signal is not proportional to the sample absorption, as is often assumed in photoacoustic spectroscopy.

The nonradial modes possess a $(2l + 1)$-fold degeneracy in an ideal sphere. Therefore, the first nonradial mode is threefold degenerate. The observed splitting of the degenerate eigenfrequency into a doublet indicates a small deviation of the resonator from the ideal spherical shape. As shown in [5.36], the observed relative deviation of the resonance frequencies is of the same order as the relative differences of the axes. Taking into account the amount of splitting, a deviation of the two axes in the percent region is estimated. Thus, an accurate determination of the split eigenfrequencies yielded detailed information on the exact ellipsoidal shape of the real sphere formed from two halves.

5.2.4 Accuracy of the Resonance Method

As mentioned before, the analysis of resonance curves yields three important quantities, namely the resonance frequency, the halfwidth of the resonance profile and the signal amplitude. These quantities are determined by fitting the resonance profiles after correction of the measured perturbed curves to a Lorentzian distribution (Fig. 5.4). The least-squares fit of the theoretical line profile to the data points gives very accurate values for the parameters of the resonance curve. The highest accuracy was found for the resonance frequency, which can be determined with an accuracy on the order of 0.01% even at low pressures around 1 torr, where the resonances are broadened substantially [5.23]. The statistical errors found for the halfwidth and the

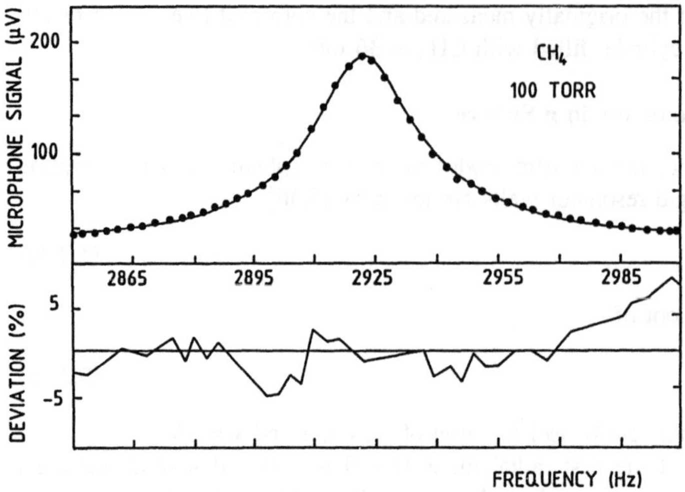

Fig. 5.4. Fit of the measured nearly undisturbed profile of the first nonradial mode in a spherical resonator to a Lorentzian profile. The lower part shows the relative deviations between data points and fitted profile

resonance amplitude were in the percent region [5.23]. These small errors clearly demonstrate the high accuracy of acoustical resonance experiments. If the behavior of several acoustic modes is studied over a large pressure and temperature range, it is possible to extract accurate information on different system properties from such a data set, especially if these properties influence the acoustic behavior in different pressure regions.

5.3 Experimental Method

5.3.1 Apparatus

Figure 5.5 is a schematic diagram of a typical experimental setup. In all experiments a modulated cw laser was used to excite the acoustic waves. Depending on the absorption characteristics of the gas under study, a 25 mW HeNe laser radiating at

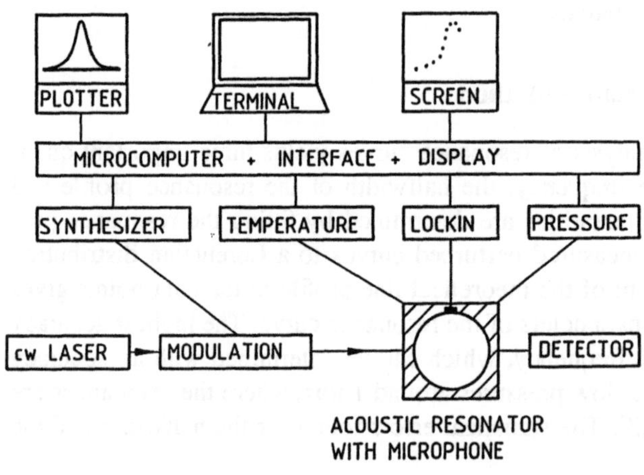

Fig. 5.5. Scheme of an experimental setup showing the photoacoustic part, electronics and computer

3.39 μm [5.27–29], a 3 W waveguide CO_2 laser [5.31] or a 30 mW argon ion laser at 488 nm [5.37] were employed as light sources. In cases of insufficient absorption of the gas at the laser wavelength available, a trace of an inert sensitizer such as SF_6 (10.6 μm) or CH_4 (3.39 μm) can be added to achieve energy deposition in the gas. The laser used for such experiments should possess a high stability and the beam diameter should be in the millimeter range.

An essential part of the setup is the modulation device and the electronics controlling the modulation frequency. If the resonance profiles are not too narrow ($>$ 10 Hz), a mechanical chopper can be used in the frequency range 0–10 kHz. A detailed description of computer-controlled stabilization and variation of the chopper frequency by means of a frequency synthesizer is given in [5.23]. With such an electronic control the frequency fluctuations caused by the chopper motor can be reduced to $\pm$ 1Hz. Most problems of poor frequency stability and limited frequency range can be avoided with a more expensive electro-optic modulation device. For example, a Pockels cell with 100% modulation has been employed in recent experiments [5.31,37]. In this case the modulation was also generated by a computer-controlled frequency synthesizer via a linear high-voltage amplifier.

Measurements have been performed with several resonators acting as acoustic cells. A cylindrical resonator, which had a length of about 10 cm and a radius of about 5 cm, was used in the energy transfer studies [5.28,29]. For the chemical relaxation experiments a smaller stainless steel cylinder with a length of about 6 cm, radius of about 2.5 cm and an inner surface polished to about 1 μm accuracy was employed [5.37]. This smaller resonator could be placed in a cryostat to vary the temperature. The spherical resonator was made in two halves from brass stock with a milling machine and had an inner diameter of about 10 cm. The windows were separated from the inner surface by holes 4 mm long and with an inside diameter of 3 mm. Two electret microphones were mounted in adapters at different positions with respect to the laser beam, allowing the investigation of the spatial distribution of nonradial modes [5.36].

The acoustic signal was detected with miniature electret microphones several millimeters in diameter and with a typical sensitivity of about 1 mV/μbar. In some cases the signal was filtered and amplified before standard two-phase lock-in amplifier processing was applied.

5.3.2 Temperature Measurement

The resonance profiles of acoustic modes are both pressure and temperature dependent, and therefore it is necessary to measure the gas temperature very accurately during a resonance experiment. Even small temperature shifts may cause a systematic deformation of the resonance curves and this leads to errors in the resonance frequency, halfwidth and signal amplitude. An accurate temperature measurement can be obtained with a quartz crystal attached to the acoustic resonator [5.23]. With such a device a temperature change of 0.001 K can be observed during an experiment, allowing an accurate correction when temperature fluctuations occur. These temperature corrections are extremely important in high-precision resonance experiments. During the recording of a resonance spectrum the temperature fluctuations are usually less than 1 K, and the data is corrected to a fixed base temperature.

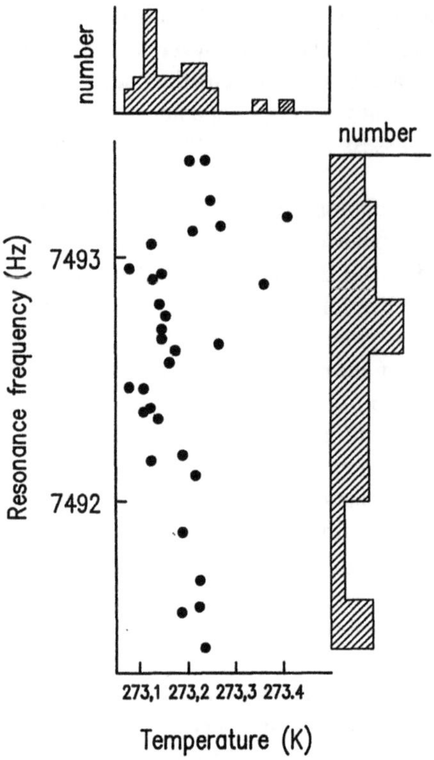

Fig. 5.6. Resonance frequencies versus temperature measured at the triple point of water for the first radial mode in a cylinder

The quartz device gives accurate information on relative temperature changes. In order to test the absolute temperature calibration and the reproducibility of the resonance experiment, measurements were performed at the triple point of water [5.37]. The cylindrical resonator was filled with argon, as the sound velocity of argon is known very precisely [5.38]. Desorption of NO_2 from the resonator walls, adsorption having occurred in previous NO_2 experiments, led to a NO_2 concentration in the ppm range. This sensitizer concentration was enough for energy deposition and excitation of acoustical modes in argon using a 488 nm argon ion laser. The results of these experiments are shown in Fig. 5.6, where the measured frequencies of the first radial mode are plotted versus temperature. The distribution of temperature values yielded 273.18 ± 0.07 K, thus, the triple points of water (273.16 K) is well reproduced. The spread in the frequencies, which is correlated to the temperature variation, was 0.6 Hz or 0.8×10^{-4}. The uncorrected theoretical value calculated for the resonance frequency was 20 Hz or 0.27% higher than the measured one. These numbers reflect the agreement between theory and experiment and the accuracy easily achieved in laser-excited acoustic spectroscopy. To improve the agreement small corrections must be introduced, which were not considered in the simple analysis. The most important corrections may be due to geometrical imperfections of the resonator symmetry such as excess volumes at the window, the gas inlet and the microphone. Additional corrections due to the small amount of NO_2 present in the gas and transport processes occurring at the resonator walls may also contribute to the observed deviations.

5.3.3 Computer Control

A first computer-controlled resonance experiment was described in detail in [5.23]. This included repeated recording of the resonator temperature with a quartz sensor, stabilization and measurement of the chopper frequency and integration of the acoustic signal detected with an electret microphone. The acoustic spectrum of several modes was measured point by point, by measuring at discrete modulation frequencies. Data analysis was started by presetting the total frequency range containing the acoustic resonances of interest. The scan through the selected range was performed by predetermining a constant interval between the frequency values to be measured. In the frequency region of an acoustic resonance the step width was shortened by an algorithm according to the increase in the signal amplitude. The analysis of the line profile could be further improved by an extension of the integration time at each frequency point. During the integration time the actual chopper frequency was compared with the respective synthesizer value and adjusted repeatedly. The signal amplitude was averaged during this time by a signal processing technique discussed in more detail in [5.23].

5.4 Theory

5.4.1 General Remarks

The investigation of processes occurring in acoustic waves and particularly in acoustical resonators has a long tradition, and just a few references are given here concerning the early theoretical developments [5.39–42]. There are several excellent books available treating different aspects of the present topic [5.43–45].

Starting around 1970 a series of papers was published dealing with optimal photoacoustic detector design [5.46–51]. Most of these papers compared the performances of a resonant and a nonresonant photoacoustic cell, to determine the ultimate detectivity of an ideal photoacoustic detector in trace analysis. The treatment of the acoustic resonances was mostly rather simple. Essentially, the homogeneous wave equation was solved to determine the eigenfrequencies of the resonator. These loss-free solutions possess a physically unreasonable infinite Q-factor at resonance. To correct for this, the solution of the homogeneous wave equation can be modified by including a mode damping [5.49]. The different loss processes occurring in the gas volume or at the resonator surface can be characterized by corresponding Q-factors. Therefore, the expression obtained for the acoustic pressure is corrected by assigning a finite Q-factor to each mode. This overall Q-factor contains all the contributions from loss processes appearing in the resonator, such as relaxation processes in the gas volume or losses caused by heat conduction and viscosity in the bulk or near the walls. Only in a few papers has the quantitative evaluation of resonance profiles been considered at all.

An improved treatment of photoacoustic signal generation was given in [5.52], where the coupled equations for sound wave propagation and thermal diffusion were solved. Relaxation effects were introduced through the expression of the source term. For a resonant cell general expressions for the shift of the resonance frequencies and for the quality factor due to volume losses were derived. The latter include bulk thermal and viscous losses and energy transfer processes.

In the following a general approach is presented to describe the processes occurring inside an acoustical resonator. This approach starts from the fundamental laws containing all important effects, rather than starting with the solved loss-free Helmholtz equation and introducing the loss effects into the solution.

5.4.2 Basic Equations

The general theoretical approach considered here is a derivation of the solutions from the fundamental equations for the conservation of mass, momentum and energy, the equation of state, the definition of density and an equation describing the kinetics.

Conservation of mass

$$\frac{\partial \varrho}{\partial t} + \varrho \nabla \boldsymbol{u} = 0 \quad , \tag{5.4.1}$$

where ϱ is the density and $\boldsymbol{u}$ is the particle velocity in a sound wave.

Conservation of momentum

$$\varrho \frac{\partial \boldsymbol{u}}{\partial t} = -\nabla p + \eta \nabla^2 \boldsymbol{u} \quad , \tag{5.4.2}$$

where p is the pressure variation caused by the acoustic wave (acoustic pressure) and η is the viscosity.

Conservation of energy

$$C_V \Delta T + P \Delta V - D\nu = \Delta Q(r,t) + \lambda \nabla^2 T \Delta t \quad , \tag{5.4.3}$$

where C_V is the specific heat at constant volume, T is the temperature, P is the total pressure at equilibrium, V is the volume, D is the dissociation energy per mole, ν is the small change of the number of moles, $Q(r,t)$ is the source function describing the amount of energy absorbed by the gas from the laser beam, λ is the thermal conductivity and t is the time.

Equation of state

$$PV = (N_1 + N_2)RT \quad , \tag{5.4.4}$$

where N_1, N_2 are the number of moles of species 1 and 2, respectively, and R is the universal gas constant.

Definition of density

$$\varrho = (N_1 M_1 + N_2 M_2)/V = NM/V \quad , \tag{5.4.5}$$

where M_1 and M_2 are the mass of one mole of species 1 and 2, respectively, N is the number of moles and M is the average mass of one mole.

Kinetic equation

$$\dot{\nu} = -\frac{1}{\tau}\nu + \left(\frac{\partial N_2}{\partial p}\right)_T \dot{p} + \left(\frac{\partial N_2}{\partial T}\right)_p \dot{T} \quad , \tag{5.4.6}$$

where τ is the relaxation time.

These six equations determine the six functions $\boldsymbol{u}, p, T, \varrho, V, \nu$, where ν is a small perturbation of the mole number N_2 from chemical equilibrium. The solution of this system of equations can be simplified considerably if only specific effects are considered.

For example, the term $\eta \nabla^2 \boldsymbol{u}$ in the Navier-Stokes equation and $\lambda \nabla^2 T \Delta t$ in the law of energy conservation describe the classical Kirchhoff losses due to transport processes in the bulk between the pressure and temperature maxima and minima. Mathematically these terms lead to complications because of their second-order derivatives, which require higher-order differentiation of other variables in order to eliminate all but one variable. This reduction can be executed, however, analytically in the case of harmonic excitation in the frequency domain. If processes in a higher pressure region are studied, where the Kirchhoff losses are negligibly small, it is reasonable to neglect these terms to simplify the mathematical analysis.

5.4.3 Kinetics in the Frequency Domain

The measurement of kinetic processes in the frequency domain is based on the excitation of a sound wave, which represents a periodic change of the state in the gas. Pressure and temperature oscillate around their equilibrium values with a very small amplitude. For example, the amplitude of a resonantly enhanced pressure wave is of the order of $0.1 - 1\,\mu$bar, whereas a typical equilibrium value for the pressure is $1 - 1000$ mbar. Each microscopic state with a given energy level is populated by energy exchange according to the Boltzmann law, depending on temperature. This energy transfer process is carried out by intermolecular collisions. The efficiency strongly depends on the amount of energy to be exchanged and on the specific properties of the system. Therefore, rotational levels are excited very effectively and in most systems only a few gas kinetic collisions are needed to equilibrate the rotational degrees of freedom with the translational degrees of freedom. This corresponds to a time scale of a few nanoseconds at atmospheric pressure. According to the Ehrenfest adiabatic principle, vibrational degrees of freedom are not equilibrated as effectively, and in fact more than 10000 collisions may be necessary to activate or deactivate a vibrational state. If the pressure is low enough and the period of the temperature variation in the sound wave becomes sufficiently short the average number of collisions during this period can no longer pass on the temperature variation to the vibrational degrees of freedom. The vibrational energy of the system is no longer in equilibrium and these degrees of freedom drop out of the distribution of excess energy. The resulting decrease of the specific heat can be detected by a change in the sound velocity, which depends on the adiabatic coefficient $\gamma = 1 + R/C_V$. Parallel to the change of the specific heat, a phase shift occurs between the local compression (density) due to the sound wave and the temperature and pressure changes. This is

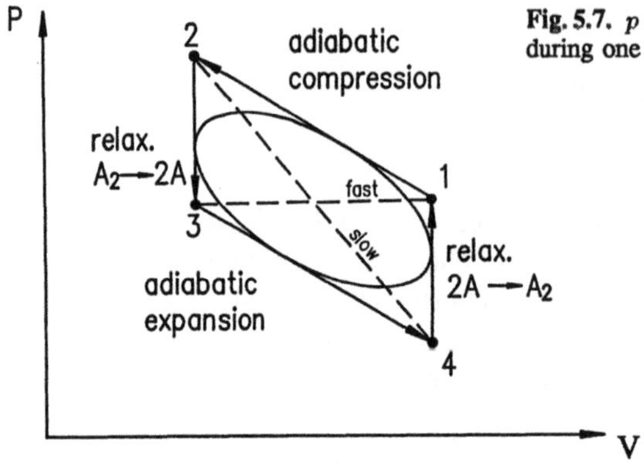

Fig. 5.7. $p - V$ diagram of the processes occurring during one cycle in a sound wave

illustrated by the idealized circle process in the $p-V$ diagram shown in Fig. 5.7. The adiabatic compression by the sound wave ($1 \rightarrow 2$) is followed by the relaxation of the system ($2 \rightarrow 3$), where the energy is equilibrated. During the second part of the period the inverse processes occur ($3 \rightarrow 4$ and $4 \rightarrow 1$). If the relaxation is fast compared to the state change in the sound wave, equilibrium is established instantaneously at any time. This means that the system moves directly between the points 1 and 3. In the other case of very slow relaxation no equilibration between the corresponding degrees of freedom can be achieved and the system oscillates between the points 2 and 4. If, however, the inverse sound frequency and the relaxation time are of the same order of magnitude, the system will move along the ellipse inside the idealized circle process and the loss is given by the area $\int p dV$ as discussed in more detail in [5.53]. These considerations are valid for all degrees of freedom that tend to follow small pertubations of the equilibrium values of pressure and temperature. This is true not only for energy exchange processes between translational and internal degrees of freedom but also for chemical reactions.

The kinetics of these relaxation processes can be described quantitatively in the following way [5.54]. If small perturbations of the equilibrium values are assumed, the concentration of the chemical species, for example, can be replaced by the equilibrium value plus the small perturbation $c(t)$:

$$C(t) = C_{\mathrm{eq}}(t) + c(t) \quad . \tag{5.4.7}$$

Nonlinear expressions of the small perturbations can be neglected in the rate equation and the contributions of the equilibrium values C_{eq} cancel. Expressing the perturbations of the concentrations of all species c_n by only one variable, say c_a, by applying the conservation of the number of atoms for each element, one obtains a first-order rate equation in c_a for the description of the kinetics:

$$\frac{\partial c_a}{\partial t} = -\frac{1}{\tau} c_a \quad . \tag{5.4.8}$$

The details of the complex reaction process are now contained in the relaxation time τ, which is a function of the rate constants, of pressure, of temperature and/or of the equilibrium concentrations. These additional complications are compensated by the extraordinary flexibility of this concept, which is not confined to the description of chemical reactions. A number of processes can formally be treated as a reaction and therefore can be analyzed by this model. An example is the excitation of a species A to an excited species A^* by absorption of light. Another example is the condensation process, where molecules "react" successively to form dimers, trimers and finally clusters.

In the frequency domain the mathematical treatment of the kinetics is simplified by the fact that the time dependence is described by the factor $\exp(i\omega t)$. Thus, time derivatives are replaced by the factor $i\omega$, and the following is obtained for the kinetics:

$$i\omega C(t) = -\frac{1}{\tau}\left(C(t) - C_{eq}(t)\right) \quad , \tag{5.4.9}$$

$$C(t) = \frac{1}{1 + i\omega\tau} C_{eq}(t) = S C_{eq}(t) \quad . \tag{5.4.10}$$

Here the equilibrium concentration $C_{eq}(t)$ is predetermined and the proportionality function describes the typical relaxation behavior. This time-independent function is called the transfer function and contains the information necessary for a determination of the relaxation time. The term $C_{eq}(t)$ may be considered as a periodic forcing function and $C(t)$ as the forced solution of the steady-state experiment with a smaller amplitude and a phase shift as shown in Fig. 5.8. The complex transfer function S can be written in the form

$$S = S_{re} + iS_{im} \quad , \quad \text{where} \tag{5.4.11}$$

$$S_{re} = \frac{1}{1 + \omega^2\tau^2} \quad . \tag{5.4.12a}$$

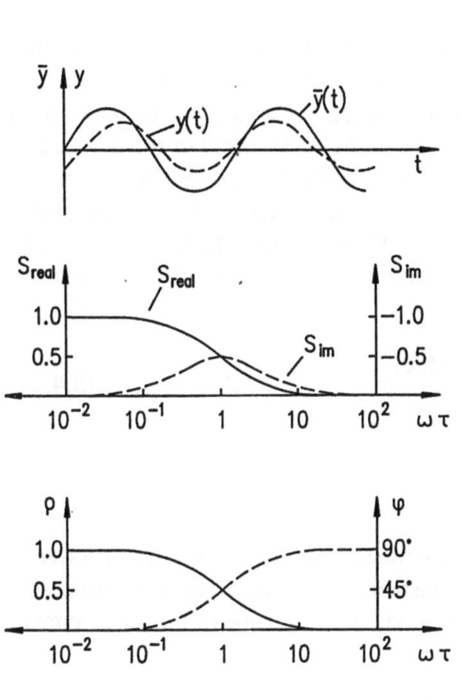

Fig. 5.8. Schematic representation of the relaxation processes in the frequency domain. *Upper part*: relaxational response $y(t)$ for a periodic forcing function $\overline{y}(t)$; *middle part*: real and imaginary parts of the transfer function S; *lower part*: effective amplitude ratio ϱ and phase angle φ

The real part of the transfer function is a step function as shown in Fig. 5.8 and describes the frequency dispersion effect in the acoustical resonator. The methods depending on the frequency behavior of S_{re} are called dispersion methods.

The imaginary part of the transfer function

$$S_{im} = -\frac{\omega\tau}{1+\omega^2\tau^2} \tag{5.4.12b}$$

shows a bell-shaped behavior around $\omega\tau = 1$ and extends over about three orders of magnitude (Fig. 5.8). The imaginary part represents the irreversible energy loss per period of oscillation as a consequence of the phase shift between the two conjugate variables density and pressure, as discussed before. All techniques based on the behavior of the imaginary part of the transfer function measure energy absorption, and therefore they are called absorption methods. The absorption effect is responsible for the irreversible loss of acoustical energy in the relaxation region $\omega\tau \approx 1$ and causes a broadening of the resonance profile.

The effective amplitude ϱ of the concentration parameters $C_{eq}(t)$ and $C(t)$ is given by

$$\varrho = \sqrt{S_{re}^2 + S_{im}^2} = \frac{1}{\sqrt{1+\omega^2\tau^2}} \tag{5.4.13a}$$

and the phase lag φ between $C_{eq}(t)$ and $C(t)$ can be determined from

$$\tan\varphi = -\frac{S_{im}}{S_{re}} = \omega\tau \quad . \tag{5.4.13b}$$

The dependence of these two properties on the value of $\omega\tau$ is also illustrated in Fig. 5.8.

This detailed discussion of the mathematics clearly shows that the relaxation phenomena observed in a photoacoustic resonance experiment, such as resonance frequency dispersion and broadening of the resonance halfwidth, can be understood on the basis of the relaxation equations considered. As an example of the description of a chemical reaction the kinetic equation is given explicitly for the simple reaction $A_2 \leftrightarrow 2A$, e.g. $N_2O_4 \leftrightarrow 2NO_2$. By defining the mole number change of the dimer by $\Delta N_2 = \nu$ and that of the monomer by $\Delta N_1 = 2\nu$ and applying the already mentioned method of variation of the equilibrium the following kinetic equation can be derived [5.37,55]:

$$\frac{(\mathrm{i}\omega + 1/\tau)\nu}{k_1} = -\frac{dK'}{dT} N_2 k_1 \Delta T - \frac{N_1^2 \Delta V k_1}{V^2} \tag{5.4.14}$$

with

$$\frac{1}{\tau} = k_2 + k_1 \frac{4N_1}{V} \quad ,$$

where k_1 and k_2 are the rate constants for dimerization and dissociation, respectively, and K' is the corresponding equilibrium constant $K' = k_2/k_1$. The right-hand side of (5.4.14) gives the thermodynamic contributions to the change of the mole number of the dimer.

5.4.4 Helmholtz Equations

The system of the fundamental equations using (5.4.14) to describe the kinetics of the reaction can be reduced to one equation for the acoustic pressure [5.55]:

$$\frac{\partial^2 p}{\partial t^2} - c^2\nabla^2 p = \Lambda \frac{\partial^2 Q(r,t)}{\partial t^2}\,\frac{1}{V} \tag{5.4.15a}$$

with

$$c^2 = \frac{RT}{M}\left(1 + \frac{k_2 A + \mathrm{i}R\omega}{k_2 B + \mathrm{i}C_V\omega}\right) \tag{5.4.15b}$$

and

$$\Lambda = \frac{R}{C_V}\left(1 - \frac{D - C_V T}{D + C_V T F(1 + \mathrm{i}\omega\tau)}\right) \quad , \tag{5.4.15c}$$

where A, B and F are real functions that depend on the mole fractions and the dissociation energy D. This is the Helmholtz wave equation including the chemical reaction $A_2 \leftrightarrow 2A$ and neglecting the classical transport losses. If the terms for viscosity and thermal conductivity are not neglected the expression contains mixed higher-order derivatives up to fourth order.

The equations obtained for the sound velocity c and the coupling function Λ will now be discussed in detail. The expression for the sound velocity was first derived by *Einstein* [5.56]. An equivalent equation which is more suitable for a qualitative discussion of the relaxation behavior is

$$c^2 = \frac{RT}{M}\left(\frac{C_P}{C_V} + \frac{A'}{B' + (1 + \mathrm{i}\omega\tau)}\right) \quad , \tag{5.4.16}$$

where A' and B' are real functions. It is interesting to consider first the two limiting cases of very high and very low frequency. For $\omega \to \infty$ the second term in the large parentheses vanishes and the expression for the sound velocity reduces to the well-known expression $c^2 = \gamma RT/M$. This case refers to the situation where the chemical reaction does not participate in the distribution of energy. The other extreme with $\omega \to 0$ describes the case where the reaction is always completely equilibrated. The value of the sound velocity is real, but different from the value without reaction and may depend on pressure, dissociation energy and the mole fraction. In between these two extreme cases the sound velocity becomes complex and depends on the frequency. The real part describes the propagation velocity of sound and yields a dispersion effect. The imaginary part is assigned to the absorption of sound and is responsible for the linewidth broadening. The expression in the large parentheses of (5.4.16) can be considered as a dynamical adiabatic coefficient $\gamma(\omega)$ that varies with frequency.

The function Λ can be considered as a coupling function describing the efficiency of the source function $Q(r,t)$ to couple energy into the sound wave. The case of $\omega \to \infty$ again refers to the limit of a negligible contribution from the chemical reaction. The value of Λ corresponds to the inverse heat capacity measured in units of the gas constant in this limit. As soon as the reaction can no longer be neglected, the second term in the parentheses tends to increase C_V, and thus to decrease the value of Λ. The reaction starts to consume energy and may be considered as an additional degree of freedom participating in the distribution of energy. For $\omega \neq 0$ the coupling function is complex and results in a phase shift between the compression by the sound wave and the response of the system. The function $\Lambda(\omega)$ describes the dynamical behavior of the specific heat.

In the following, the problem of bulk thermal and viscous losses by transport processes will be considered in more detail. For a system with one component only, a set of two coupled equations is obtained, if viscosity is taken into account in the Navier-Stokes equation and thermal conductivity is not neglected in the equation for the conversation of energy. These two equations are discussed in [5.44] and contain two variables, the acoustic pressure p and the acoustic temperature ϑ

$$\nabla^2 p = \frac{\gamma}{c^2}\left(\frac{\partial^2}{\partial t^2} - l_v c \frac{\partial}{\partial t}\nabla^2\right)(p - \alpha\vartheta) \quad , \tag{5.4.17}$$

$$l_h c \nabla^2 \vartheta = \frac{\partial}{\partial t}\left(\vartheta - \frac{\vartheta - 1}{\alpha\gamma} p\right) \quad , \tag{5.4.18}$$

where $l_h = \lambda/\varrho c C_p$ and $l_v = n/\varrho c$ are characteristic lengths typical of the transport processes and α is the pressure coefficient of thermal expansion. Since l_h and l_v possess similar values (within a factor of approximately 1.6), the mean value $(l_h + l_v)/2 = l$ may be used to reduce the set of equations to one homogeneous Helmholtz equation:

$$c^2 \nabla^2 p + \frac{\partial^2 p}{\partial t^2} + \frac{\gamma l}{c}\frac{\partial}{\partial t}\nabla^2 p = 0 \quad . \tag{5.4.19}$$

This equation contains a first-order time derivative, and thus a dissipation term. The solution of this equation, according to the boundary conditions of the resonator, contains an imaginary part in the denominator with the Q-factor due to classical bulk losses.

An approach taking into consideration bulk losses due to transport processes and in addition relaxation effects was suggested in [5.52]. In this model the two coupled equations were modified to include the rate of energy release into the translational degrees of freedom $\partial H/\partial t$,

$$\nabla^2 p = \frac{\gamma}{c^2}\left(\frac{\partial^2}{\partial t^2} - l_v c \frac{\partial}{\partial t}\nabla^2\right)(p - \alpha\Theta) \quad , \tag{5.4.20}$$

$$l_h c \nabla^2 \Theta + \frac{1}{\varrho C_p}\frac{\partial H}{\partial t} = \frac{\partial H}{\partial t}\left(\frac{C_{p0}}{C_p}\Theta - \frac{\varrho - 1}{\alpha\varrho} p\right) \quad , \tag{5.4.21}$$

where now Θ describes the fluctuations of the translational temperature and C_{p0} is the specific heat at constant pressure associated with translational and all degrees of freedom in equilibrium with translation. The information about the kinetics of the relaxation processes is contained in the source term

$$\frac{\partial H}{\partial t} = \phi_r F(k_i)\exp(\mathrm{i}\omega t) + F'(k_i)\Theta \quad , \tag{5.4.22}$$

where ϕ_r is the rate of energy deposition per unit volume due to the radiative field and $F(k_i)$ and $F'(k_i)$ are functions of the rate constants involved in the energy transfer processes. This treatment gives a general expression for the acoustic pressure that is valid for all relaxation schemes [5.52].

The reason for the volumetric or bulk relaxational losses is the phase difference between density and pressure in the relaxation region, leading to an irreversible conversion of acoustical energy into thermal energy. These processes occur in a limited pressure range determined by the $p\tau$ values of the respective relaxation processes. They usually make the main contribution to the bulk losses, for example in systems with polyatomic molecules (energy transfer) or reactive systems (chemical relaxation).

The Stokes-Kirchhoff volumetric losses due to friction connected with the compressional motion and transformation of organized energy into heat due to temperature gradients play only a minor role and can be neglected at higher pressures. The contribution of free-space viscous and thermal dissipation to the Q-factor is given by [5.44]

$$\frac{1}{Q} = \frac{\omega}{2c^2}\left(\frac{3\eta}{4\varrho} + (\gamma - 1)\frac{\lambda}{\varrho C_p} + \frac{\eta_{rel}}{\varrho}\right) \quad , \tag{5.4.23}$$

where η_{rel} is an effective bulk viscosity taking into account relaxation losses in polyatomic molecules.

5.4.5 Boundary Conditions

The Helmholtz equation for standing waves in a cylinder, for example, has to be solved by taking into account certain boundary conditions which are caused by the resonator walls. A detailed treatment of problems involving boundary conditions of sound waves in nonreacting gases can be found in [5.44]. To solve (5.4.15) the corresponding homogeneous wave equation is solved exactly by separation of the space dependence from the time dependence. The separation method yields the oscillation term exp(iωt) for the time dependence and the eigenvalue equation

$$\nabla^2 p_j(r) = k_j^2 p_j(r) \quad \text{with} \quad k_j^2 = \omega_j^2/c^2 \quad . \tag{5.4.24}$$

The boundary conditions can be satisfied only for the eigenfunctions $p_j(r)$ and the eigenvalues k_j.

In the case of a cylindrical resonator, ∇^2 is expressed in terms of cylindrical coordinates, which allow the separation of the acoustic modes into radial, longitudinal and azimuthal normal mode solutions:

$$\frac{1}{r}\frac{\partial}{\partial r}\left(r\frac{\partial}{\partial r} + \frac{\partial^2}{\partial z^2} + \frac{1}{r^2}\frac{\partial^2}{\partial \varphi^2}\right)p_j(r,z,\varphi) = -k_j^2 p_j(r,z,\varphi) \quad . \tag{5.4.25}$$

The general solution of this equation is

$$p_j = J_m(k_r r)\cos(k_z z)\begin{cases}\sin(m\varphi)\\ \cos(m\varphi)\end{cases} \quad , \tag{5.4.26}$$

where J_m is the Bessel function of order m and k_r and k_z are the wave numbers with

$$k = \sqrt{k_r^2 + k_z^2 + m^2/R_0^2} \quad . \tag{5.4.27}$$

In a photoacoustic experiment, with the laser beam travelling along the cylinder axis, azimuthal modes are not excited and therefore $m = 0$.

The boundary condition to be incorporated is that the particle velocity normal to the wall must vanish. This results in a condition for the pressure from the normal component of the Navier-Stokes equation:

$$\hat{\boldsymbol{n}} \cdot \frac{\partial \boldsymbol{u}}{\partial t} = -\hat{\boldsymbol{n}} \cdot \nabla p \tag{5.4.28}$$

where $\hat{\boldsymbol{n}}$ is the unit normal.

This means that normal to the wall the pressure gradient must be zero. If the end walls of the cylinder are positioned at $z = 0$ and $z = L$ we obtain from the resulting condition $\sin(k_z L) = 0$

$$k_z = \frac{\pi}{L} n_z \quad \text{with} \quad n_z = 0, 1, 2, \ldots \quad . \tag{5.4.29}$$

Applying the boundary conditions to the wall at $r = R_0$ leads to the result that the radial component k_r depends on the roots of the first-order Bessel function

$$\left(\frac{dJ_0(k_r r)}{dr} \right)_{R_0} = -J_1(k_r R_0) = 0 \quad , \tag{5.4.30}$$

resulting in

$$k_r = \frac{\pi \alpha_{mn}}{R_0} \quad \text{with} \quad \begin{cases} m = 0, 1, 2, \ldots \\ n = 0, 1, 2, \ldots \end{cases} \quad , \tag{5.4.31}$$

where α_{mn} is the nth root of the equation involving the mth-order Bessel function devided by π.

With the expression

$$k_j = \sqrt{k_r^2 + k_z^2} \tag{5.4.32}$$

we obtain the final result for the frequencies of the normal modes in a cylindrical cavity:

$$f_{nmn_z} = \frac{c}{2} \left[\left(\frac{\alpha_{mn}}{R_0} \right)^2 + \left(\frac{n_z}{L} \right)^2 \right]^{1/2} \quad . \tag{5.4.33}$$

In a real resonator, at least two additional boundary conditions are imposed by the interaction of the standing wave with the internal resonator surface. At the surface, the tangential component of the particle velocity is zero, whereas in the interior of the cavity it is proportional to the gradient of the acoustic pressure. Thus, viscous friction occurs in the transient region, which is responsible for dissipation of energy. Throughout the principal portion of the resonator volume, the expansion and contraction of the gas occurs adiabatically. The very small deviations from a pure adiabatic state change lead to the classical volumetric losses considered before. Near the resonator walls, however, the pressure variations become isothermal due to the high thermal conductivity of the metal wall. This leads to heat conduction, which is responsible for a thermal dissipation process. For pressures in the region of 0.1 – 100 mbar, these viscous and thermal effects take place in a very thin layer at the surface [5.49]:

$$d_\eta = \sqrt{2\eta/\varrho\omega} \quad , \quad d_\lambda = \sqrt{2\lambda M/\varrho C_p\omega} \quad , \tag{5.4.34}$$

where d_η and d_λ are the viscous and thermal boundary layers. Both boundary layer effects impose new conditions for the pressure, and the wave numbers have to be adjusted to the new conditions. For the new eigenvalues $k_j^0 + \delta k_j + \mathrm{i}\Delta k_j$ of the wave numbers, perturbation theory yields, for example for a purely radial mode in a cylindrical cavity [5.28],

$$\frac{\Delta k_j}{k_j^0} = \frac{d_\eta + (\gamma - 1)d_\lambda(1 + 1/R_0)}{L} \tag{5.4.35}$$

and

$$\delta k_j = -\Delta k_j/2 \quad , \tag{5.4.36}$$

where the absorptive contribution and the dispersive contribution depend in a non-linear way on the frequency. The contributions of nonradial modes to surface losses have been calculated in [5.28]. A quantitative theory for the frequency shift and the change of the Q-factor by viscous and thermal boundary layer effects in a cylinder was also reported in [5.57].

These losses are the dominant surface losses in an optimized acoustical resonator, where wave scattering due to surface obstructions such as gas inlet, windows and microphone is kept small. Normally, it is very difficult to take these latter effects into account quantitatively. Additional surface effects that have to be considered only in high precision experiments are the compliance of the chamber walls and dissipation at the microphone diaphragm.

5.4.6 Solutions

In the frequency domain, i.e. for a periodic source function $Q(r,t)$ and taking into account the boundary conditions, the solution of the Helmholtz equation can be found by separation of the space and time dependence, as already mentioned. For the source function one obtains with the laser power $H(r)$

$$\frac{\partial^2 Q(r,\omega)}{\partial t^2} = \mathrm{i}\omega H(r)\exp(\mathrm{i}\omega t) \quad . \tag{5.4.37}$$

The solution for the acoustic pressure is found by an infinite series expansion of the normal mode solutions p_j of the homogeneous equation with mode amplitudes A_j:

$$p(r,\omega) = \sum_j A_j(\omega)p_j(r)\exp(\mathrm{i}\omega t) \quad . \tag{5.4.38}$$

Introducing these functions into the Helmholtz equation (5.4.15) a sum of orthogonal functions is obtained. The time dependence cancels and by projection $\int p_j^* dV$ the amplitude $A_j(\omega)$ for a given mode j is determined:

$$A_j(\omega) = \frac{\mathrm{i}\omega}{(\omega^2 - c^2k_j^2)} \frac{\Lambda}{V} \int_V H(r)p_j^* dV \quad . \tag{5.4.39}$$

The first part of the right-hand side describes the frequency dependence of the solution, i.e. a Lorentzian profile with the two parameters resonance frequency and resonance width. The second part determines the amplitude value. Besides the Λ function, the geometrical coupling between the laser radiation and the acoustical mode is an important factor. This overlap between the spatial structure of the acoustical mode and the laser beam profile is described quantitatively by the overlap integral $\int H(r)p_j^* dV$. For example, this integral is zero for azimuthal modes in a cylinder with the laser beam aligned coaxial to the cylinder axis. Longitudinal modes are only weakly excited in the case of nearly homogeneous absorption of the laser radiation along the cylinder axis, because positive and negative contributions to the integral value essentially cancel. To achieve strong excitation of longitudinal modes, asymmetric excitation is needed. Radial modes possess an ideal geometry for coaxial excitation in a cylinder.

The physical quantity that is measured is the absolute magnitude of the complex pressure function at the microphone position. The complex character of the amplitude, which is caused by the phase shift with respect to the source function, is important if neighboring modes are not well separated and interferences occur. In cylindrical resonators the spacing between excited radial and longitudinal modes is relatively large and can be optimized, in addition, by choosing an appropriate ratio between radius and length. Therefore, interference effects need only be taken into consideration at low pressures or if strong relaxational losses are present that broaden the resonance profiles considerably.

The center frequency and width of the Lorentzian profile in (5.4.39) is determined by two complex numbers, namely the wave vector of mode j

$$k = k_j^0 + \delta k_j + \mathrm{i}\Delta k_j \quad , \tag{5.4.40}$$

with a dispersion and absorption term due to surface perturbations, and the velocity of sound

$$c = c^0 + c_\delta + \mathrm{i}c_\Delta \quad , \tag{5.4.41}$$

again with a dispersion and absorption term due to relaxation, e.g. chemical relaxation.

For the product of the two complex numbers we obtain

$$\begin{aligned} k_j c &= (k_j^0 + \delta k_j + \mathrm{i}\Delta k_j)(c^0 + c_\delta + \mathrm{i}c_\Delta) \\ &= k_j^0 c^0 + k_j^0 c_\delta + \delta k_j c^0 - \Delta k_j c_\Delta + \delta k_j c_\delta \\ &\quad + \mathrm{i}(\Delta k_j c^0 + k_j^0 c_\Delta + \Delta k_j c_\delta + \delta k_j c_\Delta) \quad . \end{aligned} \tag{5.4.42}$$

The resonance frequency is now given by the unperturbed value $k_j^0 c^0$ plus the surface dispersion $\delta k_j c^0$, the relaxation dispersion $k_j^0 c_\delta$ and additional dispersion contributions of higher order. The width of the resonance depends on the surface effect $\Delta k c^0$, the relaxation effect k_j^0 and second-order effects. The second-order effects may be negligible in most cases.

5.5 Applications

5.5.1 Chemical Reaction

With the argon ion laser as light source, the electro-optic modulation system and the small thermostated cylinder, the reaction $N_2O_4 \leftrightarrow 2NO_2$ was studied in the pressure range from 5 to 350 mbar and in the temperature interval of 273–322 K. The analysis of the results obtained for the resonance frequencies, line broadening and signal amplitude yield accurate values for the rate constant of the reaction and the activation energy of the dissociation process [5.37, 55].

Figures 5.9–11 show for a fixed temperature the pressure dependence of the resonance frequency (dispersion caused essentially by reaction), the width of resonance

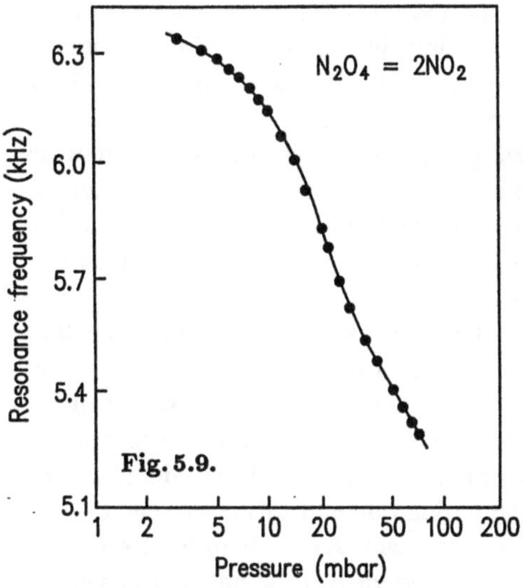

Fig. 5.9.

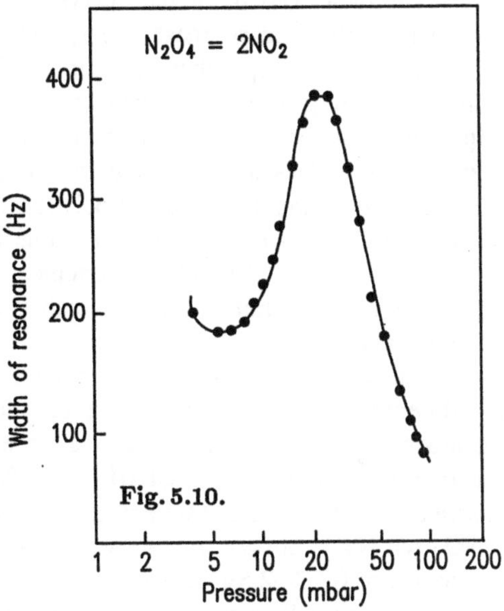

Fig. 5.10.

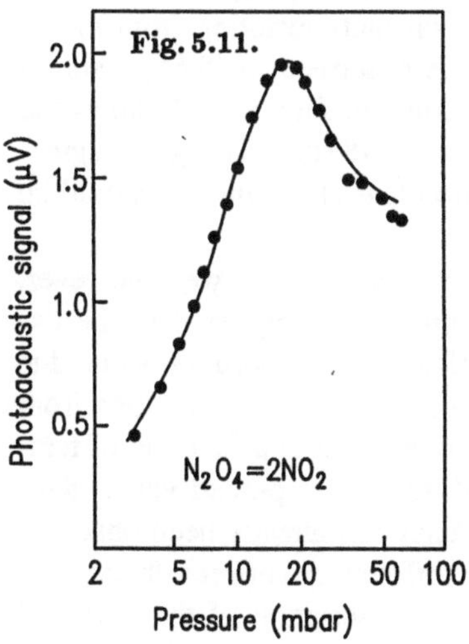

Fig. 5.11.

Fig. 5.9. The resonance frequency of the first radial mode in a cylinder as a function of pressure in the chemical relaxation zone of the system $N_2O_4 \leftrightarrow 2NO_2$

Fig. 5.10. Width of resonance of the first radial mode in a cylinder as a function of pressure in the chemical relaxation zone of the system $N_2O_4 \leftrightarrow 2NO_2$

Fig. 5.11. Photoacoustic signal of the first radial mode in a cylinder as a function of pressure in the chemical relaxation zone of the system $N_2O_4 \leftrightarrow 2NO_2$

(localized broadening caused essentially by reaction) and of the photoacoustic signal PAS = $A_{max}/Q = A_{max}\Delta f/f$, which is the maximum amplitude of the resonance divided by the resonant enhancement. The data points were fitted with a model including the effects of chemical relaxation, viscous and thermal surface losses, Stokes-Kirchhoff volume losses, and relaxation of the vibrational and rotational degrees of freedom. The latter two degrees of freedom could not be separated and an overall relaxation time of 10 ns bar at 293 K was globally assigned to the relaxation of these internal degrees of freedom. This value is not very accurate because it is extrapolated from a relatively low f/p region, however, it is in qualitative agreement with the value estimated from the Lambert Salter plot for the lowest vibration in N_2O_4 at 79 cm^{-1}, which is the torsion mode of the molecule around the N–N bond [5.58].

The data needed for the analysis of the $N_2O_4 \leftrightarrow 2NO_2$ system were taken from the following sources: The specific heat of N_2O_4 and NO_2 was calculated from the spectroscopic data given in [5.59] and the transport properties were taken from [5.60, 61]. The equilibrium constant was first determined in [5.62], whereas measurements performed later resulted in values about 2%–3% higher [5.63]. This inaccuracy creates the same uncertainty in the rate constants. The values reported by different authors for the dissociation energy D range from about 53 kJ/mol to about 61 kJ/mol with a more recent one published in [5.64]. For the modelling a mean value of 53.2 kJ/mol was assumed. An inaccuracy of 5% in D causes a 9% uncertainty in the rate constant, which is the major contribution to the error in the present results.

The rate constants resulting from the fit of the experimental data were linearly extrapolated to 1 bar pressure and their logarithm is plotted versus the inverse temperature in Fig. 5.12. The activation energy of the dissociation process can be extracted from this plot as 36.0 kJ/mol. The dissociation of N_2O_4 was also studied with a CO_2 waveguide laser and the large cylindrical resonator [5.31]. Since neither NO_2 nor N_2O_4 possess any absorption in the tuning range of this laser near 10 μm, amounts of 0.5%–2.5% of SF_6 were added as sensitizer. As no thermostat was employed in these IR experiments, the results were corrected for temperature changes as described in the experimental section. The rate constants resulting from the fit were normalized to 300 K. Similar experiments were described in [5.65], using a temperature jump method and pulsed CO_2 laser excitation. Figure 5.12 shows the corrected rate constant of the Ar^+ laser experiments at 488 nm. The good agreement indicates that photochemical effects are not important at 488 nm, as should be expected [5.66].

The theoretical model predicts the resonance frequencies very well, however, the values predicted for the sound absorption are up to 20% higher than the experimental ones. The difference between the theoretical sound absorption value due to chemical relaxation and the measured value reduced by the contributions from transport losses and vibrational and rotational relaxation losses is a function of temperature. This can be seen in Fig. 5.13, where the difference is plotted versus temperature. It is interesting to note that such a discrepancy has already been observed before, in the same order of magnitude [5.67]. The differences between theory and experiment are attributed to an incomplete theoretical description of the chemical

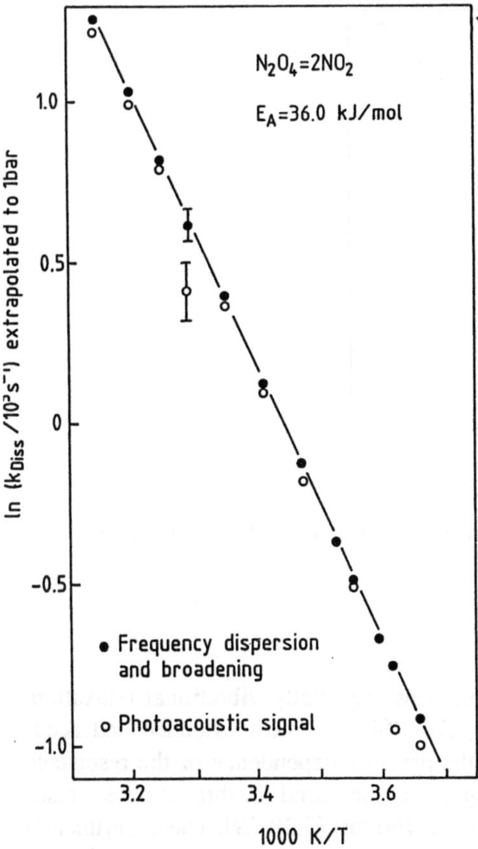

◄ **Fig. 5.12.** Arrhenius plot of the rate constants determined for N_2O_4 dissociation from the frequency dispersion and resonance broadening and the photoacoustic signal

Fig. 5.13. Deviations between theory and experiment in the case of resonance broadening as a function of pressure for the $N_2O_4 \leftrightarrow 2NO_2$ system

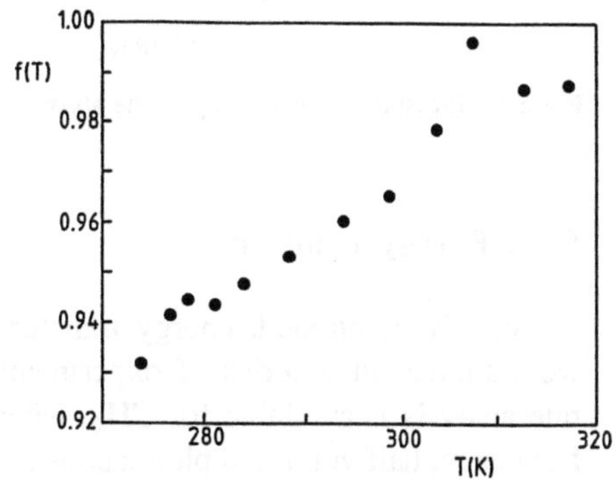

relaxation process, because deviations of the resonator from the ideal geometry tend to increase the width of the resonances. The deviations of the real resonator from the ideal geometry, although difficult to estimate quantitatively, cannot be responsible for the effect. Additional assumptions concerning the dissociation mechanism give a qualitative explanation of the effect as proposed in [5.67]. If the reaction takes place via an intermediate preactivated state, the total reaction is composed of two steps governed by two different rate constants. The two processes may possess different phase shifts as a function of temperature. With increasing separation of the relaxation times at lower temperatures, the effective sound absorption would decrease. This mechanism, however, introduces at least two additional parameters, namely the rate constant for preactivation and the energy of the preactivated state.

When inert gases are added to the N_2O_4–NO_2 mixture, other collision partners are available to activate the dimer before dissociation. Measurements with N_2 and Ar in different concentrations were performed and a shift of the position of the frequency dispersion and the absorption maximum was observed. In Fig. 5.14 the pressure where maximum absorption occurs is plotted versus the inverse mole fraction of the nitrogen oxides. Obviously the chemical relaxation time depends only on the number of collisions between nitrogen oxides, since the product of the corresponding pressure with their mole fraction is constant.

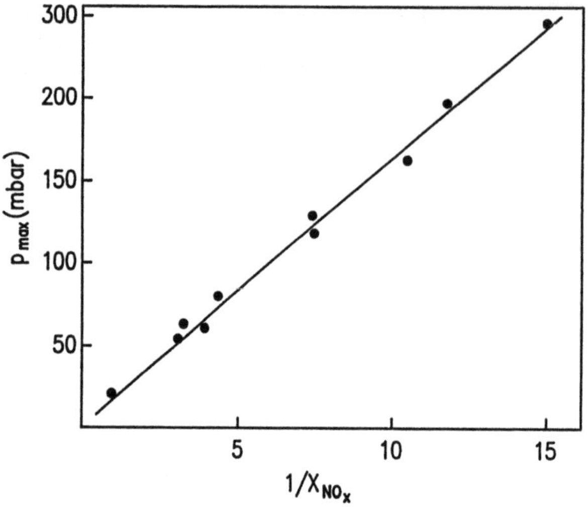

Fig. 5.14. Pressure corresponding to the absorption maximum versus inverse NO_x mole fraction

5.5.2 Energy Transfer

As already mentioned, energy transfer processes, especially vibrational relaxation, were studied in a series of experiments [5.24–30]. The most extended and accurate analysis is available for CH_4, where the pressure dependence of the resonance frequency, halfwidth and photoacoustic signal was measured for three different resonances in a cylindrical resonator between 1 and 760 torr [5.28, 29]. The experimental setup consisted of a 3.39 μm HeNe laser exciting the ν_3 mode of CH_4, a mechanical chopper and a large cylinder of about 10 cm length and diameter. In the theoretical modelling of the data obtained by computer controlled analysis, surface and volume losses due to transport processes and vibrational and rotational relaxation have been taken into account.

The values for the vibrational frequencies of CH_4 [5.68] (needed to calculate the heat capacity), the viscosity [5.69], the thermal conductivity [5.69] and the second virial coefficient [5.70] were taken from the literature and the relaxation times were determined by the fitting procedure described.

Figure 5.15 shows the pressure dependence of the resonance frequency of the first radial mode. The frequency shifts due to surface losses at low pressures and due to the real gas correction at high pressures are represented as deviations between the measured points and the ideal dispersion curve describing vibrational relaxation. Figure 5.16 presents the halfwidth of the resonance as a function of pressure for the first radial mode. The different contributions to the broadening are specified. From the bell-shaped curve it can be seen that vibrational relaxation has a large broadening effect around 40 torr, which corresponds to the center of the relaxation zone at 5.4 kHz. The broadening effect of the surface losses increases drastically at low pressures, whereas the volume losses and rotational relaxation make only minor contributions in the investigated pressure interval. Figure 5.17 exhibits the

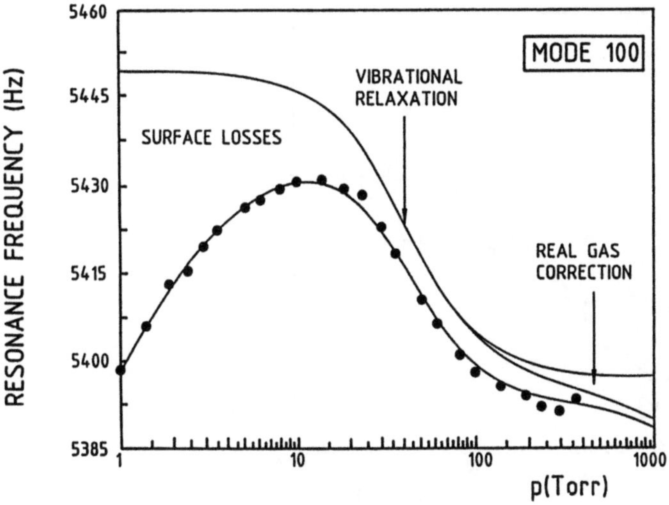

Fig. 5.15. Pressure dependence of the resonance frequency of the first radial mode for CH_4 in a cylindrical resonator. The theoretical analysis is indicated

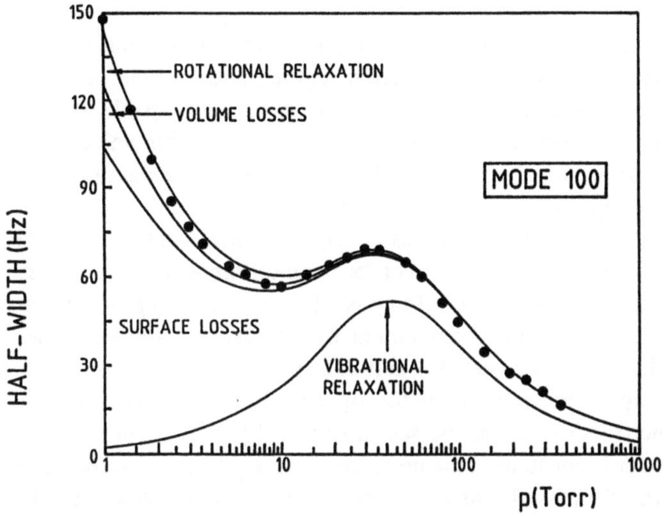

Fig. 5.16. Pressure dependence of the halfwidth of the first radial mode for CH_4 in a cylindrical resonator. The theoretical analysis is shown

photoacoustic signal observed for the three modes excited in the cylindrical resonator and the calculated signal for the first radial mode based on the vibrational relaxation time determined from the frequency dispersion and resonance broadening.

The rotational relaxation time of 1.18 ns bar was determined from the resonance broadening effect and corresponds to about 12 gas kinetic collisions needed for equilibration of the rotational degrees of freedom. Values found in the literature for this energy exchange process vary between 8.6 and 17 collisions [5.45, 58, 71]. The excellent agreement illustrates the level of quantitative understanding of the acoustic

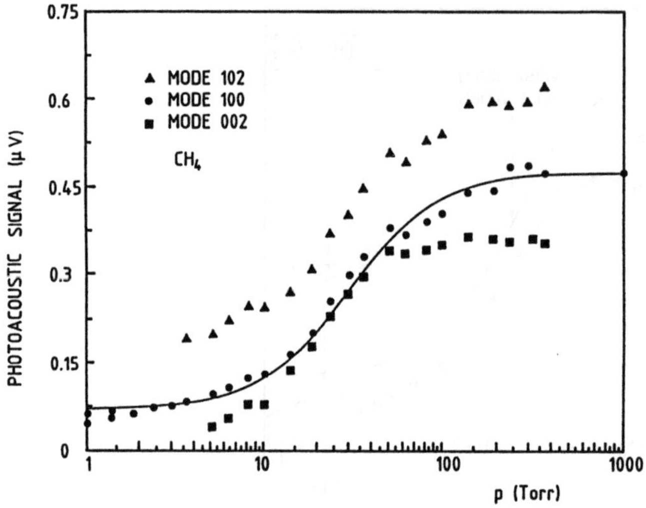

Fig. 5.17. Photoacoustic signal for the acoustic modes (002), (100) and (102) measured for CH_4 in a cylindrical resonator. The curve drawn through the data points of the (100) mode was calculated for a vibrational relaxation time of 1.74μs bar

resonance method. It is important to note that the rotational relaxation time was not determined by measurements in the main relaxation region, given by the center of the bell-shaped broadening curve, because experiments were not possible at such low pressures. From the measurement of the frequency shift no reliable value could be extracted for this relaxation time because the dispersion effect extends only over a much smaller pressure range than the broadening effect.

The vibrational relaxation of CH_4 was investigated in the whole relaxation zone by observing the dispersion and broadening effect for three different acoustic modes. The analysis yielded the accurate value of 1.74 μs bar for the vibrational relaxation time at 292 K. This result is in excellent agreement with literature values obtained by different methods, which vary between 1.4 and 2.0 μs bar [5.58, 72].

The photoacoustic signal depends upon the optical absorption coefficient in different ways for distinct modes, as can be seen in Fig. 5.17 [5.29]. For CH_4 the absorbed total optical power is constant in the pressure range 3–760 torr. Below 3 torr the absorbed optical power decreased drastically and this was taken into account by the double points in Fig. 5.17. The main pressure variation of the photoacoustic signal above 3 torr was assigned to relaxation processes. The detailed analysis of this pressure dependence yielded information on the vibrational energy distribution which is difficult to obtain by other methods such as laser-induced fluorescence.

To model the photoacoustic signal shown in Fig. 5.17 for the first radial mode, a three-level model was assumed for CH_4 relaxation following laser excitation of the ν_3 mode. This is illustrated in Fig. 5.18. One group of modes is given by the two stretching vibrations ν_1 and ν_3. The second group contains the bending vibrations ν_2 and ν_4, the first overtones $2\nu_2$ and $2\nu_4$, and the combination band ($\nu_2 + \nu_4$). The third group is given by the translational and rotational modes. The relaxation time for near-resonant vibrational to vibrational (V–V) energy exchange between ν_1, ν_3 and $2\nu_2$, $2\nu_4$, ($\nu_2 + \nu_4$) was taken from laser-induced fluorescence experiments (5.3

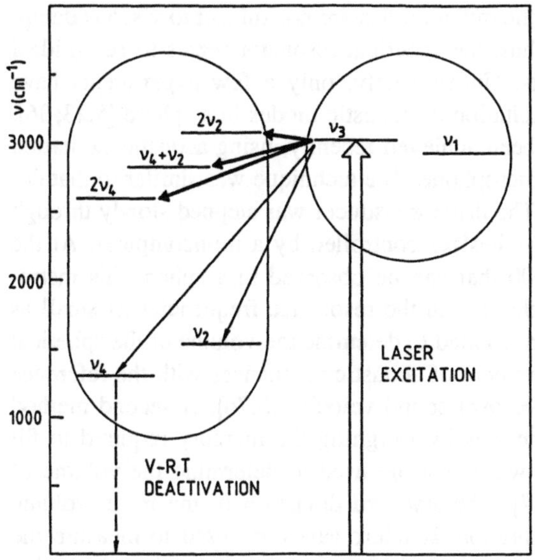

Fig. 5.18. Lowest vibrational levels of CH_4 with laser-excited mode. Different relaxation pathways are indicated. Thick arrows: double-quantum V–V transfer; thin arrows: single-quantum V–V transfer; dashed arrow: V–R, T transfer

ns bar) [5.73,74]. The relaxation time for vibrational to rotational and translational (V–R,T) energy exchange from the ν_4 mode was taken from the present dispersion and absorption experiments (1.74 μs bar).

The fact that a finite photoacoustic signal is observed at low pressures around 1 torr proves that very fast near-resonant V–V exchange processes contribute to the signal. The ratio of this fast translational energy release to the slow energy release thus contains additional information about the internal vibrational energy distribution in the molecules. Application of the three-level model leads to the conclusion that not only does near-resonant V–V transfer [$\nu_1, \nu_3 \rightarrow 2\nu_2, 2\nu_4, (\nu_2 + \nu_4)$] occur, but nonresonant V–V transfer ($\nu_1, \nu_3 \rightarrow \nu_2, \nu_4$) must also contribute to a small but measurable extent to the equilibration process between stretching and bending modes. The quantitative analysis yields a relaxation time of 35 ns bar for this latter nonresonant V–V exchange process [5.29].

5.5.3 Thermophysical Properties and Fundamental Constants

The resonant frequencies of an acoustic resonator are determined by the dimension of the cell and by the physical properties of the gas that determine the sound velocity, such as heat capacity, virial coefficient, temperature and molecular mass. If the geometry of the resonator is known with high accuracy the measurement of acoustic resonance frequencies yields values for the sound velocity and the corresponding properties with high precision.

Of special interest in this respect is the sphere, because it is the highest symmetry resonator and has a surface-to-volume ratio which is smaller than in any other cavity of the same volume. This property leads to a reduction of surface losses. In addition, the resonance frequencies of the radially symmetric modes of a gas-filled sphere are not affected by volume-preserving deformations of the resonator's shell in first-order perturbation theory, and the radial modes are not subject to viscous damping at the cavity wall [5.75–77]. Thus, the spherical resonator seems to be an ideal tool for high-precision experiments. Unfortunately, only a few experiments have been performed employing laser excitation of acoustic modes in a sphere [5.23, 36]. However, substantial progress has been achieved recently, using acoustic excitation of resonator modes with a second microphone. The technique was similar to that described before for laser excitation. The drive transducer was stepped slowly through the resonance using a frequency synthesizer controlled by a minicomputer. At the very high Q-factors of about 10 000 that can be observed in a sphere this instrumentation is capable of detecting changes in the resonance frequencies as small as 10^{-7} [5.76]. Different methods were applied to detemine the volume of the spherical cavity. One method was the measurement of acoustic resonances with the reference gas argon, which possesses a well-known sound velocity [5.76]. A second method for the determination of the volume was by weighing the mercury required to fill the resonator [5.35]. Finally, microwaves can be used to determine the volume of a spherical acoustic resonator [5.77]. The standard deviation of the mean volume was 0.29 ppm for the mercury calibration. If microwaves are used to measure the resonator volume, an acoustic experiment with the same resonator can determine the ratio of the sound velocity to the velocity of light. This ratio can be measured with an accuracy approaching a few parts per million, thus, microwave resonance experiments are a promising alternative to the mercury procedure [5.77].

Precision acoustic measurements studying radial resonances in a spherical resonator were performed for C_2H_4 in the temperature range 0°–100° C and the pressure range 0.15–1.0 MPa [5.78]. In this high-pressure region surface losses are small and also vibrational relaxation has only a minor effect, because the relaxation zone is located at pressures about two orders of magnitude lower. These low losses lead to very high Q-factors, typically 2000–10 000. In the case of these narrow resonance profiles with a halfwidth of near 1 Hz, the nonselective acoustic excitation of modes with a microphone causes no difficulty. The acoustic sources excite all cavity modes, radial and nonradial, with comparable amplitudes. The values obtained for the sound velocity had a precision of 0.003%. The measurements yielded accurate data for the virial coefficient and the vibrational relaxation time of C_2H_4, despite the fact that the analysis of the relaxation process was performed far away from the relaxation zone.

Recently, the value of the universal gas constant was redetermined by sound velocity measurements in argon at the triple point of water (273.16 K). The resonance frequencies of the first five radial modes of a sphere were measured as a function of the frequency of the drive transducer in the frequency range 2.5–9.5 kHz and the pressure range 25–500 kPa. The resonance frequencies were converted to sound velocity values using the volume determined by the mercury weight method. The data analysis was based on a theoretical model of the spherical resonator, taking into account the following phenomena: (1) the thermal boundary near the resonator wall, including imperfect thermal accommodation (2), bulk dissipation, and (3) the coupling of shell motion and the gas motion. Perturbation theory was used to account for surface roughness, effects of transducers, and imperfections in geometry.

The value obtained for the sound velocity in the zero-pressure limit was

$$c^2 = 94756.178 \pm 0.065\,\mathrm{m^2/s^2} \quad .$$

With this value, a new value for the universal gas constant was determined from $R = c^2 M/\gamma T$:

$$R = 8.314471 \pm 0.000014\,\mathrm{J/mol\,K} \quad .$$

With this measurement the standard error of R has been reduced by a factor of 5 to 1.7 ppm [5.35, 78].

The generally accepted value of the gas constant, 8.31441 J/mol K, which is still given in most textbooks, was determined by the conventional method of limiting densities [5.79]. This value was improved by measurements using a high-frequency acoustic interferometer. The length changes in a variable-length cylindrical cavity were measured with an accuracy of two parts per million, yielding a value of 8.31448 J/mol K with an accuracy of 25 ppm [5.80]. The recent most accurate measurements with a spherical resonator, discussed in detail above, correspond to a low-frequency interferometer with fixed path length [5.78]. The improvement achieved by the acoustic resonance techniques is shown in Fig. 5.19.

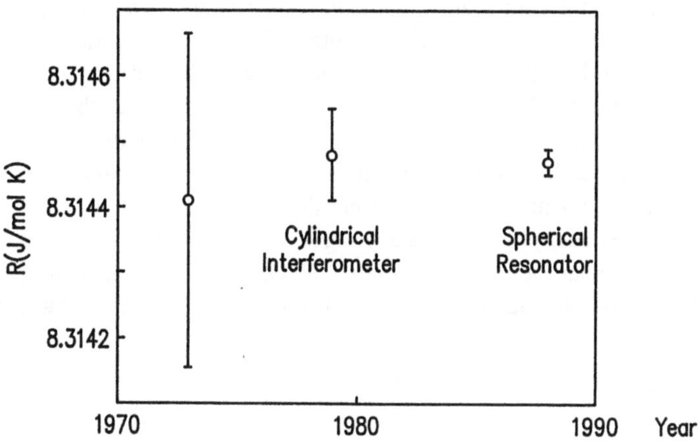

Fig. 5.19. Improvements in the accuracy of the universal gas constant by acoustic techniques; 1979: variable path length cylindrical resonator; 1988: fixed volume spherical resonator

5.5.4 Condensation Effects

In sound velocity experiments with nitrogen a sharp decrease in the apparent value, which was as large as 1%, was observed when the two-phase boundary was approached at low temperatures [5.81]. The observation of this anomaly led researchers to perform similar sound velocity measurements on propane under suitable conditions of temperature and pressure in spherical and cylindrical resonators [5.82]. The anomalous decrease of the sound velocity was modelled on the basis of a pre-condensation effect occurring at the walls when the pressure is increased toward the saturated vapor pressure. The authors showed that the rate of change of the

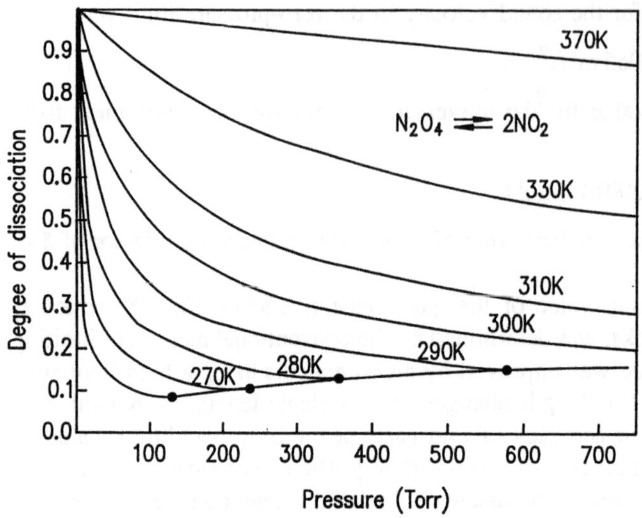

Fig. 5.20. Degree of dissociation versus pressure in the system $N_2O_4 \leftrightarrow 2NO_2$ for different temperatures. The points indicate phase transitions

thickness of the film coating the wall makes a major contribution to the real and imaginary parts of the acoustic admittance. This change of thickness results from condensation and evaporation during the course of an acoustic cycle and may occur also at pressures much lower than the thermodynamic vapor pressure. The theoretical model explains at least qualitatively several features of the experimental results and predicts additional resonator losses if the precondensation effect appears at low pressures and in films only a few molecular diameters thick.

Similar effects near the two-phase boundary were observed for the system $N_2O_4 \leftrightarrow 2NO_2$. The relative amount of dissociated N_2O_4 is plotted as a function of pressure in Fig. 5.20 for various temperatures. If the system moves along an

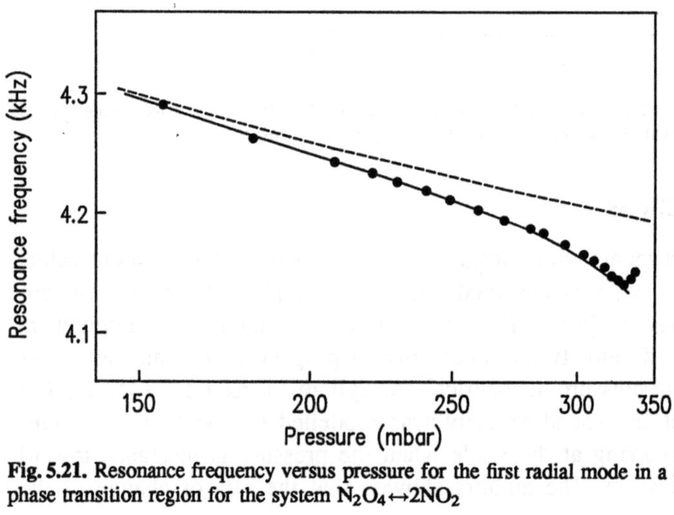

Fig. 5.21. Resonance frequency versus pressure for the first radial mode in a cylindrical resonator in the phase transition region for the system $N_2O_4 \leftrightarrow 2NO_2$

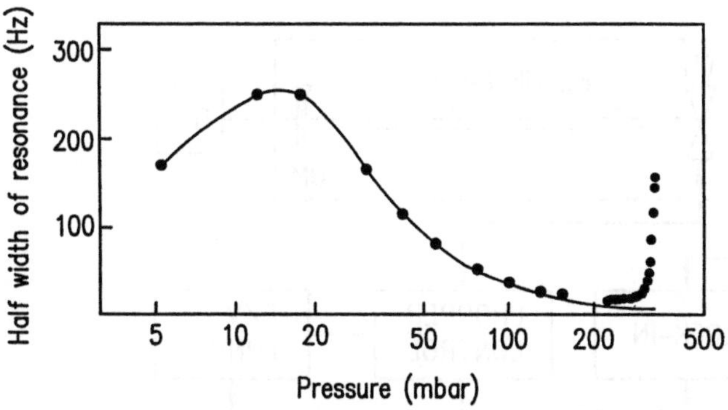

Fig. 5.22. Halfwidth of resonance versus pressure in the chemical relaxation and phase transition region for the system $N_2O_4 \leftrightarrow 2NO_2$

isotherm in the direction toward higher pressures it approaches the intersection between the isotherm and the vapor pressure curve indicating the equilibrium between the condensed phase and the vapor phase. Near these conditions the resonance frequencies deviate considerably from the ideal behavior of the gas mixture, as can be seen in Fig. 5.21. Even more pronounced is the sharp increase of the halfwidth of the resonance profiles in this pressure region, which is compared with the corresponding broadening effect of the chemical relaxation process in Fig. 5.22. It is interesting to note that the relative size of the effect obviously correlates with the energetics of the corresponding effect, namely the dissociation energy of N_2O_4 (53.2 kJ/mol) and the heat of evaporation (38 kJ/mol). The behavior of the resonance frequency and halfwidth can be understood at least qualitatively in terms of a simple model assuming the formation of clusters in the resonator volume. Even just the simple assumption of an average cluster size in dynamical equilibrium with the molecules explains qualitatively the resonance broadening and decrease of the resonance frequency. Of course, there is an additional decrease due to the higher average molecular mass of the species in the mixture. In a quantitative theory the cluster distribution has to be taken into account.

5.5.5 Intracavity Experiments

In many previous photoacoustic experiments resonance cells were placed inside the optical laser resonator ("intracavity"), where a higher laser power can be obtained, to improve the signal. The goals of these intracavity experiments was either to increase the sensitivity of trace gas detection or to monitor high overtone transitions with weak line strength. In these investigations the microphone signal was considered as proportional to the sample absorption without discussing the theory of acoustic resonances in detail.

Figure 5.23 shows a setup with a cylindrical cell inside the laser cavity used to study the intracavity photoacoustic signal quantitatively and to make a comparison with extracavity operation. Measurements were performed for C_2H_4, CH_3Cl, CH_3F, C_2H_6 and CH_4, where the optical absorption coefficients determined by transmission experiments increase from C_2H_4 to CH_4, employing excitation with a 3.39μm HeNe

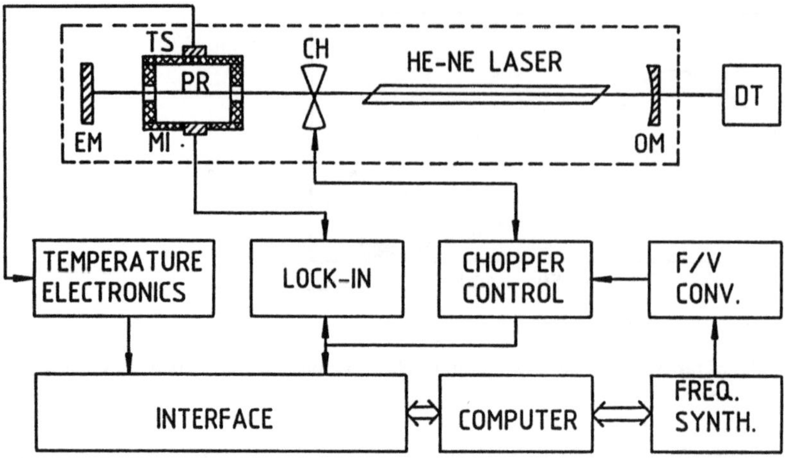

Fig. 5.23. Scheme of experimental setup of intracavity experiments; EM: end mirror; TS: temperature sensor; MI: microphone; PR: photoacoustic resonator; CH: mechanical chopper; OM: output mirror; DT: thermopile detector

laser [5.34]. The mechanical chopper operated in the laser cavity in this experiment modulated the laser field with one of the resonance frequencies of the cylinder. The photoacoustic cell acted as an intracavity power detector and also as a source of additional losses for the laser activity, depending on the pressure and the absorption coefficient of the gas. The signal strength and Q-factor of the first radial mode of the cylinder were determined. Good agreement between experiment and theory was observed for all gases in the case of low optical losses in the laser resonator. Large deviations between experimentally determined signals and theory were found for high losses, e.g. near laser threshold.

Figure 5.24 gives a comparison between the extracavity and intracavity signal for C_2H_4, which is a weak absorber at 3.39 μm [5.32]. Threshold for laser operation was not reached in this case, even at the highest pressures studied. The intracavity

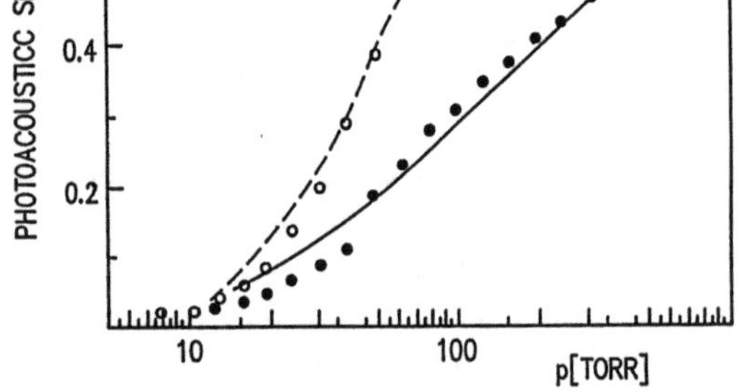

Fig. 5.24. Comparison of the photoacoustic signal of the first radial mode in a cylinder filled with C_2H_4 for extracavity and intracavity operation. Solid line: calculated extracavity signal. Dashed line: intracavity signal calculated assuming that the intracavity power is $2/T$ times the output of the laser

power was about a factor of two higher than the extracavity power. This can be explained by the transmittance T of the output mirror, which was about 36% at 3.39μm. At the small losses introduced in these experiments the intracavity power can be regarded as constant and estimated to be $2/T$ times the extracavity power. As expected, the resulting intracavity enhancement of the photoacoustic signal varies between 1.5 and 2.5. Theory gives a good description of the extracavity signal and also of the intracavity results as long as the intracavity losses are small. The deviations at C_2H_4 pressures below 150 torr indicate that the intracavity losses are small. The deviations at C_2H_4 pressures above 150 torr indicate that the intracavity losses are no longer small and that the $2/T$ approximation is no longer valid.

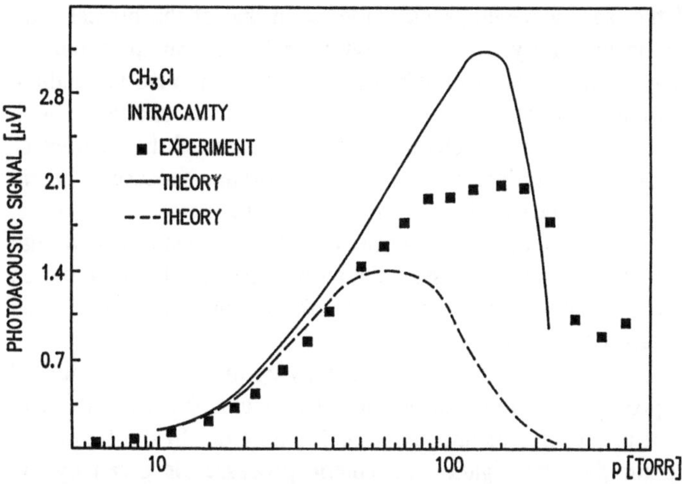

Fig. 5.25. Intracavity photoacoustic signal versus pressure for CH_3Cl. Squares: experimental data. Solid line: calculated photoacoustic signal assuming exponential gain. Dashed line: calculated photoacoustic signal assuming linear gain

Figure 5.25 shows the pressure dependence of the intracavity signal of the first radial mode of the cylinder for CH_3Cl, which possesses a higher absorption than C_2H_4 [5.33, 34]. One would expect the laser oscillation to stop completely as soon as the losses become larger than the laser gain. This is indeed observed. No output at 3.39 μm was measured above 250 torr and the photoacoustic signal was not altered if the end mirror of the laser cavity was blocked. The still observable photoacoustic signal is probably due to amplified spontaneous emission, a phenomenon commonly encountered in high gain lasers. For CH_3Cl two approaches were developed to describe the signal. The first one, represented by the upper line, assumed exponential gain as a function of path length through the active laser medium. This is the usual case for small losses. The lower, dashed line shows the prediction of the theory in the case of a linear gain per unit path length in the light- amplifying medium. Both models describe the low-loss output, but for high losses near laser threshold the reality lies in between the limiting situations of exponential and linear response. The breakdown of laser activity occurring as a consequence of the destroyed feedback inside the cavity is described reasonably well by the models considered.

These results illustrate possible ways to study the laser behavior by intracavity experiments. The ability of the photoacoustic resonator to detect the deposited power and to vary the internal losses can be used to determine laser properties such as saturation intensities, small-signal gain or optimum output coupling.

5.6 Conclusions

In photoacoustic resonance spectroscopy the acoustic resonances of a cavity such as a sphere or cylinder are studied. These acoustic modes can be excited with a modulated cw laser if the laser wavelength coincides with one of the optical transitions of the molecules in the cavity. By doping with a relatively small amount of an optically absorbing gas, properties of nonabsorbing species can also be studied. Compared with conventional excitation methods, such as acoustic excitation with a microphone, laser excitation is mode selective. The reduction of the number of excited modes by symmetry constraints may be very important, because coherent superposition of neighboring modes leads to a distortion of the resonance profiles, which must be corrected in high precision experiments. Thus, in systems with high losses, due, for example, to relaxation processes, the more expensive laser excitation may be advantageous. On the other hand, redundant information is obtained by the investigation of several acoustic modes.

The information is obtained in acoustic spectroscopy by fitting the undisturbed or corrected resonance profiles to Lorentzian profiles. This yields the resonance frequency, the halfwidth of the resonance and the signal amplitude. In the frequency domain the time scale for the investigation of kinetic processes is given by the frequency-to-pressure ratio. The frequency is fixed to certain values by the eigenfrequencies of the cavity. Therefore, the pressure is varied over a large range. At low pressures the physical limits of acoustic resonators are given by the breakdown of the kinetic gas theory. This limit is reached if the transport processes cause boundary layers of a thickness comparable to the dimensions of the resonator or acoustical wavelength. This is usually the case at pressures below 0.1 mbar. The lowest pressure where a reasonable signal-to-noise ratio can be achieved determines the minimum molecular collision frequency and thus the time resolution in collision processes such as energy transfer and chemical reactions. In reality one has the advantage of long tails in the localized dissipation zones, especially for sound absorption. This allows the determination of relaxation times at pressures one or two orders of magnitude away from the pressure characterizing the dissipation center. Usually, the width of the resonances decreases with increasing pressure, because the surface losses become smaller. At high pressures the real gas correction increases, and finally the useful pressure range is limited by the condensation process, which can also be studied by the photoacoustic technique. If miniature microphones are used to detect the resonances, their response limits the frequency range to values below 20 kHz.

The high information content of the resonance method is extracted from the pressure dependence of the eigenfrequencies, the resonance halfwidth and the signal amplitude. The resonance frequencies can be determined with the highest precision. If the geometry of the resonator is known, the most accurate value of the sound velocity can be obtained. The recent redetermination of the universal gas constant clearly shows that with low-frequency fixed volume resonators a higher accuracy can be achieved than with high-frequency acoustic interferometers of variable length. The frequency dispersion due to energy transfer and classical relaxation processes can be described quantitatively.

The halfwidth of the resonance depends on irreversible energy dissipation and essentially measures sound absorption. The values obtained are not as accurate as for the resonance frequencies. Nevertheless, this is an important quantity, because the broadening effect due to relaxation processes extends over many orders of magnitude, and therefore allows the determination of relaxation times in a large pressure range. In the case of energy transfer processes a quantitative description of the broadening effect seems to be possible, however, for chemical reaction processes deviations between theory and experiment are observed for the broadening of the acoustic resonance profile.

The pressure dependence of the signal amplitude yields additional information on the optical absorption process and on energy exchange between the excited state and the other vibrational degrees of freedom. To obtain the photoacoustic signal without resonance amplification the Q-factor of the resonance must be measured. For this corrected signal a quantitative theory is available, e.g. for the laser-excited acoustic modes in a cylinder. This is the main advantage of resonant operation of a photoacoustic cell. Nonresonant operation of a small cell may give a higher signal, however, a theoretical description is no longer possible.

The photoacoustic resonance method yields detailed information on a variety of system properties such as optical, thermophysical and kinetic properties. If the measurements are perfomed in the ideal pressure range, e.g. for transport coefficients at low pressures, for relaxation processes such as energy exchange or chemical reaction in the main relaxation zone, or for deviations from the ideal gas behavior in the high pressure region, a very high accuracy can be achieved, which can compete with any other technique.

Acknowledgements. Financial support of our research work in the field of photoacoustics, mainly reviewed in this article, by the Deutsche Forschungsgemeinschaft, the European Community and the Fonds der Chemischen Industrie is gratefully acknowledged.

References

5.1 A.G. Bell: Am. J. Sci. **20**, 305 (1880)
5.2 W.C. Röntgen: Philos. Mag. **11**, 308 (1881)
5.3 J. Tyndall: Proc. R. Soc. London **31**, 307 (1881)
5.4 A.G. Bell: Philos. Mag. **11**, 510 (1881)
5.5 C.F. Dewey, Jr., R.D. Kamm, C.E. Hackett: Appl. Phys. Lett. **23**, 633 (1973)
5.6 E. Max, L.-G. Rosengreen: Opt. Commun. **11**, 422 (1974)
5.7 P.D. Goldan, K. Goto: J. Appl. Phys. **45**, 4350 (1974)
5.8 A.M. Angus, E.E. Marinero, M.J. Colles: Opt. Commun. **14**, 223 (1975)

5.9 S. Shtrikman, M. Slatkine: Appl. Phys. Lett. **31**, 830 (1977)
5.10 R. Gerlach, N.M. Amer: Appl. Phys. Lett. **32**, 228 (1978)
5.11 L.J. Thomas III, M.J. Kelly, N.M. Amer: Appl. Phys. Lett. **32**, 736 (1978)
5.12 K.P. Koch, W. Lahmann: Appl. Phys. Lett. **32**, 289 (1978)
5.13 N. Ioli, P. Violino, M. Meucci: J. Phys. E. **12**, 168 (1979)
5.14 D.H. Leslie, G.L. Trusty: Appl. Opt. **20**, 1941 (1981)
5.15 O. Poizat, G.H. Atkinson: Anal. Chem. **54**, 1485 (1982)
5.16 M. Seaver, T.J. Manuccia, J.R. McDonald: Chem. Phys. Lett. **123**, 164 (1986)
5.17 G.L. Loper, J.A. Gelbwachs, S.M. Beck: Can. J. Phys. **64**, 1124 (1986)
5.18 M.W. Sigrist: In *Photoacoustic and Photothermal Phenomena*, ed. by P. Hess, J. Pelzl, Springer Ser. Opt. Sci., Vol.58 (Springer, Berlin, Heidelberg 1988) p. 114
5.19 P. Hess: Top. Curr. Chem. **111**, 1 (1983)
5.20 J. Gelfand, W. Hermina, W.H. Smith: Chem. Phys. Lett. **65**, 201 (1979)
5.21 W.H. Smith, J. Gelfand: J. Quant. Spectrosc. Radiat. Transfer **24**, 15 (1980)
5.22 K.V. Reddy, D.F. Heller, M.J. Berry: J. Chem. Phys. **76**, 2814 (1982)
5.23 A. Karbach, J. Röper, P. Hess: Rev. Sci. Instrum. **55**, 892 (1984)
5.24 K. Frank, P. Hess: Chem. Phys. Lett. **68**, 540 (1979)
5.25 K. Frank, P. Hess: Ber. Bunsenges. Phys. Chem. **84**, 724 (1980)
5.26 V.P. Zharov, S.G. Montanari: Opt. Spectrosc. (USSR) **51**, 66 (1981)
5.27 A. Karbach, J. Röper, P. Hess: Chem. Phys. **82**, 427 (1983)
5.28 A. Karbach, P. Hess: J. Chem. Phys. **83**, 1075 (1985)
5.29 A. Karbach, P. Hess: J. Chem. Phys. **84**, 2945 (1986)
5.30 R.H. Johnson, R. Gerlach, L.J. Thomas III, N.M. Amer: Appl. Opt. **21**, 81 (1982)
5.31 M. Fiedler, P. Hess: In *Photoacoustic and Photothermal Phenomena*, ed. by P. Hess, J. Pelzl, Springer Ser. Opt. Sci., Vol. 58 (Springer, Berlin, Heidelberg 1988) p. 79
5.32 J. Röper, G. Chen, P. Hess: Appl. Phys. B **43**, 57 (1987)
5.33 A. Neubrand, J. Röper, P. Hess: In *Photoacoustic and Photothermal Phenomena*, ed. by P. Hess, J. Pelzl, Springer Ser. Opt. Sci., Vol. 58 (Springer, Berlin, Heidelbeg 1988) p. 84
5.34 J. Röper, A. Neubrand, P. Hess: J. Appl. Phys. **64**, 2838 (1988)
5.35 M.R. Moldover, J.P.M. Trusler, T.J. Edwards, J.B. Mehl, R.S. Davis: Phys. Rev. Lett. **60**, 249 (1988)
5.36 A. Karbach, P. Hess: J. Appl. Phys. **58**, 3851 (1985)
5.37 M. Fiedler, P. Hess: To be published
5.38 A.R. Colclough, T.J. Quinn, T.R.D. Chandler: Proc. R. Soc. London A **368**, 125 (1979)
5.39 G. Damköhler: Z. Elektrochem. **48**, 62 (1942)
5.40 G. Damköhler: Z. Elektrochem. **48**, 116 (1942)
5.41 L. Fritsche: Acustica **10**, 189 (1960)
5.42 L. Fritsche: Acustica **10**, 199 (1960)
5.43 K.F. Herzfeld, T.A. Litovitz: *Absorption and Dispersion of Ultrasonic Waves* (Academic, New York 1959)
5.44 P.M. Morse, K.V. Ingard: *Theoretical Acoustics* (McGraw-Hill, New York 1968)
5.45 A.J. Matheson: *Molecular Acoustics* (Wiley-Interscience, London 1971)
5.46 L.B. Kreuzer: J. Appl. Phys. **42**, 2934 (1971)
5.47 L.G. Rosengren: Appl. Opt. **14**, 1960 (1975)
5.48 R.D. Kamm: J. Appl. Phys. **47**, 3550 (1976)
5.49 L.B. Kreuzer: In *Optoacoustic Spectroscopy and Detection*, ed. by Yoh-Han Pao (Academic, New York 1977) p 1
5.50 E. Nodov: Appl. Opt. **17**, 1110 (1978)
5.51 E. Kritchman, S. Shtrikman, M. Slatkine: J. Opt. Soc. Am. **68**, 1257 (1978)
5.52 M.R. Da Silva: Can. J. Phys. **64**, 1098 (1986)
5.53 H.O. Kneser: Ergeb. exakten Naturwiss. **22**, 121 (1949)
5.54 M. Eigen, L. de Mayer: In *Techniques of Organic Chemistry*, Vol. 8, Part 2, ed. by A. Weissberg (Interscience, New York 1963) p. 895
5.55 M. Fiedler: Dissertation, Universität Heidelberg (1989)
5.56 A. Einstein: Sitzungsber. Dtsch. Akad. Wiss. Berlin, Kl. Math. Phys. Tech. 380 (1920)
5.57 R. Gerlach, R.H. Johnson, N.M. Amer: Tech. Digest, Fourth Int. Topical Meeting on Photoacoustic, Thermal and Related Sciences, Montreal (1985)
5.58 J. Lambert: *Vibrational and Rotational Relaxation in Gases* (Clarendon, Oxford 1977) p. 66
5.59 R.F. Holland, W.B. Maier: J. Chem. Phys. **78**, 2928 (1983)
5.60 R. Brokaw, R. Svehla: J. Chem. Phys. **44**, 4643 (1966)
5.61 H. Sonntag, A.C. Tam, P. Hess: J. Chem. Phys. **86**, 3950 (1987)

5.62 M. Bodenstein: Z. Phys. Chem. **100**, 68 (1922)
5.63 L. Harris, K.L. Churney: J. Chem. Phys. **47**, 1703 (1967)
5.64 D.R. Powell, E.T. Adams: J. Chem. Phys. **82**, 1947 (1978)
5.65 P. Gozel, B. Calpini, H. van den Bergh: Isr. J. Chem. **24**, 210 (1984)
5.66 G.H. Myers, D.M. Silver, F. Kaufmann: J. Chem. Phys. **44**, 718 (1966)
5.67 G. Sessler: Acustica **10**, 44 (1960)
5.68 G. Herzberg: *Infrared and Raman Spectra of Polyatomic Molecules* (Van Nostrand, New York 1945)
5.69 R.C. West (ed.): *Handbook of Chemistry and Physics*, 61st ed. (Chemical Rubber Co., Cleveland 1980)
5.70 J.H. Dymond, E.B. Smith: *The Virial Coefficients of Gases and Mixtures* (Clarendon, London 1980)
5.71 R. Holmes, G.R. Jones, N. Pusat: Trans. Faraday Soc. **60**, 1220 (1964)
5.72 E. Avramides, T.F. Hunter: Chem. Phys. **57**, 441 (1981)
5.73 P. Hess, C.B. Moore: J. Chem. Phys. **65**, 2339 (1976)
5.74 P. Hess, A. Kung, C.B. Moore: J. Chem. Phys. **72**, 5525 (1980)
5.75 M.R. Moldover, M. Waxman, M. Greenspan: High Temp. – High Pressures **11**, 75 (1979)
5.76 J.B. Mehl, M.R. Moldover: J. Chem. Phys. **74**, 4062 (1981)
5.77 J.B. Mehl, M.R. Moldover: Phys. Rev. A. **34**, 3341 (1986)
5.78 M.R. Moldover, J.P.M. Trusler, T.J. Edwards, J.B. Mehl, R.S. Davis: J. Res. Natl. Bur. Stand. **93**, 85 (1988)
5.79 T. Batuecas: In *Atomic Masses and Fundamental Constants*, Vol. 4, ed. by J.H. Sanders, A.H. Wapstra (Plenum, London 1972) p.534
5.80 A.R. Colclough, T.J. Quinn, T.R.D. Chandler: Proc. R. Soc. London A **368**, 125 (1979)
5.81 B.A. Younglove, R.D. McCarty: J. Chem. Thermodyn. **12**, 1121 (1980)
5.82 J.B. Mehl, M.R. Moldover: J. Chem. Phys. **77**, 455 (1982)

6. Application of the Photoacoustic Effect to Studies of Gas Phase Chemical Kinetics

G. J. Diebold

With 23 Figures

Over the last decade, a number of applications of the photoacoustic effect to studies of chemical kinetics have appeared in the literature. From this work it is clear that the photoacoustic effect, perhaps as a result of its inherent high sensitivity and selectivity, is a welcome addition to the rather limited number of diagnostic methods chemists have for unravelling complex chemical reaction mechanisms. In addition to analytical applications, a few experimental methods based on the response of the photoacoustic effect to evolved heat have been developed. In pulsed experiments the rate of heat evolution determines the shape of the acoustic waveform; equally, in a modulated, continuous wave experiment the pathways for heat liberation determine the amplitude and phase of the sound wave. Perhaps the response that is most characteristic of chemical reactions vis-a-vis purely physical processes is the liberation of an amount of heat by the absorption of a photon that can exceed the energy of the photon itself. In such cases of the "chemically amplified" photoacoustic effect, the properties of the chemical reaction become paramount in determining the amplitude and phase of the acoustic wave. The object of this review is to summarize photochemical energy release mechanisms insofar as they relate to sound wave production as well as to recount the analytical applications of the photoacoustic effect to gas phase chemical kinetics; the wide range of phenomena and novel effects uncovered over the last several years make the subject of this review pleasing from a scientific viewpoint and suggest this area for further inquiry.

6.1 Pulsed Excitation

6.1.1 Signal Description

In general, the absorption of short-wavelength radiation results in the production of an excited molecular electronic state. Eventually the excitation energy is lost as a result of fluorescence, internal energy conversion to other states, energy transfer through collisions with other molecules, or molecular photodissociation. Only the last of these is an inherently chemical process in that a bond is broken. Although the process of photodissociation itself produces sound waves as a result of a local increase in mole number and through thermalization of the recoil energy of the fragments, the more interesting possibility is the initiation of chemical reactions, which release their own characteristic energy. The time scales for photofragmentation and chemical reaction are generally quite disparate. Photodissociation of a diatomic molecule, for instance, may take place on a subpicosecond time scale; the rate of chemical reaction, on the other hand, depends on the pressure of the reactants and the reactivity of the species themselves, and may proceed at anywhere from the gas kinetic rate to an infinitesimally slow rate.

The production of sound waves by the evolution of heat, whether it be through chemical or physical processes, is a problem in fluid dynamics where the excess pressure $p(\mathbf{r},t)$ is found (for negligible heat conduction) as a solution to the wave equation [6.1],

$$\left(\nabla^2 - \frac{1}{c^2}\frac{\partial^2}{\partial t^2}\right)p(\mathbf{r},t) = \frac{-\beta}{C_p}\frac{\partial}{\partial t}H(\mathbf{r},t)\,, \tag{6.1.1}$$

where c is the sound speed, β and C_p are the isobaric volume expansion coefficient and heat capacity, respectively, of the gas, and $H(\mathbf{r},t)$ is the rate of heat evolution per unit volume at a point $\mathbf{r}$ at time t. For a perfect gas, β is equal to $1/T$, the inverse of the temperature. By making use of the relation [6.2]

$$\gamma - 1 = \frac{T\beta^2}{\kappa_s C_p \rho}\,, \tag{6.1.2}$$

where κ_s is the isentropic compressibility and ρ the density, the proportionality constant β/C_p on the right-hand side of (6.1.1) can be written as $(\gamma-1)/c^2$, where γ is the ratio of the heat capacity at constant pressure to that at constant volume. When the heat evolution is defined by only the temporal and spatial profile of the laser, in a cylindrically symmetric geometry the right hand-side of Eq. (6.1.1) reduces to $-(\alpha\beta/C_p)\partial I(r,t)/\partial t$, where I is the laser intensity, r is the radial coordinate, and α is the absorption coefficient.

Sound production by a cylindrically symmetrical laser beam exciting an optically thin liquid has been discussed by *Bebchuk* et al. [6.3], *Lai* and *Young* [6.4], *Heritier* [6.5] and *Brueck* et al. [6.6], and has been studied experimentally by *Sullivan* and *Tam* [6.7], and *Sigrist* and *Kneubuhl* [6.8]. It should be noted that the calculations for liquids given by these authors apply equally to gases (unless unusually low pressures are employed) except that the contribution from electrostriction can be expected to be comparatively small. The formulation of the problem by *Heritier* will be outlined here, where a Gaussian heating function of the form

$$H(r,t) = \frac{2\alpha\bar{E}_0}{\pi^{3/2}w_0^2\tau_l}\exp\left(-\frac{2r^2}{w_0^2}-\frac{t^2}{\tau_l^2}\right) \tag{6.1.3}$$

is considered. Here, $\bar{E}_0$ is the total energy in the pulse, w_0 is the beam radius, and τ_l is the pulse width; that is, the e^{-1} points are defined as τ_l and $w_0/\sqrt{2}$. To solve the wave equation for the pressure as given by (6.1.1) the time-dependent quantities are written in terms of their Fourier transforms:

$$\begin{aligned} p(r,t) &= \int_{-\infty}^{\infty} d\omega\, p(r,\omega)e^{-i\omega t}\,,\\ p(r,\omega) &= \frac{1}{2\pi}\int_{-\infty}^{\infty} dt\, p(r,t)e^{i\omega t}\,. \end{aligned} \tag{6.1.4}$$

(Heritier evidently uses the complex congugate of these transforms. His derivation is modified here to conform with this convention.) Substitution of the first of these as well as the corresponding Fourier transform of the heating function $H(r,\omega)$ into (6.1.1) gives

$$\left(\nabla^2 + \frac{\omega^2}{c^2}\right)p(r,\omega) = \frac{i\omega\beta}{C_p}H(r,\omega)\,, \tag{6.1.5}$$

which has a Green's function solution of the form

$$\begin{aligned} p(r,\omega) &= \frac{-i\pi}{2}H_0^{(1)}(kr)\int_0^r J_0(k\rho)S(\rho,\omega)\rho d\rho \\ &- \frac{i\pi}{2}J_0(kr)\int_r^\infty H_0^{(1)}(k\rho)S(\rho,\omega)\rho d\rho\,, \end{aligned} \tag{6.1.6}$$

where J_0 and $H_0^{(1)}$ are *Bessel* and *Hankel* functions respectively, $S(r,\omega)$ is the right-hand side of (6.1.5), and k is the wave vector ($k = \omega/c$). The second integral can be neglected far from the source ($r >> 2w_0$), and the remaining integral can be evaluated using standard integrals to give the pressure spectral density as

$$p(r,\omega) = \left(\frac{\alpha\beta\bar{E}_0}{8\pi C_p}\right)\omega H_0^{(1)}(kr)\exp\left(-\frac{\omega^2\tau_l^2}{4} - \frac{\omega^2 w_0^2}{8c^2}\right)\,. \tag{6.1.7}$$

This expression can then be Fourier transformed into the time domain to give

$$p(r,t) = \frac{\alpha\beta\bar{E}_0 c^2}{8\pi C_p w_0^2}\int_{-\infty}^{\infty} dq\ qH_0^{(1)}(q\hat{r})\exp\left(-\frac{q^2}{8} - \frac{q^2\hat{\tau}_l^2}{4} - iq\hat{t}\right)\,, \tag{6.1.8}$$

where the following dimensionless quantities have been defined:

$$q = \frac{\omega w_0}{c}\,,\quad \hat{r} = \frac{r}{w_0}\,,\quad \hat{\tau}_l = \frac{c\tau_l}{w_0}\,,\quad t = \frac{ct}{w_0}\,.$$

This expression can be simplified by using asymptotic forms for $H_0^{(2)}(q\hat{r})$ and the integral can then be expressed in terms of hypergeometric functions. Equally, the integral in (6.1.8) can be computed using the fast Fourier transform method. A representative waveform computed from (6.1.8) is shown in Fig. 6.1. It is of note that when (6.1.8) is expressed in terms of hypergeometric functions, the waveform becomes a function of the parameter

$$\frac{(t - r/c)}{(\tau_l^2 + w_0^2/2c^2)^{\frac{1}{2}}}\,,$$

which shows the signal to be a function of the retarded time $(t - r/c)$ and to have an effective pulse width of $(\tau_l^2 + w_0^2/2c^2)^{\frac{1}{2}}$, so that the greater of the laser pulse width or the acoustic transit time across the beam determines the observed pulse width. *Sullivan* and *Tam* [6.7] have shown that there is agreement between experiments and the theoretical results given by *Heritier*, and by *Lai* and *Young*.

When the absorbing gas is optically thick, as is often the case in the ultraviolet spectral region, the beam is strongly attenuated as it traverses the

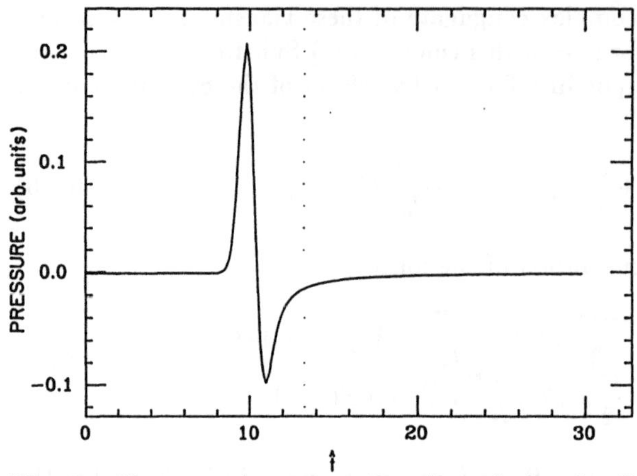

Fig. 6.1. Photoacoustic pressure versus dimensionless time for rapid energy release, computed from (6.1.8) with $\hat{r} = 10$ and $\hat{\eta} << 1$

gas. In Cl_2, for instance, at several mm Hg a 355 nm laser beam with a total pulse energy of 10 mJ is totally absorbed within a few centimeters, making the above calculation inappropriate. *Karcher* et al. [6.9] have investigated the production of sound waves in highly absorbing media and have calculated the shape of the waveform expected for an exponentially absorbed light beam. They consider an infinite absorbing medium a extending above the xy plane bounded by another medium b below the xy plane. The heating function for a sinusoidally modulated beam can be written as

$$H(\mathbf{r}, t) = \alpha I_0 e^{-i\omega t - \alpha z} , \tag{6.1.9}$$

where α is the absorption coefficient, I_0 is the beam intensity, and ω is the modulation frequency. Substitution of this expression into (6.1.1) gives a differential equation in the spatial variable whose solution is

$$\tilde{p}_\alpha(\omega) = -\frac{iI_0\beta c_a}{C_p}\left(\frac{q_\alpha e^{(-\alpha z - i\omega t)}}{1+q_\alpha^2}\right) , \tag{6.1.10}$$

where

$$q_\alpha = \omega/\alpha c_a$$

is the frequency parameter and c_a is the sound speed in medium a, the gas. The total pressure p_a in medium a is the sum of the driving pressure $\tilde{p}_\alpha$ and a travelling wave solution to the homogeneous wave equation; in medium b, the pressure p_b is simply a travelling wave. The acoustic pressure can thus be written as

$$\begin{aligned} p_a &= \tilde{p}_\alpha + \hat{P}_a e^{i(k_a z - \omega t)} , & z > 0 , \\ p_b &= \hat{P}_b e^{-i(k_b z + \omega t)} , & z < 0 , \end{aligned} \tag{6.1.11}$$

where k_a and k_b are wave vectors in media a and b, respectively, and $\hat{P}_a$ and $\hat{P}_b$ are complex amplitudes.

The amplitude factors are found by equating the pressures and accelerations of the two media at their interface. This gives the frequency domain solution for the sound wave in medium a as

$$p_a(\omega) = \frac{i\beta I_0 c_a}{C_p}\left(\frac{q_\alpha e^{(-iq\hat{t}_\alpha - \hat{z})}}{1+q_\alpha^2} - \frac{(q_\alpha\hat{\rho}\hat{c}+i)e^{-iq\hat{\tau}_\alpha)}}{(1+\hat{\rho}\hat{c})(1+q_\alpha^2)}\right), \tag{6.1.12}$$

where the dimensionless time $\hat{t}_\alpha$ and retarded time $\hat{\tau}_\alpha$ are defined as

$$\hat{t}_\alpha = c_a\alpha t\,, \qquad \hat{\tau}_\alpha = c_a\alpha\left(t - \frac{z}{c_a}\right),$$

and where the dimensionless sound speed $\hat{c}$ and density ratio $\hat{\rho}$ are defined as

$$\hat{c} = c_a/c_b \qquad \hat{\rho} = \rho_a/\rho_b\,,$$

where c refers to a sound speed, ρ refers to a density and the subscripts a and b refer to the respective properties in these two media. This expression can be Fourier transformed to give the time domain acoustic response to delta function heating. Evaluation of the Fourier integrals gives the far field pressure in medium a as

$$p_a(t) = \frac{\alpha\beta E_0 c_a^2}{C_p}\begin{cases} \frac{1}{2}e^{\hat{\tau}_\alpha}\,, & \hat{\tau}_\alpha < 0\,, \\ \frac{1}{2}\left(\frac{1-\hat{\rho}\hat{c}}{1+\hat{\rho}\hat{c}}\right)e^{-\hat{\tau}_\alpha}\,, & \hat{\tau}_\alpha > 0\,, \end{cases} \tag{6.1.13}$$

where E_0 is the energy fluence (energy per area) of the beam. The signal thus has a risetime and falltime of $c_a\alpha$. The first terms containing $\hat{z}$ in (6.1.12) can be seen to decay rapidly with distance and have been ignored in the far field solution. Note also that the factor in parenthesis containing $\hat{\rho}$ and $\hat{c}$ is the reflection coefficient $\tilde{R}$ for a plane wave striking an infinite plane boundary at normal incidence. For large $\hat{\rho}\hat{c}$, the value of $\tilde{R}$ approaches -1, for small $\hat{\rho}\hat{c}$ then $\tilde{R}$ becomes 1, and for $\hat{\rho}\hat{c}$ equal to unity $\tilde{R}$ is zero. The results are in agreement with the results given by *Carome* et al. [6.10] and *Burmistrova* et al. [6.11], which represent limiting cases of (6.1.13) in the sense that only extreme values of $\hat{\rho}\hat{c}$ were considered by these authors.

Perhaps the simplest photochemical reaction is molecular photodissociation, where radiant energy $h\nu$ breaks a chemical bond in the process

$$A_2 + h\nu \rightarrow 2A\,. \tag{6.1.14}$$

The photoacoustic signal arises from two sources, the evolution of heat through rapid equilibration of the translational recoil energy of the fragments (a process that usually requires only a few collisions) and a pressure increase owing to an increase in the mole number. For a perfect diatomic gas where every photon absorbed produces two atoms, the gas pressure P is given by

$$P = (N_0 + N_\nu)\frac{k_B T}{V}\,, \tag{6.1.15}$$

where N_0 is the original number of molecules in the gas volume V, N_ν is the number of photons absorbed, T is its temperature and k_B is Boltzmann's constant. The excess pressure p in a photodissociation reaction taking place at constant volume is then

$$p = \left(\frac{\partial P}{\partial N}\right)_{V,T} dN + \left(\frac{\partial P}{\partial T}\right)_{N,V} dT\ . \tag{6.1.16}$$

The contribution from the second term leads to the usual wave equation for the photoacoustic effect, (6.1.1.), see [6.12]. The first term describes a radially directed force $\mathbf{F}(x,t)$ generated inside the irradiated region and gives an additional pressure of the form

$$p(t) = -\frac{1}{4\pi}\int \frac{\nabla \cdot \mathbf{F}(\mathbf{x}', t - |\mathbf{x} - \mathbf{x}'|/c)}{|\mathbf{x} - \mathbf{x}'|} dV' . \tag{6.1.17}$$

Park and *Diebold* [6.13] have shown that if the gas is only partially dissociated and the light beam intensity is constant in space, the acoustic signal is proportional to a quantity p' given by

$$p' = \alpha E_0 \left[\frac{k_B T}{h\nu} + (\gamma - 1)\left(\frac{h\nu - D_0}{h\nu}\right)\right] , \tag{6.1.18}$$

where D_0 is the bond energy of the molecule (expressed in units of energy per atom) and E_0 is the energy fluence of the light beam. The second term represents the contribution to the acoustic signal from equilibration of the translational energy of the photofragments, while the first term describes the pressure signal from the mole number increase caused by photodissociation. The relative magnitude of the two terms depends on the magnitude of γ, and the energy of recoil of the fragments compared with $k_B T$. Since both processes represent essentially instantaneous processes on the time scale of the photoacoustic effect (which is a few microseconds for a laser beam diameter on the order of 1 mm), the description of the waveform given by either (6.1.8) or (6.1.13) is valid for this particular photochemical means of sound production. Note that the reverse of reaction (6.1.14), molecular recombination, can often be neglected since it is a three-body process, and, depending on the pressure, is too slow to contribute to the signal on a microsecond time scale. Short pulse excitation effectively freezes out this process. The problem of calculation of the pressure by low-frequency modulation of a continuous light beam, on the other hand, would normally require inclusion of the heat release arising from the recombination reactions.

If molecular photodissociation is followed by chemical reaction, additional heat release is possible. In this case, the evolution of heat and the subsequent production of sound may be governed by the temporal profile of the reaction rather than that of the light pulse. For instance, in the photodissociation of a molecule A_2 in the presence of a reactant BC, the reactions

$$\begin{aligned} A_2 + h\nu &\to 2A , \\ A + BC &\xrightarrow{k_0} AB + C \ , \end{aligned} \tag{6.1.19}$$

are characterized by two sources of heat. The first arises from the translational recoil energy of the photofragments, which, as discussed above, is expected to be governed by the temporal profile of the laser or the transit time of sound across the laser beam. The second source of energy is that liberated by the reaction; however, since the rate of disappearance of A is given by

$$[A] = [A]_0 \, \mathrm{e}^{-k_0[BC]t} , \tag{6.1.20}$$

where $[A]_0$ is the concentration of A immediately after the firing of the laser and where $[BC]$ is assumed to be far in excess of $[A]_0$, the heat release will follow an exponential law. (This is valid under the assumption of a light pulse whose duration is far shorter than the time constant for the reaction.)

Another kinetic mechanism following molecular photodissociation of a molecule AB involves reaction of the photoproducts with the parent molecule,

$$\begin{aligned} AB + h\nu &\rightarrow A + B \,, \\ A + AB &\underset{k_r}{\overset{k_f}{\rightleftarrows}} A_2 + B \,. \end{aligned} \tag{6.1.21}$$

If some A_2 is initially present, the reaction can proceed in the reverse direction as well as the forward direction, which, depending on the rate constant and reaction enthalpy, can lead to the absorption of energy from the system, giving an outwardly propagating rarefaction wave. The rate equation for the $[A]$ when $[AB]$ and $[A_2]$ are far greater than $[A]_0$ can be written as

$$\frac{d[A]}{dt} + \gamma_r[A] - 2k_f[AB][A]_0\mu = 0 \,, \tag{6.1.22}$$

where the parameters γ_r and μ are defined as

$$\gamma_r = k_f[AB] + k_r[A_2]$$

$$\mu = k_r[A_2]/k_f[AB] \;.$$

Solution of this differential equation gives the $[A]$ as

$$[A] = [A]_0\left[\frac{2\mu}{1+\mu} + \left(\frac{1-\mu}{1+\mu}\right)e^{-\gamma_r t}\right] \,. \tag{6.1.23}$$

The rate of heat evolution is proportional to the product of time rate of change of the $[A]$ and the enthalpy of reaction ΔH_r (a negative quantity for an exothermic reaction). The heating function in (6.1.1), neglecting the spatial factor, thus becomes

$$H(t) = -k_f[AB][A]_0\Delta H_r\theta(t)(1-\mu)e^{-\gamma_r t} \quad , \tag{6.1.24}$$

where $\theta(t)$ is the Heaviside function. (Note that all concentrations and enthalpies should be expressed in terms of particle densities.) Equation (6.1.24) shows either heat evolution or absorption depending on the magnitude of μ. There is also the possibility of no heat liberation: by adjustment of the ratio $[A_2]/[AB]$ in an experiment it is possible, depending on the magnitude of the equilibrium contant, to determine a null or "isophotoacoustic" point. (A similar effect has been found in a liquid phase reaction, see [6.14].) For small $[A_2]$, the parameter μ can be made to be less than 1, so that the reaction liberates heat; for large $[A_2]$, μ will be greater than 1 and the reaction will consume energy, reversing the sign of the acoustic signal. It is thus evident that measurement of such an isophotoacoustic point permits a determination of the equilibrium constant for reaction (6.1.21).

If the experimental conditions are such that the reaction proceeds too rapidly to be resolved on the time scale of the experiment (i.e. $\gamma_r^{-1} << w_0/\sqrt{2}c$ or τ_l) then the acoustic wave contains a contribution from the total heat per volume evolved in the reaction $\bar{H}$ given by

$$\bar{H} = (-\Delta H_r)\Big(\frac{1-\mu}{1+\mu}\Big)[A]_0 \ , \tag{6.1.25}$$

where $[A]_0$ can be expressed in terms of the incident optical absorption coefficient of the gas if desired. The temporal profile of the acoustic wave is then described by (6.1.8) or (6.1.13), but the amplitude of the wave must include contributions from the mole number increase, the heat corresponding to translational energy of the photofragments, and the enthalpy of reaction.

A quantitative description of the acoustic waveform produced by a Gaussian light beam initiating reaction sequences (6.1.19 and 21) requires a solution to the wave equation (6.1.1) with a heating function of the form

$$H_{ch}(r,t) = \frac{2\alpha\bar{E}_{ch}}{\pi w_0^2}k_f[AB](1-\mu)\theta(\gamma_r t)\exp\Big(-\frac{2r^2}{w_0^2}-\gamma_r t\Big) \ , \tag{6.1.26}$$

where $\bar{E}_{ch} = -\Delta H_r \bar{E}_0/h\nu$. The pressure spectral density is found by substituting $H_{ch}(r,\omega)$, the Fourier transform of (6.1.26), into (6.1.5) as above, giving

$$p_{ch}(r,\omega) = \frac{\alpha\beta\bar{E}_{ch}}{8\pi C_p}\Big(\frac{1-\mu}{1+\mu}\Big)H_0^{(1)}(kr)\Big(\frac{\omega}{1-i\omega/\gamma_r}\Big)\exp(-k^2w_0^2/8) \ . \tag{6.1.27}$$

The functional form of the acoustic signal generated by a chemical reaction can be determined by a solution of the wave equation (6.1.1). If $p_{ch}(r,w)$ is transformed into the time domain, the result is

$$p_{ch}(t) = \frac{\alpha\beta\bar{E}_{ch}c^2}{8\pi w_0^2 C_p}\Big(\frac{1-\mu}{1+\mu}\Big)\int_{-\infty}^{\infty}dq\ H_0^{(1)}(q\hat{r})\Big(\frac{q}{1-iq/\hat{\gamma}}\Big)\exp(-\frac{q^2}{8}-iq\hat{t}) \ , \tag{6.1.28}$$

where a dimensionless decay contant $\hat{\gamma} = \gamma_r w_0/c$ has been defined. This expression can be evaluated using the fast Fourier transform, however, the limiting cases of large $\hat{\gamma}$ can be found quite simply from inspection of the Fourier integral.

It is evident that the exponential function in q^2 acts as a cutoff function so that the other factors in the integrand contribute to $p_{ch}(t)$ only if they are nonzero within the e^{-1} point, $q = \sqrt{8}$, of the exponential function. Thus, if the magnitude of $\hat{\gamma}$ is so large that $1 << \hat{\gamma}/\sqrt{8}$ (or $2\sqrt{2}c/w_0 << \gamma_r$), then the denominator of the factor in parenthesis never alters the value of the integrand within the cutoff point of the exponential function, and the waveform given by *Heritier*, [(6.1.8), but with $\tau_l = 0$], is regained. This, of course refers to the case where chemical energy is released on a time scale short compared with the acoustic transit time across the the beam. The effective pulse width is then determined by the acoustic transit time across the laser beam. In general, the integral must be evaluated numerically; no simple functional dependence on time is apparent even when $\hat{\gamma} << 1$. A plot of the pressure for two different values of $\hat{\gamma}$ is given in Fig. 6.2.

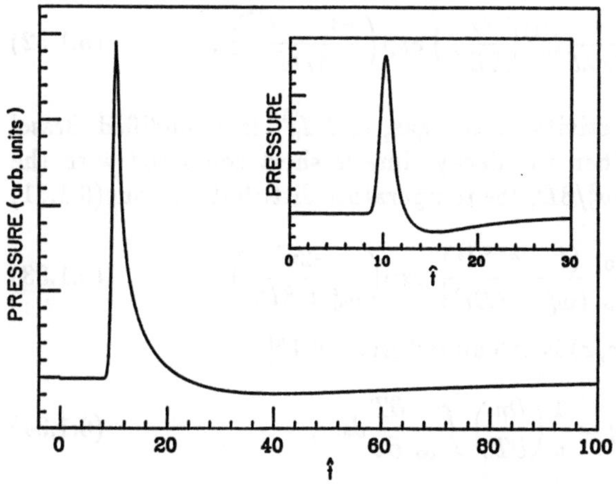

Fig. 6.2. Pressure versus dimensionless time for exponential energy release, as calculated from (6.1.28). The dimensionless rate parameter $\hat{\gamma}$ is 0.05. Inset: pressure versus time with $\hat{\gamma} = 0.33$

Generally speaking, photoacoustic monitoring of reactions in the gas phase is analogous to the liquid-phase experiments described by *Braslavsky* et al. [6.15], *Peters* [6.16], and *Rothberg* [6.17] where the enthalphy of reaction is determined by the magnitude of the photoacoustic signal. In the case of liquids, though, the reactions occur rapidly so that they are not easily time resolved.

It is also possible to monitor chemical reactions initiated by pulsed laser beams by probe beam deflection techniques. Here, the temperature rise in the irradiated region results in a spatial gradient in the index of refraction of the gas which deflects a probe laser beam. As described by *Sontag* et al. [6.18], for a Gaussian pump beam that causes heat to be released exponentially in time with a decay constant τ_h, the temperature distribution in the gas can be described as a solution to the heat equation,

$$\frac{\partial^2 T}{\partial r^2} + \frac{1}{r}\frac{\partial T}{\partial r} - \frac{1}{D}\frac{\partial T}{\partial t} = \frac{-N(r,t)h\nu}{D\rho C_p \tau_h} e^{-t/\tau_h} , \qquad (6.1.29)$$

where the radial distribution of excited molecules is given by

$$N(r,t) = \frac{2\alpha \bar{E}_0}{\pi w_0^2 h\nu} e^{-2r^2/w_0^2} \qquad (6.1.30)$$

and where $\bar{E}_0$ is the energy in the pump beam, D is the thermal diffusivity of the gas, and $h\nu$ is the amount of energy liberated by absorption of a photon. The temperature distribution is found as an integral of the driving function $Q(r,t)$ on the right-hand side in (6.1.29) over a Green's function,

$$T(r,t) = 2\pi \int_0^t dt' \int_0^\infty r' Q(r,' t') G(r, r'; t - t') dr' , \qquad (6.1.31)$$

where the Green's function is given by

$$G(r, r', \delta) = \frac{1}{4\pi\lambda\delta} I_0^{(1)}\left(\frac{rr'}{2D\delta}\right) \exp\left(\frac{-r^2 - r'^2}{4D\delta}\right) , \tag{6.1.32}$$

where λ is the thermal conductivity of the gas, and $I_0^{(1)}$ is a modified Bessel function of the first kind. When the decay time is short compared with the thermal diffusion time, $\tau_h << w_0^2/8D$, the temperature distribution from (6.1.31) becomes

$$T(r, t) = \frac{2\alpha \bar{E}_0}{\pi \rho C_p} \frac{(1 - e^{-t/t_h})}{(w_0^2 + 8Dt)^{\frac{3}{2}}} \exp\left(\frac{-2r^2}{w_0^2 + 8Dt}\right) . \tag{6.1.33}$$

The probe beam deflection $\phi(r, z)$ is calculated from [6.19]

$$\phi(t, z) = \frac{1}{n}\left(\frac{\partial n}{\partial T}\right) \int_{-\infty}^{\infty} \frac{\partial T}{\partial z} dx , \tag{6.1.34}$$

where n is the index of refraction of the gas. Substitution of (6.1.33) into (6.1.34) gives

$$\phi(t, z) = \frac{1}{n}\left(\frac{\partial n}{\partial T}\right) \frac{(-8\alpha \bar{E}_0 z)}{\sqrt{2\pi} \rho C_p} \frac{(1 - e^{-t/t_h})}{(w_0^2 + 8Dt)} \exp\left(\frac{-2z^2}{w_0^2 + 8Dt}\right) . \tag{6.1.35}$$

This expression, which includes the effects of both heat release and thermal diffusion, can be used for any chemical or physical process where heat release follows an exponential law. The photothermal signals from, for instance, reactions (6.1.19) and (6.1.21) are thus described by (6.1.35) (under the restrictions of the derivation). The signals predicted by (6.1.35) generally show a rapid pressure increase followed by slow decay, as shown in Fig. 6.3.

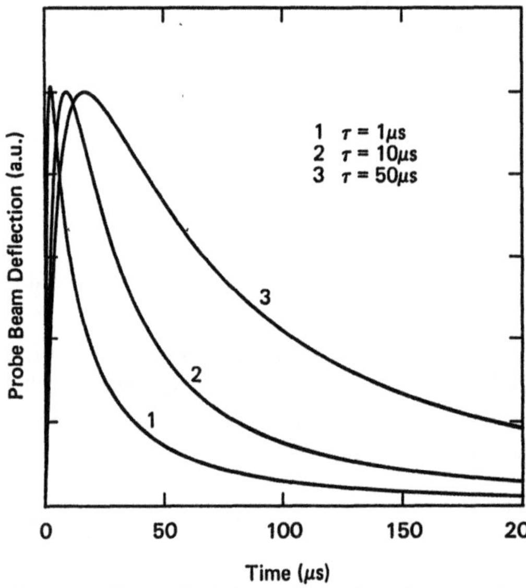

Fig. 6.3. Beam deflection versus time for several values of τ_l. [6.18]

6.1.2 Experimental Results

Koseki et al. [6.20] have studied the reaction of Br atoms with NOBr through photodissociation of NOBr at 532 nm. The reaction sequence is of the general form of reactions (6.1.21) discussed above, that is,

$$\begin{aligned} NOBr + h\nu &\rightarrow Br + NO \, , \\ Br + NOBr &\underset{k_r}{\overset{k_f}{\rightleftarrows}} Br_2 + NO \, , \end{aligned} \qquad (6.1.36)$$

where ΔH_r is -17.4 kcal mole^{-1}. A pulsed Nd:YAG laser beam was used to initiate the reaction. The evolution of heat was monitored using the thermal lens method with a HeNe laser beam propagating collinearly with the 532 nm beam. Two decays were reported in the observed signal, the first one originating from the chemical reaction (6.1.36) followed by a slower heat release attributed to thermal diffusion. By adding known amounts of NO to the initial mixture, the authors found that the reverse reaction had no influence on the observed rate. Thus the parameter μ in (6.1.28) was shown to be so small as to be essentially negligible. The authors measured a rate constant of $(5.3 \pm 0.7) \times 10^{-12}$ cm^3 molecule^{-1} s^{-1} for the forward reaction at 298K, in excellent agreement with previous measurements, and found the activation energy to be 1.3 ± 0.4 kcal^{-1}mole. Figure 6.4 shows a plot of the observed rate constant as a function of [NOBr], from which the rate constant at 298 K was determined.

Diebold and coworkers [6.9, 13] have studied the photodissociation of Cl_2in the near ultraviolet using the output from an excimer-laser-pumped dye laser and a harmonic output of a Nd:YAG laser at 355 nm. Signals were recorded using

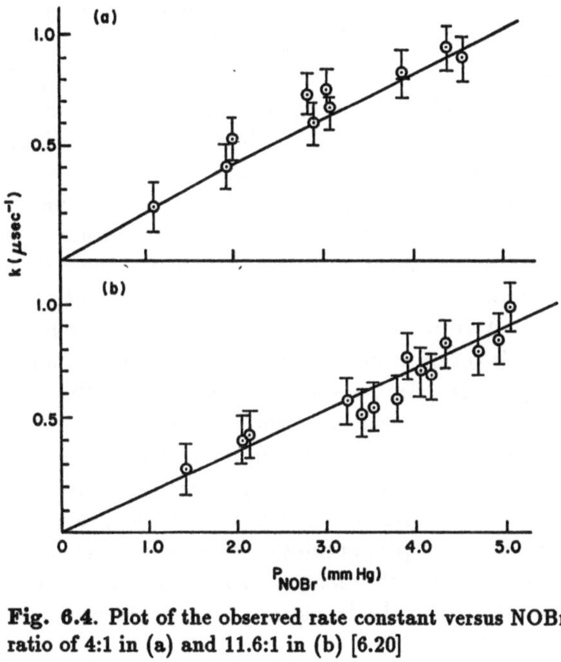

Fig. 6.4. Plot of the observed rate constant versus NOBr pressure at an initial [NO] to [Br_2] ratio of 4:1 in (a) and 11.6:1 in (b) [6.20]

the beam deflection technique where a HeNe laser beam is passed through the gas mixture and is directed onto a split photodiode [6.19, 21]. The exciting laser beam was expanded to fill the entrance window of the cell so that the acoustic wavelength was small compared with the dimensions of the laser beam. The signal from the split photodiode is proportional to the density gradient *along* the propagation direction of the ultraviolet laser beam. The high extinction coefficient of Cl_2, characteristic of most gases in the ultraviolet region, causes the gas to be optically thick even at moderate pressures. The waveforms recorded in various Cl_2 - Ar mixtures clearly showed the exponential waveform predicted by (6.1.13). Since gases at moderate pressures have such low densities relative to Pyrex (the window material),the parameter $\hat{\rho}\hat{c}$ is small compared with 1 so that the pressure rises and falls exponentially. The beam deflection signal responds to the time *derivative* of the pressure so that the waveform reverses sign at $\hat{\tau}_a = 0$.

When the Cl_2 partial pressure is reduced so that the gas mixture is optically thin, the acoustic signal is described by (6.1.8) and (6.1.18). In this case a cylindrical acoustic wave is generated and the photodiode is used in the conventional beam deflection geometry. In the experiments reported in [6.13], a focused HeNe beam was directed at perpendicular incidence to a 355 nm laser beam. The beam deflection signal produced by irradiating a 5% Cl_2 in Ar mixture at 200 mm Hg with 355 nm radiation shows qualitative agreement with the derivative of the signal shown in Fig. 6.1. This gives some evidence that all sound generating processes take place on a time scale short compared with the acoustic transit time across the laser beam and that the amplitude of the signal is proportional to p' given in (6.1.18).

The authors also showed that at very low pressures of Cl_2, an anisotropic pressure wave is produced by photodissociation. This arises as a consequence of the preferred direction of recoil of Cl atoms with respect to the electric vector of the 355 nm beam. Pressure signals were recorded with an electret microphone that responded to the burst of recoiling Cl atoms with an exponentially damped oscillatory signal. The amplitude of the microphone signal recorded as a function of the angle of rotation of the polarization vector of the laser beam showed an asymmetry characteristic of a perpendicular transition in Cl_2.

Note that dissociation of a molecule into electronically excited fragments does not alter the analysis given above provided the energy in the fragments is rapidly converted to heat. In addition, if absorption takes place from the excited vibrational or rotational levels of the molecular ground state, the additional recoil energy gained from the excited vibrational-rotational level cancels out in the acoustic signal, provided the relaxation time of the molecular ground state is fast compared with the acoustic transit time across the laser beam. That is, the extra translational energy gained by dissociating a molecule through a "hot band" is exactly compensated for by the energy the ensemble of molecules in a Boltzmann distribution of translational and internal energy states is required to give up in order to maintain thermal equilibrium. Since both of these processes are generally rapid at atmospheric pressure (depending on the species present), and since the signal from the mole number increase is usually negligible, the photoacoustic effect becomes proportional to the translation energy release $h\nu - D_0$, and acts effectively as a calorimeter.

Barker et al. [6.22] have used this property of the photoacoustic effect to study the exothermic reaction sequence

$$\begin{aligned} N_2O_5 + h\nu &\longrightarrow NO_2 + NO_3 \, , \\ NO_3 + NO &\longrightarrow 2NO_2 \, . \end{aligned} \tag{6.1.37}$$

A microphone was placed in the reaction chamber and the gas mixture was irradiated with the frequency-doubled output of a dye laser pumped by a pulsed excimer laser. The acoustic signal originates from the total energy release $\hbar_0$ from reactions (6.1.37). That is, $\hbar_0$ has a contribution from the photodissociation reaction of $hv - D_0$, and a contribution of $-\Delta H_0$ from the bimolecular reaction so that the total energy released can be expressed as

$$\hbar_0 = h\nu - D_0 - \Delta H_0 \, . \tag{6.1.38}$$

The authors follow *Rothberg* et al. [6.23], who have derived expressions for the photoacoustic signal arising from chemical reactions in solution. First, the response of the detector is modeled as an underdamped oscillator. The impulse response function is then convoluted with the forcing function obtained from solution of the heat equation to obtain the voltage output from the transducer. *Barker* et al. further introduce a chemical lifetime parameter and perform a spatial integration to account for the extended line source of their beam geometry. A typical microphone response, shown in Fig. 6.5, can be fit to the numerically simulated waveform. The authors find a clear increase in the photoacoustic signal as [NO] is increased. The bimolecular reaction of NO_3 with NO is too rapid to be resolved with the microphone and hence appears as a signal amplitude increase. The authors also conclude that the photodissociation reaction proceeds with a quantum yield of nearly unity at 290 nm.

Sontag et al. [6.18, 24] have used the beam deflection technique to study heat liberation in an equilibrium mixture of NO_2 with N_2O_4, and in CS_2 – both irradiated with the 337 nm output of a N_2 laser. The experimental apparatus, shown in Fig. 6.6, consists of an N_2 laser whose output is focused into a stain-

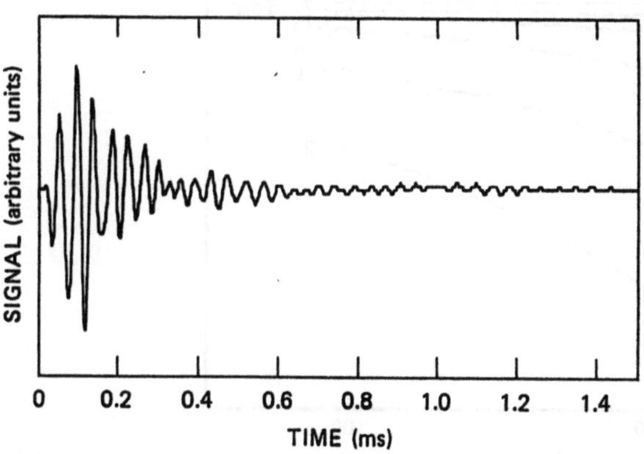

Fig. 6.5. Microphone signal from irradiation of N_2O_5 with the frequency-doubled output of a pulsed dye laser at 290 nm. [6.22]

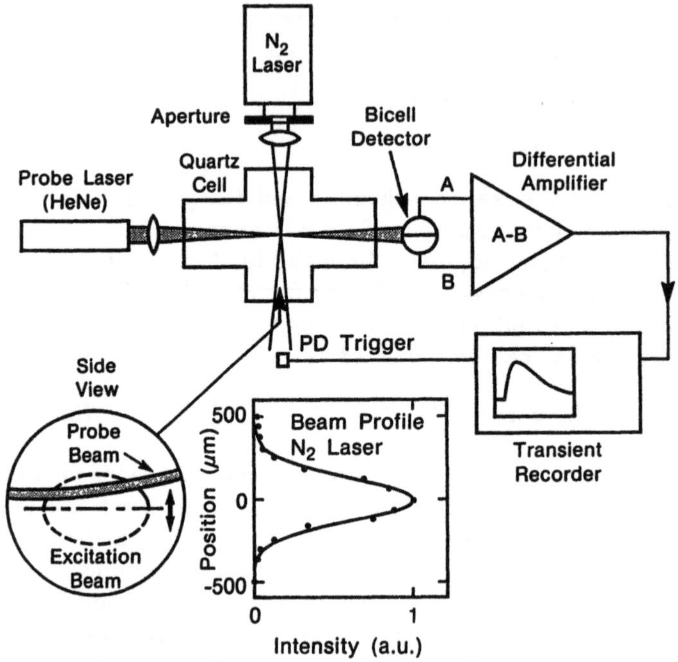

Fig. 6.6. Experimental apparatus for recording photothermal signals. The square inset shows the intensity distribution of the N_2 laser beam perpendicular to its propagation direction. [6.18]

less steel reaction cell equipped with quartz windows, and a probe HeNe laser beam focused in the excitation region. The deflection of the HeNe laser beam is detected with a split photodiode whose output is fed to a differential amplifier, the output of which is displayed on an oscilloscope. Transmission measurements at 337 nm in CS_2 showed a linear pressure dependence over the range 1-200 mm Hg. As shown in Fig 6.7, the beam deflection signal at pressures less than 10

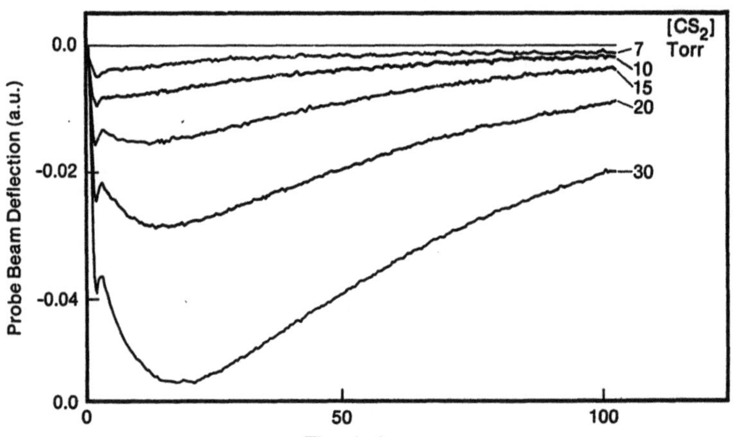

Fig. 6.7. Photothermal beam deflection signal versus time in CS_2 vapor. [6.18]

mm Hg displayed a fast rise followed by a slow decay arising from thermal diffusion. At higher pressures, chemical effects occur that change the character of the waveform altogether. The authors conclude that immediately after the laser is fired, quenching of the excited electronic state, or internal energy conversion takes place. Based partly on previous work, the initial step leading to energy release appears to be

$$\begin{aligned} CS_2 + h\nu &\longrightarrow CS_2^* \,, \\ CS_2^* + CS_2 &\longrightarrow 2CS + S_2 \,, \end{aligned} \tag{6.1.39}$$

which is energetically allowed in the wavelength region below 360 nm. This sequence is possibly followed by the reactions

$$\begin{aligned} 4S_2 &\longrightarrow S_2 \,, \\ nCS &\longrightarrow (CS)_n \,, \end{aligned} \tag{6.1.40}$$

both of which are exothermic. The authors also find particulate formation in the irradiated region, as evidenced by observation of light scattering in the irradiated region and particulate buildup on surfaces.

Similar beam deflection experiments were done in $NO_2 - N_2O_4$ mixtures. Absorption coefficient measurements gave $\alpha = 0.5\ \mathrm{cm^{-1}bar^{-1}}$ for NO_2 and $\alpha = 26\ \mathrm{cm^{-1}bar^{-1}}$ for N_2O_4 at 337 nm, so that both species can be excited by the N_2 laser. Irradiation of NO_2 at this wavelength is known to result in the formation of atomic O, which can undergo reaction through the sequence

$$\begin{aligned} NO_2 + h\nu &\longrightarrow NO + O \,, \\ O + NO_2 &\longrightarrow NO + O_2 \,. \end{aligned} \tag{6.1.41}$$

In addition, a reaction similar to (6.1.39) can take place,

$$NO_2^* + NO_2 \longrightarrow 2NO + O_2 \,. \tag{6.1.42}$$

The experimental data however could be fit to the thermal diffusion function (6.1.35) given above, and there appears to be no evidence of reactions (6.1.41 and 42) taking place. Such reactions can be present; the data indicate only that if they do take place then they proceed on a time scale short compared with the resolution of the apparatus.

Lewis and *Zeltmann* [6.25] have used a pulsed dye laser in the range 360-410 nm to investigate the decomposition of UF_6. Absorption of light in this region is known to decompose UF_6 according to

$$UF_6 + h\nu \longrightarrow UF_5 + F \,. \tag{6.1.43}$$

The photoacoustic effect thus is proportional to the difference between the photon energy and the $F - UF_5$ bond energy described by (6.1.18) if the effect of the mole number increase is negligible. This is a reasonably good assumption provided that the internal energy of the fragments is rapidly thermalized on the time scale of the measurement. The authors recorded photoacoustic signals with a fast microphone in a brass cell with UF_6 at a pressure of 20 mm Hg over the above wavelength range using a series of laser dyes. They found that the photoacoustic signal could be accurately modeled from a knowledge of the optical absorption coefficient. The UF_5 – F bond energy was determined to be 60 ± 1 kcal $mole^{-1}$, in agreement with previous measurements. This finding basically corroborated the assumption of no post-photolysis chemistry taking place and a unit quantum yield for the photodissociation. The addition of H_2 to UF_6 gave an increase in the signal amplitude of a factor of 1.8, but the effect was attributed to an enhancement in the energy relaxation rate rather than chemical reaction between the photofragments and H_2.

6.2 Continuous Excitation

6.2.1 Theory

Photoacoustic measurements can be carried out in the time domain by using a pulsed light source and recording the acoustic waveform in time. Alternately, in a frequency domain experiment a continuous light beam can be amplitude modulated and the phase shift and amplitude of the acoustic signal recorded as a function of the modulation frequency. From a purely mathematical point of view the information obtained in these two kinds of measurements is identical. Clearly, chemical effects can be investigated in the frequency domain using the photoacoustic effect.

The starting point for analysis of such experiments is a mathematical description of the energy relaxation rates for the system shown in Fig. 6.8. This diagram shows in a very general way the energy level diagram of a typical large organic molecule with a singlet level S_1 that is excited optically from the ground state S_0, and a triplet state T. The singlet level is allowed to relax to the ground state at a rate k_S and to the triplet state at a rate k_{ST}. Both of these processes contribute to give the singlet state a lifetime of τ_S. The triplet state with energy

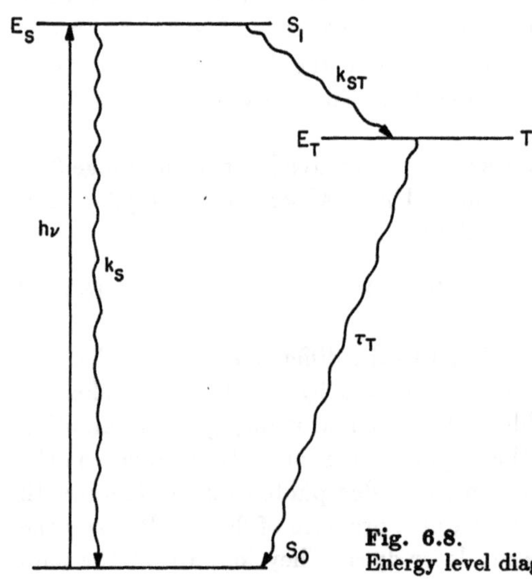

Fig. 6.8.
Energy level diagram for heat flow in a large molecule

E_T (which is normally metastable) relaxes to the ground state with a lifetime τ_T. Although such a kinetic scheme was originally used by *Hunter* et al. [6.26] and *Hunter* and *Stock* [6.27] to model the decay of energy in organic molecules excited by ultraviolet light, in fact, this model has been used to interpret experimental data when a number of chemical and physical processes involving both fast and slow energy release mechanisms are present.

As described by *Hunter* and co-workers, the rate equations for the populations of the singlet n_S and triplet n_T populations are given by

$$\begin{aligned} \frac{d}{dt}n_S &= \frac{\alpha I_0}{h\nu}e^{i\omega t} - \frac{n_S}{\tau_S} , \\ \frac{d}{dt}n_T &= k_{ST}n_S - \frac{n_T}{\tau_T} , \end{aligned} \tag{6.2.1}$$

where $h\nu$ is the energy of the incident photon, and I_0 is the intensity of the radiation modulated at a frequency ω. (Thus $\alpha I_0/h\nu$ represents the number of photons absorbed per unit volume.) Solution of these equations gives

$$\begin{aligned} n_S &= \frac{(\alpha I_0 \tau_S/h\nu)}{(1+\omega^2\tau_S^2)^{\frac{1}{2}}}e^{i(\omega t-\phi_S)} , \\ n_T &= \frac{(\alpha I_0 \tau_S k_{ST}\tau_T/h\nu)}{(1+\omega^2\tau_S^2)(1+\omega^2\tau_T^2)^{\frac{1}{2}}}e^{i(\omega t-\phi_S-\phi_T)} , \end{aligned} \tag{6.2.2}$$

where the phase angles ϕ_S and ϕ_T are defined through the relations

$$\tan\,\phi_S = \omega\tau_S , \qquad \tan\,\phi_T = \omega\tau_T .$$

For a uniformly irradiated ideal gas, the rate of change of the pressure is given by

$$\frac{dP}{dt} = \frac{R}{C_v'}H(t) , \tag{6.2.3}$$

where C_v' is the molar heat capacity of the gas and R is the gas constant. The heat per unit volume liberated is given by a vector sum of the contributions from each pathway as

$$H(t) = (k_S E_S + k_{ST}E_{ST})n_S + (E_T/\tau_T)n_T , \tag{6.2.4}$$

where E_S and E_T are the energies of the singlet and triplet states, respectively, and E_{ST} is the singlet-triplet energy gap. Substitution of (6.2.2, 4) into (6.2.3) gives the acoustic signal in the limit $\tau_S << \tau_T$ as

$$p_{cw} = \frac{(\alpha I_0/h\nu)R}{C_v'\omega}\left(\hat{H}^2 + \frac{(\hat{L}^2+2\hat{H}\hat{L})}{(1+\omega^2\tau_T^2)}\right)^{\frac{1}{2}} e^{i(\omega t-\phi-\frac{\pi}{2})} , \tag{6.2.5}$$

where the following parameters are defined:

$$\begin{aligned} \hat{H} &= \tau_S(k_S E_S + k_{ST}E_{ST}) , \\ \hat{L} &= \tau_S k_{ST} E_T , \\ \tan\,\phi &= \frac{\omega\tau_T}{1+(\hat{H}/\hat{L})(1+\omega^2\tau_T)} . \end{aligned}$$

From this result, the well-known $\pi/2$ phase shift at low frequency and a prefactor proportional to $\gamma - 1$ can be seen. When $\hat{H}$ is zero, (6.2.5) reduces to the expression for the photoacoustic effect for a two-level system [6.28, 29] with $\tan\phi = \omega\tau_T$. By measuring either the amplitude or the phase shift as a function of frequency, both τ_T and the ratio $\hat{H}/\hat{L}$ can be determined. Since this result was derived under the assumption that τ_S is much smaller than τ_T, the quantity $\hat{H}$ represents that part of the absorbed energy that relaxes faster than the modulation period, and $\hat{L}$ the part involved in the slow energy relaxation of the triplet state. A plot of the phase lag versus modulation frequency for several values of $\hat{H}/\hat{L}$ is given in Fig. 6.9, showing the relatively pronounced effect of the $\hat{H}/\hat{L}$ ratio on the shape of the curve.

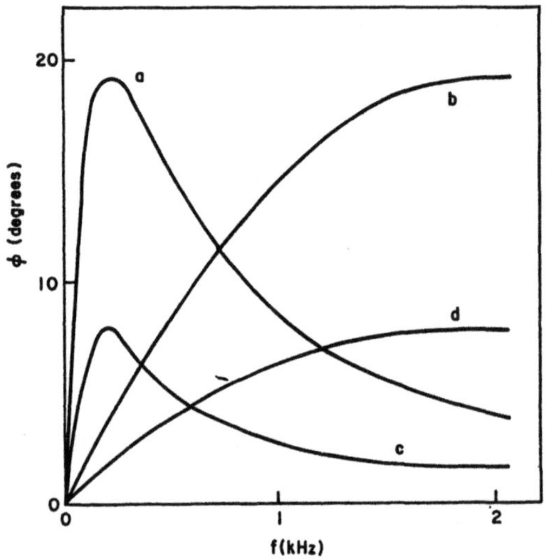

Fig. 6.9. Phase lag versus modulation frequency for (a) $\tau_T = 10^{-3}$s, $\hat{H}/\hat{L} = 1.0$; (b) $\tau_T = 10^{-4}$s, $\hat{H}/\hat{L} = 1.0$; (c) $\tau_T = 10^{-3}$s, $\hat{H}/\hat{L} = 3.0$; (d) $\tau_T = 10^{-4}$s, $\hat{H}/\hat{L} = 3.0$. [6.27]

The influence of an increase in mole number has been considered by *Hunter* and *Kristjansson* [6.30]. A derivation of the properties of the acoustic signal for the process

$$\begin{aligned} A + h\nu &\longrightarrow A^* \,, \\ A^* &\xrightarrow{k_d} B + C \,, \end{aligned} \tag{6.2.6}$$

where k_d is the rate constant for decomposition of A^*, proceeds from consideration of the linear differential equations describing the four species in (6.2.6) for sinusoidal modulation of the incident radiation,

$$\begin{aligned} \frac{d[A]}{dt} &= -\frac{\alpha I_0}{h\nu} e^{i\omega t} \,, \\ \frac{d[A^*]}{dt} &= \frac{\alpha I_0}{h\nu} e^{i\omega t} - k_d[A^*] \,, \\ \frac{d[B]}{dt} &= \frac{d[C]}{dt} = k_d[A^*] \,. \end{aligned} \tag{6.2.7}$$

Use of the ideal gas law together with the elementary solution to these equations gives an expression for the pressure signal of the form

$$p_d(t) = \frac{\alpha I_0}{\omega}\left(\frac{k_B T}{h\nu}\right)\frac{e^{i(\omega t-\phi_d-\pi/2)}}{(1+\omega^2\tau_d^2)^{\frac{1}{2}}}, \tag{6.2.8}$$

where $\tau_d = k_d^{-1}$ and $\tan\,\phi_d = \omega\tau_d$. The authors note that the mole number increase generally produces a small signal compared with the equilibration of the translational energy of the fragments. For instance, I_2 excited at 546 nm gives $h\nu - D_0 = 5878\ \mathrm{cm}^{-1}$. For a molar heat capacity of 28 $\mathrm{JK^{-1}mole^{-1}}$ at room temperature, the mole number increase is responsible for less than 10% of the acoustic signal. This figure, of course, increases as $h\nu - D_0$ is made smaller.

In a simple molecular photodissociation process

$$\begin{aligned} AB &\rightarrow A + B^*\,, \\ B^* + M &\xrightarrow{k_q} B + M\,, \end{aligned} \tag{6.2.9}$$

the mole number increase and the translational energy act to give a prompt heat release. The quenching of the internal energy of the fragment B^* in the second step of (6.2.9) often takes place at a rate slow enough to be resolved in a phase modulation experiment. Thus, the first two processes can be identified with $\hat{H}$ and the quenching of the internal energy of the fragment can be identified with $\hat{L}$, so that the phase lag information can be analyzed according to (6.2.5) with τ_T associated with k_q^{-1}.

An important application of this analysis is the determination of the branching ratio q for formation of an excited fragment versus a ground state fragment [6.31]. For the fraction of the products giving ground state atoms, the heat liberated is $(1-q)(h\nu - D_0)$; for the fraction giving excited fragments the energy liberated is $q(h\nu - D_0 - E_f)$ where E_f is the internal energy of the fragment. These two sources make contributions to $\hat{H}$, while $\hat{L}$ has a contribution proportional to qE_f. Thus, the $\hat{H}/\hat{L}$ ratio is given by

$$\hat{H}/\hat{L} = [(1-q)(h\nu - D_0) + q(h\nu - D_0 - E_f)]/qE_f\,, \tag{6.2.10}$$

which can be solved for q as

$$q = \frac{h\nu - E_f}{E_f[(\hat{H}/\hat{L}) + 1]}\,.$$

For photodissociation of molecular species leading to excited atomic states, E_f is known; determination of the $\hat{H}/\hat{L}$ ratio as a function of exciting wavelength in a photoacoustic experiment thus gives q, a quantity of great interest from the point of view of optimizing the output of a photodissociation laser and which can be obtained only with difficulty using optical techniques.

6.2.2 Experimental Results

In addition to using the formal description of the photoacoustic effect given in the previous section for a number of studies of vibrational relaxation, *Hunter*

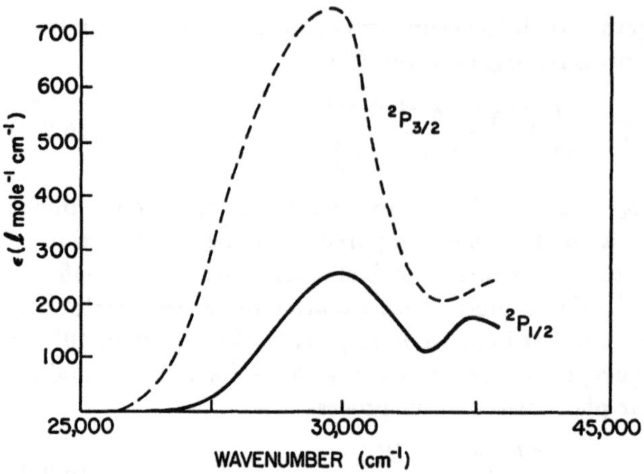

Fig. 6.10. Absorption versus wavelength for production of I $^2P_{\frac{3}{2}}$ and $^2P_{\frac{1}{2}}$ atoms from photodissociation of CH_2I_2 obtained from photoacoustic phase lag measurements. [6.32]

and co-workers [6.31-35] have studied the photodissociation of several iodine-containing polyatomic molecules. The experimental apparatus consisted of a super-pressure Hg arc lamp whose output was passed through a monochromator and amplitude modulated between 0.1 and 2 kHz. The cell was constructed of stainless steel and was equipped with a Bruel and Kjaer model 4144, one inch (2.5 cm) diameter, capacitance microphone. The output of the microphone was amplified, filtered, and detected with a lock-in amplifier. Particular attention was given to eliminating Helmholtz resonances in the cell. In CH_2I_2, *Hunter* and *Kristjansson* [6.32] were able to determine q at several values of the incident photon energy, as shown in Fig. 6.10. The experimental data were found to fit the functional dependence given by (6.2.5), which indicated that the simple dissociation and energy relaxation scheme depicted in Fig. 6.8 for heat release was valid. Measured values for the quenching time were found to agree with existing literature values. The data obtained over the range 247 - 365 nm showed q to vary from a high value of 0.41 to less than 0.04. Similar measurements [6.33] in I_2 taken at several Ar ion laser output wavelengths showed q to range between 0.62 and 0.27.

In the photofragmentation of CH_3I, a more complicated scheme of energy liberation was required to explain the photoacoustic phase lag data [6.31, 34]. At 247 nm the photolysis of CH_3I at 1 mm Hg in 100 mm Hg of Kr gives photoacoustic signals characterized by a fast component and two slow components of heat evolution for modulation frequencies in the range 0.2 - 3 kHz. The fast processes include equilibration of the recoil energy of the photofragments (both CH_3 and I), and the vibrational relaxation of the excited CH_3. The slow reactions are quenching of the $^2P_{\frac{1}{2}}$ state of atomic I,

$$I^* + M \rightarrow I + M \; , \tag{6.2.11}$$

where M is a third body, as well as reaction of the photofragments, according to

$$CH_3 + I + M \xrightarrow{k_t} CH_3I + M \; , \tag{6.2.12}$$

or $$CH_3 + I_2 \xrightarrow{k_b} CH_3I + I\,, \tag{6.2.13}$$

where the I_2 comes as a byproduct of the photolysis. By varying the light intensity in a series of experiments the rate of the slowest process was found to vary, indicating that it was a reactive process [through either (6.2.12) or (6.2.13)]. On changing to CD_3I, which is known to relax I^* much more slowly than CH_3I, the second slow process was identified as being a quenching process of the form of (6.2.11). In addition to measuring q as a function of wavelength for CH_3I, CD_3I, and CF_3I, the authors investigated the question of physical relaxation versus chemical reaction of I $^2P_{\frac{1}{2}}$ in CD_3I through the mechanisms

$$I^* + CD_3I \rightarrow I + CD_3I^* \tag{6.2.14}$$

$$I^* + CD_3I \rightarrow I_2 + CD_3\,. \tag{6.2.15}$$

On addition of n-hexane, which dominates the relaxation rate, it was found that the $\hat{H}/\hat{L}$ ratio was unchanged (to within 4%), indicating that the chemical quenching mechanism is not the dominant relaxation mechanism.

In a separate study [6.35], reactions of the methyl radical with I and I_2 were investigated. Here, (6.2.5) was used to analyze the data with $\hat{L}$ being the combination of the enthalpies of reactions (6.2.12 and 13) and $\hat{H}$ being the enthalpy of all processes fast compared with modulation period. The triplet relaxation time τ_T in (6.2.5) was taken as a reaction time constant defined by

$$\tau_T^{-1} = k_t[I] + k_b[I_2]\,. \tag{6.2.16}$$

The pressure of the alkyl iodide was adjusted to be high enough so that vibrational relaxation took place on a time scale fast compared with the reactions; thus, only the chemical reactions were affected by changing the modulation frequency. The authors were able to determine rate constants from plots such as shown in Fig. 6.11 not only for reactions (6.2.12 and 13) but also for the C_2H_5, CD_3 and CH_2I analogs of these reactions.

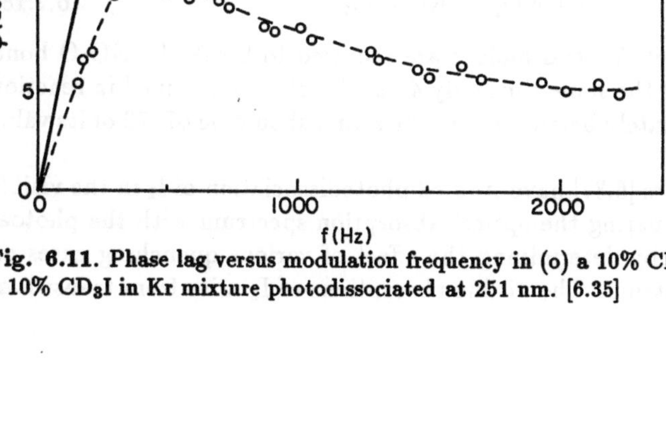

Fig. 6.11. Phase lag versus modulation frequency in (o) a 10% CH_3I in Kr mixture, and (x) a 10% CD_3I in Kr mixture photodissociated at 251 nm. [6.35]

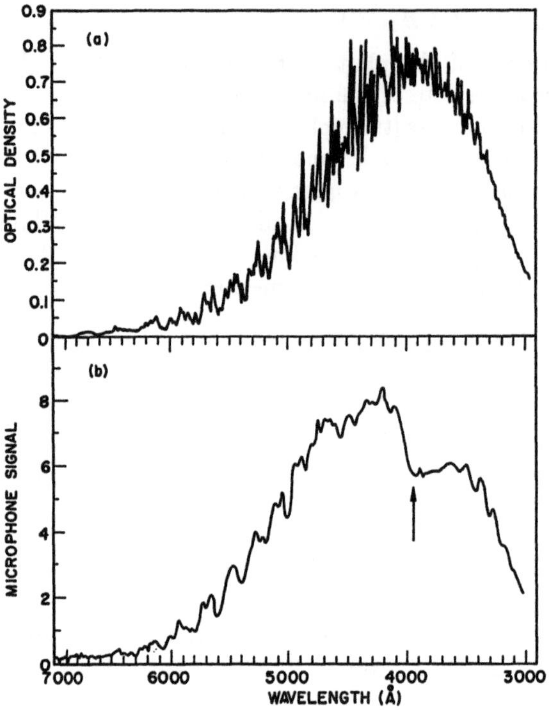

Fig. 6.12. (a) Optical density versus wavelength, (b) photoacoustic signal versus wavelength for NO_2 at 10 mm Hg. The dissociation limit is indicated by the arrow. [6.36]

Harshbarger and *Robin* [6.36] have studied the photodissociation of NO_2 in the visible and near ultraviolet spectral region by taking the absorption spectrum of the NO_2 and comparing it with the photoacoustic spectrum. The experimental apparatus was a conventional arc lamp whose output was dispersed in a monochromator and directed into a reaction cell equipped with a microphone. Typical spectra are shown in Fig. 6.12. The most striking result is the dip in the spectrum that takes place at about 400 nm. The authors interpret the falloff in the heat release as arising from the onset of the photodissociation of NO_2,

$$NO_2 + h\nu \rightarrow NO(^2\Pi) + O(^3P)\ , \tag{6.2.17}$$

followed by the exothermic (ΔH = -47 kcal mole^{-1}) reaction

$$O + NO_2 \rightarrow NO + O_2\ . \tag{6.2.18}$$

At the dissociation limit, 72 kcal mole^{-1} are required to break the NO-O bond in reaction (6.2.17). At the same time only 47 kcal mole^{-1} are gained in reaction (6.2.18). Thus, immediately below 400 nm the signal should be 47/72 of its value at 400 nm.

The same authors [6.37] have studied photodissociation of I_2 in the visible spectral region by comparing the optical absorption spectrum with the photoacoustic spectrum, and by determining the effect of various quenching gases on the photoacoustic spectrum. The visible absorptions of I_2 arise from transitions to (1) the $^3\Pi_{1u}$ state, which dissociates into two ground state $^2P_{\frac{3}{2}}$ atoms, (2) the repulsive $^1\Pi_{1u}$ state, which also dissociates into two ground state atoms, or (3) the $^3\Pi_{0u^+}$ state, which dissociates into one ground state and one excited state $^2P_{\frac{1}{2}}$ atom. The authors find that with neat I_2, and I_2 in a number of gases including NO, He, N_2, SO_2, CH_3I, and C_3H_8, the photoacoustic spectra are all essentially identical. These spectra are also similar to the optical absorption spectrum. However, the photoacoustic spectrum of I_2 in O_2 shows a distinct drop in amplitude near 520 nm. This is interpreted as arising from the onset of predissociation of the $^3\Pi_{0u^+}$ state with the production of excited I $^2P_{\frac{1}{2}}$. This state of I lies 7603 cm^{-1} above the $^2P_{\frac{3}{2}}$ ground state, and is close to resonance with the excited electronic $^1\Delta_g$ state of O_2 at 7886 cm^{-1}; hence the energy transfer can take place through the collisional process

$$I^*(^2P_{\frac{1}{2}}) + O_2(^3\Sigma_g^-) \rightarrow I + O_2^*(^1\Delta_g) , \qquad (6.2.19)$$

which is endothermic by only 284 cm^{-1}. The dramatic drop in the photoacoustic signal amplitude comes about in that O_2^* has a lifetime on the order of 0.4 s, which differs markedly from the lifetime (estimated to be less than 10^{-6}s) for the energy relaxation processes that are operative at wavelengths longer than 520 nm. Based on a simple two-level model with $\hat{H} = 0$ in (6.2.5), the transition from where $\omega\tau << 1$ with no O_2 to $\omega\tau >> 1$ in the presence of O_2 at least qualitatively explains the drop in signal level. In fact, the $^3\Pi_{0u^+}$ level responsible for part of the absorption in this region is not totally predissociated; dissociation is estimated to give 50%-70% I with the remaining I* so that the excess translational energy from the ground state fragments still gives a significant acoustic signal. To bolster this case further, the authors also carried out analogous experiments in Br_2, where the excited Br atom has an energy mismatch with O_2 of 4201 cm^{-1} (far greater than k_BT). No corresponding effect was observed at the dissociation limit of Br_2 in the presence of O_2.

The photoacoustic spectrum of SO_2 as studied by *Harshbarger* and *Robin* [6.36] shows a decreased amplitude resulting from dissociation as in the case of NO_2, but the onset of the dissociation reaction,

$$SO_2 + h\nu \longrightarrow SO + O , \qquad (6.2.20)$$

does not appear to be so pronounced. There are a number of chemical and physical processes that take place in electronically excited SO_2 to quench the incident energy; however, the rates of these processes are too rapid to be resolved with modulation frequencies in the kilohertz range.

A few high-resolution experiments have been reported where chemical reactions are induced by the absorption of light from a continuous dye laser. *Colles* and *Marinero* [6.38] (see also [6.39]) have measured the fluorescence and photoacoustic spectrum of I_2 as a function of wavelength in the 16-2, 14-1, and 12-0 bands of the $B - X$ transition. The 14-1 band, whose rotational structure is resolved by the narrow linewidth of the dye laser shows a photoacoustic signal that increases with rotational quantum number. The lowering of both the fluorescence and photoacoustic signals in the 12-0 band gives strong evidence of predissociation.

There are a number of polyatomic molecules that have been studied photoacoustically using short-wavelength light where both chemical and physical energy release mechanisms are possible. As reviewed by *Robin* [6.40], a number of these molecules give a photoacoustic effect that can be explained primarily in terms of physical processes [6.41]. Such molecules include biacetyl [6.27, 42, 43], benzene [6.42, 44, 45], azulene [6.46], pyridine [6.42, 43], pyrizine [6.43, 47], 2-6-dimethylpyrazine [6.43], pyrimidine [6.43], sym-triazine [6.43], disulfur monoxide [6.48], thioformaldehyde [6.48], and benzaldehyde [6.49]. (The last of these can be photodissociated but no further reaction ensues as the photoproducts are closed-shell molecules.) In a few cases photochemical reactions take place following absorption of ultraviolet light — these are discussed below.

6.3 Nonlinear Effects

As was discussed above, when a diatomic molecule is photodissociated to give two atoms, the process of recombination must be included if the photoacoustic effect is recorded at low modulation frequency. *Diebold* [6.50, 51] has considered the effect of termolecular recombination in the reaction sequence

$$\begin{aligned} A_2 + h\nu &\longrightarrow 2A \,, \\ A + A + M &\underset{\tilde{k}_d}{\overset{\tilde{k}_r}{\rightleftharpoons}} A_2 + M \,. \end{aligned} \tag{6.3.1}$$

The time rate of change in the [A] is given by

$$\begin{aligned} \frac{d[A]}{dt} &= 2\tilde{k}_d[A_2][M] - 2\tilde{k}_r[A]^2[M] \\ &+ 2[A_2]\rho_\nu B(1 + d\ \sin\ \omega t) \,, \end{aligned} \tag{6.3.2}$$

where ρ_ν is the radiation density, B is the Einstein coefficient for absorption and photodissociation of A_2, and d is the modulation depth of the radiation. It is assumed that there is no relaxation of the excited electronic state of A_2. Equation (6.3.2) can be written in terms of a dimensionless concentration x as

$$\frac{dx}{dt} + ax^2 = b + \delta \sin\ \omega t \,, \tag{6.3.3}$$

where

$$x = [A]/[A_2] \,, \qquad a = 2\tilde{k}_r[M][A_2] \,,$$
$$b = 2(\rho_\nu B + \tilde{k}_d[M]) \,, \qquad \delta = 2\rho_\nu B d \,.$$

In an experiment it is easy to make the radiation density large enough so that $\tilde{k}_d[M]$ can be neglected relative to $\rho_\nu B$; the quantity δ/b then approximately equals d, the modulation depth of the radiation. Unfortunately (6.3.3), referred to as a generalized form of the Riccati equation, is a *nonlinear* first-order differential equation. Three different approaches to obtaining a perturbation solution to (6.3.3) are given in [6.50]; here, however, only the lowest-order solution will be obtained. First, a new variable w is defined as $w = x - (b/a)^{\frac{1}{2}}$. [It is easy to see that $w = 0$ corresponds to the steady-state solution of (6.3.3) with no modulation.] Then (6.3.3) becomes

$$\frac{dw}{dt} + 2(ab)^{\frac{1}{2}} w + aw^2 = \delta \sin \omega t \, . \tag{6.3.4}$$

According to the method of successive approximation [6.52], the lowest-order solution to this equation is found by ignoring the nonlinear term; this gives x as

$$x = \left(\frac{b}{a}\right)^{\frac{1}{2}} \left(1 + \frac{(\delta/b)}{2(1+\lambda^2)^{\frac{1}{2}}} \sin(\omega t - \phi_R)\right) , \tag{6.3.5}$$

where

$$\tan \phi_R = \lambda = \frac{\omega}{2(ab)^{\frac{1}{2}}} \cong \frac{\omega}{4\sqrt{\rho_\nu B \bar{k}_r [M][A_2]}} \, .$$

The recombination reaction (6.3.1) gives a highly vibrationally excited A_2 molecule, which, provided relaxation takes place rapidly compared with the modulation period, means that the bond energy of A_2 is given to translational heating of the gas. The rate of heat deposition is then

$$H(t) = D_0 \bar{k}_r [A]^2 [M] \, , \tag{6.3.6}$$

which can then be substituted into (6.2.3) along with (6.3.5) to give the fundamental frequency component of the pressure as

$$p_r^{(1)}(t) = -\left(\frac{R}{C_v'}\right) \rho_\nu B [A_2] D_0 \left(\frac{\delta}{b}\right) \frac{\cos(\omega t - \phi_R)}{\omega(1+\lambda^2)^{\frac{1}{2}}} \, . \tag{6.3.7}$$

The next contribution to the pressure comes from considering the nonlinear term in (6.3.4). This gives a pressure at the overtone frequency 2ω of

$$\begin{aligned} p_r^{(2)}(t) \;=\; & \left(\frac{R}{C_v'}\right) \rho_\nu B [A_2] D_0 \left(\frac{\delta}{b}\right)^2 \left(\frac{\sin(2\omega t - 2\phi_R - \phi_R')}{16\omega(1+\lambda^2)(1+\lambda'^2)^{\frac{1}{2}}} \right. \\ & \left. - \frac{\sin(2\omega t - 2\phi_R)}{16\omega(1+\lambda^2)} \right) , \end{aligned} \tag{6.3.8}$$

where $\lambda' = \lambda/2$ and $\tan \phi_R' = \lambda'$. In fact, an infinite number of overtones are found in the perturbation solution of (6.3.4). It can be shown, however, that the overtone series can be expressed as a power series in the modulation depth (δ/b). Provided this quantity is small, the intermodulation terms at the fundamental frequency will not contribute to the fundamental frequency component of the pressure and the approximate solution (6.3.5) can be used. A further solution to this problem is given for square wave modulation in [6.50]. The results can be seen to be qualitatively similar to the case for sinusoidal modulation.

There are two noteworthy features of this solution for the pressure. The first is the existence of overtones, which is clearly a feature of the nonlinear response, and the second is the unusual intensity dependence of the phase lag given in (6.3.5), which shows the phase lag to decrease with increasing light intensity as $\rho_\nu^{-\frac{1}{2}}$.

O'Connor and *Diebold* [6.53] have included the effect of depletion of the A_2 concentration by intense light, and the effects of the mole number increase and the thermalization of the recoil energy of the fragments on the photoacoustic signal. The effect of the last two phenomena is to change the signal amplitude and phase, giving the measured phase lag Φ_r as

$$\tan\ \Phi_r = \frac{\lambda}{1+\hat{\alpha}(1+\lambda^2)}\ , \tag{6.3.9}$$

where $\hat{\alpha}$ is, as in (6.2.5), a ratio of the sum of the recoil and equivalent mole number increase energies divided by the heat released by photodissociation [6.54]. This expression, of course, reduces to (6.3.5) when $\hat{\alpha} = 0$.

O'Conner and *Diebold* have also reported experimental measurements of phase lags and amplitudes from the termolecular recombination of Cl radicals. The authors used an Ar ion laser at 488 nm to excite the $B\ 0_u^+$ state, which predissociates to give two ground state $^2P_{\frac{3}{2}}$ atoms. The laser beam was expanded in a Galilean telescope so that only the central position of the beam filled a 1.75 cm diameter by 3.8 cm long Teflon cell, as shown in Fig. 6.13. The laser beam was sinusoidally amplitude modulated with an acousto-optic light modulator operating in zeroth order. The modulation depth was adjusted to be 0.1. Data were taken at fixed frequencies by measurement of both the amplitude and phase of the acoustic signal as a function of laser intensity and modulation frequency. Plots of the phase lag versus laser power and modulation frequency are shown in Fig. 6.14. The authors measured $2(ab)^{\frac{1}{2}}$ as 238 ± 10 rad s^{-1} at 100 mW, in reasonably good agreement with existing data for Cl_2.

The somewhat unusual dependence of the phase lag on light intensity given by (6.3.5) can be understood readily in terms of a mean radical lifetime τ_R

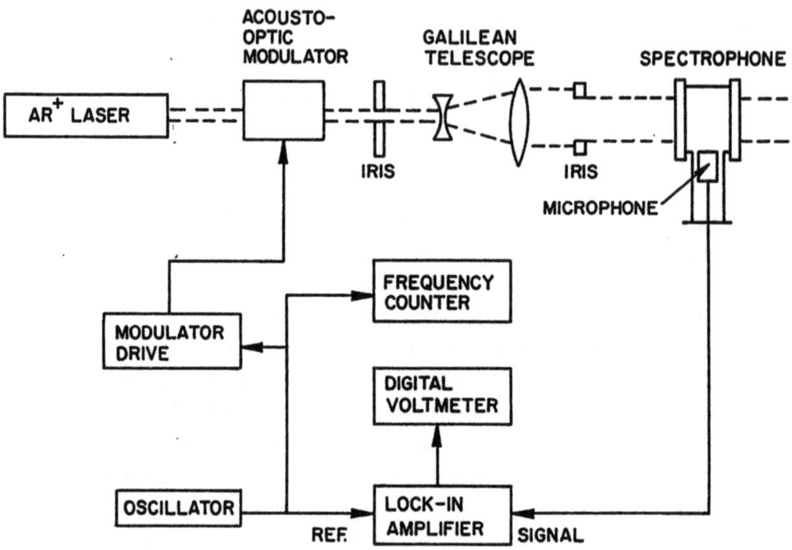

Fig. 6.13. Experimental apparatus for measuring the rate of termolecular recombination of Cl radicals. [6.53]

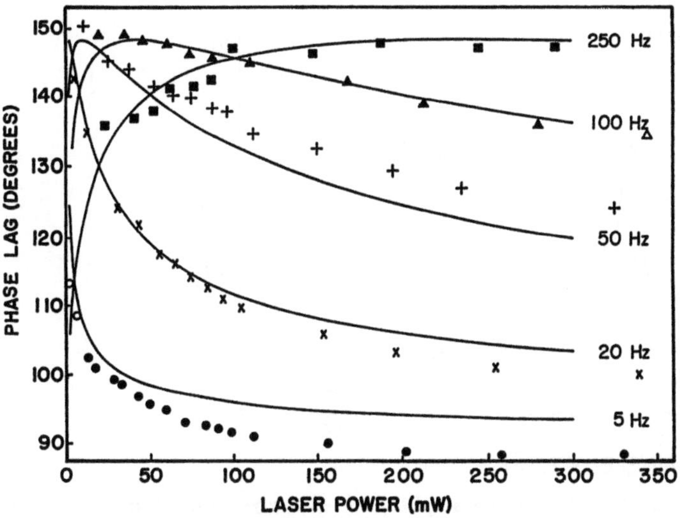

Fig. 6.14. Phase lag versus laser power in Cl_2 photodissociated at 488 nm. [6.53].

defined as the concentration of radicals divided by twice their rate of removal. From (6.3.2 or 3), τ_R is seen to be

$$\tau_R = \frac{x}{2ax^2} = \frac{1}{2(ab)^{\frac{1}{2}}} , \tag{6.3.10}$$

which shows that the phase lag expression in (6.3.5) can be written as

$$\tan \phi_R = \omega \tau_R , \tag{6.3.11}$$

which, again, is the limiting form of (6.2.5) for a two-level system. The shortening of τ_R by an increase in light intensity comes as a consequence of a recombination rate bimolecular in [A]: the more radicals per unit volume present, the faster recombination takes place, which, in turn, reduces the radical lifetime.

The change in slope of the curves of phase lag versus modulation frequency can be understood in terms of a competition between the prompt signal, which always has a $\pi/2$ fixed phase lag with respect to the modulation and an ω^{-1} amplitude dependence, and the signal from recombination, which has a variable phase lag and whose amplitude falls off as ω^{-1} at low frequencies and ω^{-2} at high frequencies. At low frequency (or high light intensity) the signal from recombination dominates the sound production so that an arctangent curve is traced out by the phase versus frequency curve. However, as the frequency becomes large, $\omega\tau_R$ become greater than 1 and the amplitude of this effect becomes smaller in relation to the signal produced by the mole number increase and the recoil energy of the fragments. The phase thus comes to be controlled by the last two effects, and tends to a value of $\pi/2$.

Nonlinear effects have also been observed by *Robin* et al. [6.55] in glyoxal excited to its first singlet state. When the (0,0) band of the $S_1 \longleftarrow S_0$ transition of glyoxal was excited the photoacoustic signal was found to exhibit a quadratic

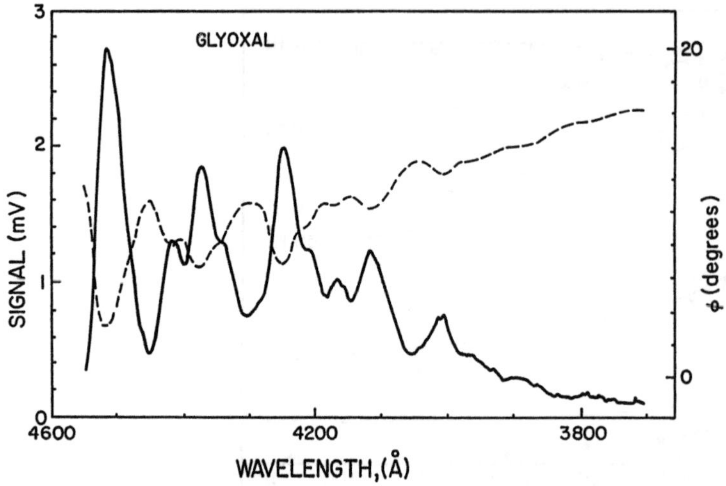

Fig. 6.15. Photoacoustic phase (- - -) and amplitude (—) in glyoxal at 10 mm Hg. The change in phase with absorbed energy is indicative of a nonlinear heat evolution mechanism. [6.55]

dependence on light intensity. The absorption spectrum (Fig. 6.15) was also found to exhibit a phase lag decrease wherever the absorption increased. The authors interpreted their data as being consistent with a reaction scheme where radiation excites glyoxal from the singlet ground state S_0 to the first excited singlet state S_1, which transfers its energy to a triplet state T_1, which, in turn, reacts to give an excited species E. The species E reacts in an exothermic bimolecular reaction to give products (CO, H_2, a polymer, and possibly H_2CO) in the reaction sequence

$$\begin{aligned} S_0 &+ h\nu \longrightarrow S_1 \,, \\ S_1 &+ S_0 \longleftrightarrow T_1 + S_0 \,, \\ T_1 &+ S_0 \longrightarrow E \,, \\ E &+ E \longrightarrow \text{Products} \,. \end{aligned} \tag{6.3.12}$$

If equilibrium is maintained in the second step, then the rate equation for [E] is found to be

$$\frac{d[E]}{dt} + 2k_G[E]^2 = K[S_0]^2\rho_\nu B(1 + d\,\sin\,\omega t)\,, \tag{6.3.13}$$

where k_G is the rate constant for the last reaction and K is a constant. This equation can be reduced to a form identical with (6.3.3), used to describe termolecular recombination of Cl radicals. Hence, the nonlinear effects found in Cl_2, namely a phase lag that decreases with an increase in light intensity, are expected. Clearly, a heat release (at least) bimolecular in the excited species concentration follows from the phase lag dependence shown in Fig. 6.15. This system would appear to merit further investigation to confirm the mechanism given in (6.3.12), and to identify the various species present.

The photolysis of acetaldehyde has been investigated by *deGroot* et al. [6.56], *Emeis* et al. [6.57] and *Harshbarger* and *Robin* [6.36]. The presence of nonlinear heat-generating mechanisms is indicated by the presence of an acoustic signal at twice the modulation frequency as well as by an intensity-dependent signal amplitude. Two spectral regions were identified, one for wavelengths greater than 320 nm where the radiationless transitions govern the release of heat, and a second one between 270 and 320 nm, where the heat release is second order in the radical concentration. The mechanism for sound production was found to be

$$\begin{aligned} A(S_0) + h\nu &\longrightarrow A(S_1)\,, \\ A(S_1) &\longrightarrow A(T)\,, \\ A(T) + B(S_0) &\longrightarrow A(S_0) + B(T)\,, \\ B(T) &\longrightarrow 2R\,, \\ 2R &\longrightarrow B\,, \end{aligned} \tag{6.3.14}$$

where the second reaction is fast intersystem crossing, A is acetaldehyde, B is biacetyl formed in the reaction vessel over a period of time, and R is the acetyl radical. This mechanism, deduced with the help of data from phosphorescence measurements [6.58] and additional photoacoustic measurements carried out with acetaldehyde – biacetyl mixtures, also leads to a Riccati equation of the form (6.3.3) for the [R].

In certain spectral regions, biacetyl [6.27,42,43], acetophenone [6.49] and *p*-trifluoromethylacetophenone [6.49] exhibit complicated photochemical effects as a result of photodecomposition of the parent molecule into reaction products. In the case of biacetyl, *Hunter* and *Stock* [6.27] were able to determine the triplet-triplet quenching rate. In general, the capability of the photoacoustic technique of determining kinetic mechanism or rate parameters in reacting systems depends on the simplicity of the pathways for energy release: only in those cases where the mechanisms of energy evolution are straightforward and take place on experimentally accessible time scales can rate parameters be obtained. From the work reported here it is clear that valuable kinetic information can be obtained on a large number of gas phase species using the photoacoustic effect. These experiments serve as well to point up a unique feature of the photoacoustic effect: its response to evolved energy.

6.4 Chemical Amplification

In certain chemical systems there exist propagation mechanisms for radicals where the introduction of a single radical causes a product molecule to be produced and one or more radicals to be regenerated as well. Such regenerative sequences of chemical reactions are referred to as "chain reactions", which are characterized by quantum yields of products greater than 1. The existence of quantum efficiencies or "chain lengths" that exceed unity, and which have been calculated [6.59] to be as large as 10^9, combined with reaction enthalpies that are on the order of magnitude of a diatomic bond energy, suggest that chain reactions can be expected to give a sizeable amplification of the magnitude of the photoacoustic effect.

One such system [6.59, 60] is the two-center reaction between a halogen X_2 and H_2, where the introduction of an X radical through photodissociation (the initiation reaction) results in the production of two HX product molecules (through the chain propagation steps) while at the same time regenerating an X radical. The photochemical reaction sequence can be written in terms of initiation, propagation, and termination steps as

$$\begin{array}{lll} (a) & X_2 + h\nu \rightarrow 2X, & \text{Initiation} \\ (b) & X + \mathrm{H}_2 \underset{k_{-X}}{\overset{k_X}{\rightleftharpoons}} \mathrm{H}X + H\,, & \text{Chain} \\ (c) & \mathrm{H} + X_2 \xrightarrow{k_H} \mathrm{H}X + X\,, & \text{Propagation} \\ (d) & 2X + M \underset{\bar{k}_d}{\overset{\bar{k}_r}{\rightleftharpoons}} X_2 + M\,, & \text{Termination}\,. \end{array} \quad (6.4.1)$$

Newly produced X radicals repeat the propagation steps (6.4.1b and c) until they encounter other X radicals and are consumed in reaction (d), referred to as the termination reaction. Because the termination reaction is bimolecular in [X], and is consequently slow, the number of cycles of propagation before termination takes place is generally large.

In their analysis of reactions (6.4.1), *Diebold* and *Hayden* [6.61] note that the propagation reactions taken together do not alter the [X]. That is, provided reaction (c) is fast compared with (b) then the introduction of H_2 into a spectrophone containing X_2 does not alter the photochemically produced [X]. This can be put on a quantitative basis by considering the rate equation for the radical concentrations,

$$\begin{aligned} d[X]/dt &= 2[X_2]\rho_\nu B(1 + d\sin\omega t) - k_X[X][\mathrm{H}_2] + k_{-X}[\mathrm{H}X][\mathrm{H}] \\ &+ k_H[\mathrm{H}][X_2] - 2\bar{k}_r[X]^2[M] + 2\bar{k}_d[X_2][M] \end{aligned} \quad (6.4.2)$$

and

$$d[\mathrm{H}]/dt = k_X[X][\mathrm{H}_2] - k_{-X}[\mathrm{H}X][\mathrm{H}] - k_H[\mathrm{H}][X_2]\,. \quad (6.4.3)$$

Provided a stationary state for [H] is maintained during a modulation period, a Riccati equation is obtained for $[X]$.

The rate of heat release in the system is the sum of the contributions from the propagation reactions and can be expressed as

$$H(t) = \left[\frac{k_X[\mathrm{H}_2][X_2](-\Delta H_{rx})}{1 + (k_{-X}[\mathrm{H}X]/k_H[X_2])}\right] x(t), \quad (6.4.4)$$

where ΔH_{rx} is the total enthalpy for the propagation reactions and $x(t)$ is the solution to (6.3.3), as given by (6.3.5). Substitution of $H(t)$ into (6.2.3) gives the pressure signal from the chain reacting system as

$$p_{rx}(t) = \frac{k_X[\mathrm{H}_2][X_2]}{1 + k_{-X}[\mathrm{H}X]/k_H[X_2]}\left(\frac{\Delta H_{rx} R}{C_v'}\right)\left(\frac{(\delta/b)(b/a)^{\frac{1}{2}}}{2\omega(1+\lambda^2)^{\frac{1}{2}}}\cos(\omega t - \phi_R)\right), \quad (6.4.5)$$

where a, b, λ, δ, and ϕ_R are given by (6.3.3-5) above, and where only the fundamental frequency component of $x(t)$ has been included. There are, of course, contributions to the pressure from the mole number increase, the recoil energy of the photofragments, and the recombination reaction; however, these effects are comparatively small in systems with large chain lengths and can be neglected.

It is possible to express this result in terms of the photochemical chain length Φ_{ph} defined as the number of product molecules HX produced per X radical introduced into the system by the initiation reaction. According to *Benson* [6.59], the photochemical chain length is given as

$$\Phi_{ph} = \frac{2k_X[\mathrm{H}_2]}{1 + (k_{-X}[\mathrm{H}X]/k_H[X_2])}(ab_{ph})^{-\frac{1}{2}} , \qquad (6.4.6)$$

where $b_{ph} = 2\rho_\nu B$. This result, which can be found by assuming the stationary state for $[X]$ and [H] is derived under the assumption that $\rho_\nu B/\bar{k}_d[M] >> 1$, that is, the reactions are photochemical rather than thermal in origin. If (6.4.6) is substituted into (6.4.5) the acoustic signal is found to be

$$p_{rx}(t) = \left(\frac{\Delta H_{rx} R}{2C_v'}\right)\rho_\nu B[X_2]\, \Phi_{ph}\left(\frac{\delta}{b_{ph}}\right)\frac{\cos(\omega t - \phi_R)}{\omega(1+\lambda^2)^{\frac{1}{2}}} , \qquad (6.4.7)$$

which shows the signal amplitude to be directly proportional to the chain length.

There is a fundamental difference in the response of the fundamental frequency component of the acoustic signal given above, and the corresponding signal from photodissociation alone with respect to changes in $\rho_\nu B$, the steady part of the incident radiation intensity. At low frequency ($\lambda << 1$) and for a fixed alternating component of the radiation intensity δ, as $\rho_\nu B$ increases, the amplitude of the acoustic signal from photodissociation alone is unchanged; the amplitude of the acoustic signal produced by the chain reactions on the other hand decreases as $(\rho_\nu B)^{-\frac{1}{2}}$, inversely proportional to the square root of the steady component of the radiation intensity. The physical basis for this dependence is that as the steady component of the radiation increases in intensity, the $[X]$ increases, thereby raising the three-body recombination rate by an amount proportional to the square of $[X]$. Since the rate of HX production is only linear in $[X]$, the ratio of the termination to chain reaction rates increases, thereby reducing the chain length. At the opposite extreme, when the modulation frequency is high or the light intensity low ($\lambda >> 1$) the acoustic signal at fixed δ becomes independent of $\rho_\nu B$ and decreases with frequency as ω^{-2}. Here, the modulation period is shorter than the radical lifetime, which, given the response $x(t)$ to the linearized Riccati equation, results in an additional factor of ω in the denominator of the expression for $p_{rx}(t)$.

O'Connor and *Diebold* [6.62] have reported chemical amplification of the photoacoustic effect in Cl_2 - H_2 - N_2 mixtures using the experimental apparatus shown in Fig. 6.13. The 476 nm line from an Ar ion laser was used to photodissociate Cl_2 in a spectrophone equipped with a corrosion resistant microphone. The microphone was a Bruel and Kjaer Model 4125 condenser microphone whose diaphragm was replaced with a 1.3 x 10^{-2} cm thick Ni foil. The foil was then coated with a chlorosulfonated polyethylene paint. Typical experimental data taken with H_2(20%) - Cl_2(70%) - N_2(20%) mixtures are shown in Fig. 6.16.

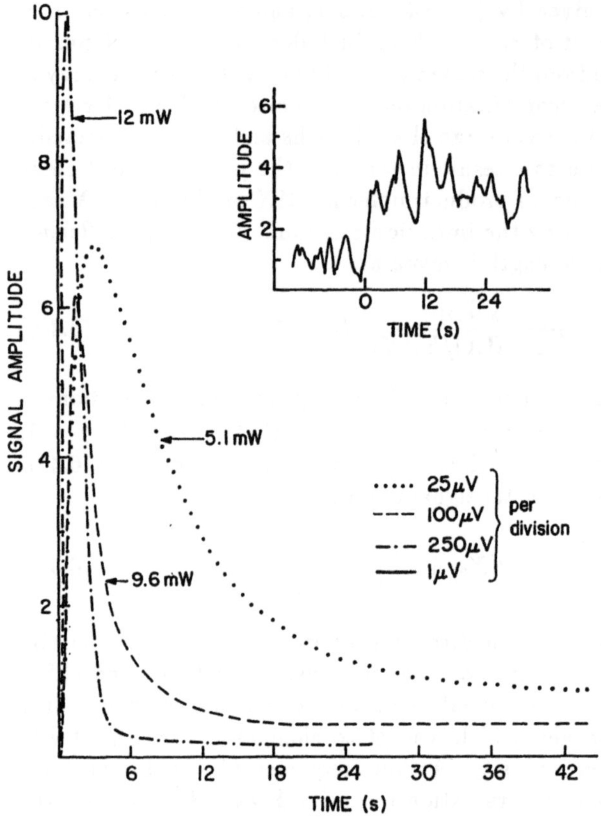

Fig. 6.16. Photoacoustic signals versus time in static H_2 - Cl_2 - N_2 mixtures (1 atm total pressure) at several laser powers. The lock-in amplifier time constant is 1s. Inset: Photoacoustic signal amplitude versus time in a Cl_2-N_2 mixture with a lock-in amplifier time constant of 3s. [5.62]

For comparison, the signal from a Cl_2(85%) - N_2(15%) mixture irradiated with a 75 mW laser beam is shown in the inset. The small signal from Cl_2 is characteristic of the relative insensitivity of this particular microphone, which was designed to withstand both corrosion from Cl_2 and the explosions which take place readily at powers above 12 mW. The photoacoustic signal follows the rate of chemical reaction — the higher the laser power, the more rapid the consumption of reactants. As can be seen from (6.4.5), the photoacoustic effect acts as a monitor of the concentrations of reactants through the product $[H_2][Cl_2]$. From a broad perspective, the photoacoustic effect used in this way acts as the frequency domain analog of flash photolysis. In contrast with flash photolysis, the high sensitivity of the photoacoustic effect permits experiments to be carried out at low light intensities, hence, the effects of small amounts of impurities on the overall reaction rate can be determined. The photoacoustic effect thus possesses some advantages over flash photolysis, where relatively insensitive techniques such as atomic or molecular absorption are used to monitor concentration.

Isotope effects on the rate of reaction can be studied using the photoacoustic effect. *Wyers* and *Diebold* [6.63] have investigated the photoacoustic effect in stoichiometric H_2 - Cl_2 mixtures and compared the signal amplitudes to those generated in D_2 - Cl_2 mixtures. If (6.4.5) is evaluated for these two systems under identical experimental conditions early in the reaction, when [HCl] = [DCl] $\cong$ 0, then the ratio of the peak photoacoustic pressures for these two systems is given by

$$\frac{P_H}{P_D} = \frac{k_{Cl}^H \Delta H_H}{k_{Cl}^D \Delta H_D} , \tag{6.4.8}$$

where H and D refer to quantities obtained with H_2 and D_2 respectively.

The photoacoustic data exhibited high signal-to-noise ratios and gave a value of P_H/P_D in fair agreement with previously measured rate constants. It is worth noting at this point that the photoacoustic effect in chain reacting systems is extremely sensitive to trace impurities — the longer the chain length, the more sensitive the product yield is to the presence of chemical species that terminate the reactions. In fact, it is the comparatively low reaction rate of the termolecular recombination reaction that is responsible for the remarkably long length of the chains. It is known, for instance, that the addition of only 1% O_2 to a H_2 - Cl_2 mixture can change the rate of reaction by a factor of 10^3 [6.59]. Thus, the accuracy of such photoacoustic measurements is contingent on gas purity. It would appear, for this reason alone, that the value of the photoacoustic effect for measurements in chain reacting systems lies in its capability for in situ studies of the influence of slight changes in parameters such as light intensity, chemical composition, or impurities on overall reaction rate more than in its ability to determine rate parameters.

O'Connor et al. [6.64] (see also [6.65]) have investigated the effect of NO on the H_2-Cl_2 reaction. The effect of NO was modeled by considering the reaction sequence

$$\begin{aligned} \mathrm{Cl} + \mathrm{NO} + M &\xrightarrow{k_{NO}} \mathrm{NOCl} + M , \\ \mathrm{Cl} + \mathrm{NOCl} &\xrightarrow{k_{NOCl}} \mathrm{NO} + \mathrm{Cl}_2 , \end{aligned} \tag{6.4.9}$$

where the rate constant for the second reaction is known to be 70 times faster than the first. This pair of reactions acts to consume Cl and produce Cl_2, leaving the NO concentration unchanged. Thus NO can be said to act as a homogeneous catalyst for the removal of Cl radicals, which, depending on NO concentration, can completely dominate the chain termination process. Under the assumption of a steady state concentration of NOCl, the rate equation for the dimensionless [Cl] is

$$\frac{dx}{dt} + 2k_{NO}[\mathrm{NO}]x = 2\rho_\nu B(1 + d \sin \omega t) . \tag{6.4.10}$$

The steady-state solution to this equation is

$$x(t) = \frac{\rho_\nu B d}{k_{NO}[\mathrm{NO}](1 + \lambda_{NO}^2)^{\frac{1}{2}}} \sin(\omega t - \phi_{NO}) , \tag{6.4.11}$$

where

$$\tan \phi_{NO} = \frac{\omega}{2k_{NO}[\mathrm{NO}]} = \lambda_{NO} .$$

Now, provided the termination step is dominated by reactions (6.4.9) (rather than termolecular recombination) and the [NO] is not so great as to eliminate the chain reactions as the dominant source of heat release in the system, then the energy per unit volume released into the system is again given by (6.4.4) and the acoustic signal is given by

$$p_{NO}(t) = \left(\frac{\Delta H_{rx} R}{C_v'}\right) \frac{\rho_\nu B d k_{Cl}[H_2][Cl_2]}{k_{NO}[NO]\omega(1+\lambda_{NO}^2)^{\frac{1}{2}}} \cos(\omega t - \phi_{NO}) , \qquad (6.4.12)$$

where k_{Cl} refers to the forward reaction rate constant in (6.4.1b). This equation differs from the others in this section that describe chain reactions in that it represents *linear* response — there is no dependence of the phase lag on light intensity and the signal amplitude increases linearly with the radiation density.

Diebold and co-workers carried out measurements in a Teflon cell equipped with an electret microphone. The microphone was protected from the corrosive gas mixtures by a thin Teflon diaphragm. High-purity gas mixtures were made up, introduced into the cell, and irradiated with the 488 nm output of an Ar ion laser. Data were taken immediately after the photoacoustic signals reached a stable amplitude. The plot of photoacoustic amplitude versus mole fraction of NO shown in Fig. 6.17 clearly shows the $[NO]^{-1}$ dependence given in (6.4.12). The inset shows the insensitivity of the phase lag to changes in radiation density, in striking contrast to Fig. 6.14.

Choi and *Diebold* [6.66] have used chemical amplification of the photoacoustic effect in conjunction with a gas chromatograph for trace detection. In

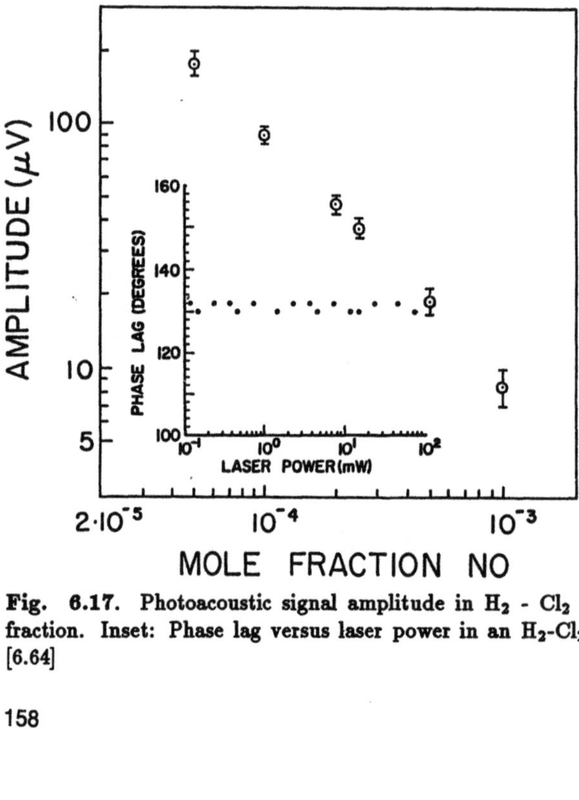

Fig. 6.17. Photoacoustic signal amplitude in H_2 - Cl_2 mixtures versus added NO mole fraction. Inset: Phase lag versus laser power in an H_2-Cl_2 mixture containing 1 ppm NO. [6.64]

the first set of experiments, N_2 was used as the carrier gas. The eluant from the chromatograph was fed to a spectrophone along with a small flow of Cl_2. An Ar ion laser at 488 nm photodissociated the Cl_2 to give a photoacoustic effect. When a hydrocarbon was injected onto the chromatograph it was carried at its characteristic elution time into the acoustic cell, giving a chemically amplified photoacoustic effect. A dramatic increase in the magnitude of the lock-in amplifier signal versus time was recorded. Detection limits on the order of a few tenths of a nanomole were found.

When H_2 was used as the carrier gas in the chromatograph a large chemically amplified signal was generated in the acoustic cell. Introduction of small amounts of gases that act to terminate chain reactions such as NO or O_2 resulted in a large decrease in the signal amplitude. The authors reported that the photoacoustic detector has a sensitivity of a few picomoles for NO, and that it is insensitive to species such as CO, N_2O, and CO_2 that do not quench chain reactions.

In the experiments with N_2 as the carrier gas, a gain factor can be calculated by dividing the chemically amplified signal by the signal present when only Cl_2 produces a photoacoustic effect. From (6.4.7) it can be seen that the gain, which is directly proportional to the photochemical chain length, increases as the laser power decreases. A plot of gain versus laser power shown in Fig. 6.18, taken by injecting fixed quantities of the various species onto the column, clearly shows this effect. Of course, some of these species undergo complex chain reactions which involve several chlorination steps. The gain does not increase without limit as the laser power is reduced, but reaches a limit determined by the thermal chain length.

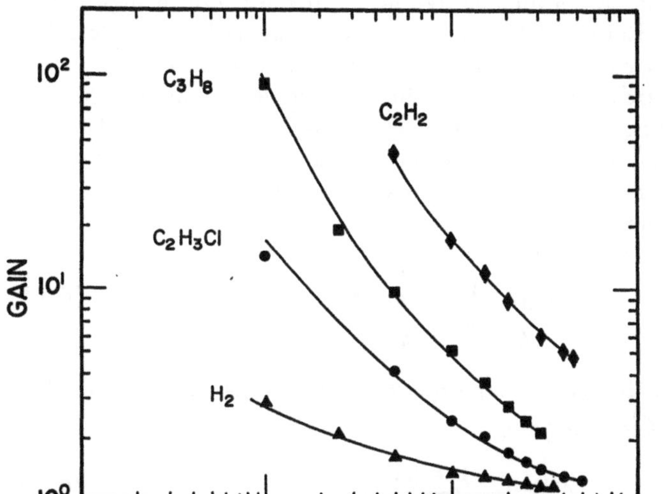

Fig. 6.18. Photoacoustic gain versus laser power for 200 nmole of several gases injected onto a chromatography column. The increase in the photochemical chain length as the laser power is reduced causes the gain to increase. [6.66]

Generally speaking, the photoacoustic effect acts as an analog of the method of intermittent activation (also called the rotating sector method) [6.67], a technique employed some time ago where photochemical reactions were studied by illuminating a reaction vessel through a hole in a rotating wheel. The reaction rate was studied (generally by a titration at the end of the experiment) as a function of the rotation rate of the wheel, from which the order of the reaction could be determined. The photoacoustic effect has in common with the method of intermittent activation a measurement of reaction rate as a function of modulation frequency; the photoacoustic method, however, directly monitors the heat evolution in the system and has the additional advantages of fast response and high sensitivity.

6.5 Unimolecular Reactions

When certain molecules are excited to high vibrational levels of the ground state, or an electronically excited state, there is the possibility of an isomerization reaction taking place. If the isomerization reaction is carried out in a vessel where collisional deactivation of the parent species can also take place, then there is a competition between collisional deactivation of the initially prepared state and isomerization. Since collisional deactivation is a pressure-dependent process and unimolecular isomerization is pressure independent, the former can be used as an internal "clock" for calibration of the rate of unimolecular isomerization by measuring the isomerization rate as a function of pressure. While there exist other methods for studying unimolecular reactions, the photoacoustic technique used in conjunction with a continuous dye laser provides both a high degree of specificity in the excitation of the initial excited state and a high detection sensitivity for recording changes in reactant or product concentration.

According to *Reddy* and *Berry* [6.68-70], the simplest model for this process considers a molecule A photoactivated at a rate k_a by radiation $h\nu$. The excited species A^* can be collisionally deactivated at a rate k_M by an inert species M, or it can isomerize at a rate $k(\epsilon)$ to a new chemical species $\tilde{A}$ according to the Stern-Volmer scheme,

$$\begin{aligned} A + h\nu &\xrightarrow{k_a} A^* , \\ A^* + M &\xrightarrow{k_M} A + M , \\ A^* &\xrightarrow{k(\epsilon)} \tilde{A} . \end{aligned} \tag{6.5.1}$$

Under steady-state conditions, the time rate of change in $[A^*]$ is zero. Solution of the rate equation for $[A^*]$ gives

$$[A^*] = \frac{\rho_\nu k_a [A]}{k_M[M] + k(\epsilon)} , \tag{6.5.2}$$

which on substitution into the rate equation for [A], yields

$$\frac{d[A]}{dt} = -\left(\frac{\rho_\nu k_a k(\epsilon)}{k_M[M] + k(\epsilon)}\right)[A] . \tag{6.5.3}$$

This shows that [A] decreases exponentially in time and that the quantity in round brackets can be regarded as an apparent rate constant k_{app}, which is most conveniently expressed as

$$\frac{1}{k_{app}} = \frac{1}{\rho_\nu k_a} + \left(\frac{k_M}{k(\epsilon)k_a\rho_\nu}\right)[M] \,. \tag{6.5.4}$$

It is thus clear that if experimental values of k_{app}^{-1} versus [M] are plotted, $(\rho_\nu k_a)^{-1}$ represents the intercept at $[M] = 0$, and that the slope of the plot, along with the intercept, yields a value for the unimolecular rate constant.

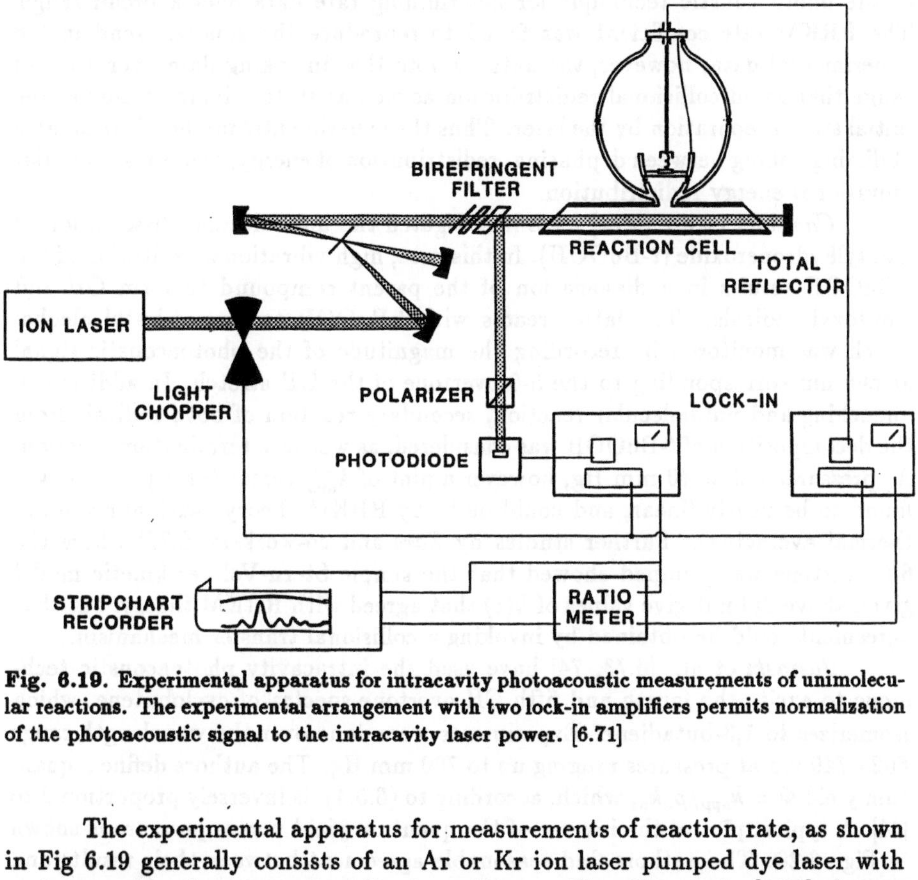

Fig. 6.19. Experimental apparatus for intracavity photoacoustic measurements of unimolecular reactions. The experimental arrangement with two lock-in amplifiers permits normalization of the photoacoustic signal to the intracavity laser power. [6.71]

The experimental apparatus for measurements of reaction rate, as shown in Fig 6.19 generally consists of an Ar or Kr ion laser pumped dye laser with an intracavity photoacoustic cell. According to *Chandler* et al. [6.71] the use of the intracavity cell in a dye laser with totally reflecting mirrors results in a continuous intracavity power of 120 W when the laser is pumped with a 15 W ion laser. Such high average powers combined with the high sensitivity of the photoacoustic effect permit recording of weak overtone spectra. Since excitation of high overtones is requisite for producing a molecular state with sufficient energy to cause isomerization, every effort is made to obtain as high a sensitivity as possible. The experiments are carried out by recording the photoacoustic signal amplitude versus time at a fixed wavelength so that the rate of disappearance of the parent compound is monitored. Measurements are repeated at various pressures to generate a plot that corresponds to (6.5.4). The dependence of the unimolecular decay rate on the energy deposited in the parent molecule can also be determined by repeating the experiment at different excitation wavelengths. This dependence is predicted by RRKM theory.

Reddy and *Berry* [6.68] have conducted experiments on the isomerization of methyl isocyanide (CH_3NC) to methyl cyanide (CH_3CN). The absorption spectra of these molecules were assigned in the region of 890 - 540 nm and values of $k(\epsilon)$ were obtained over much of the same region. Values for $k(\epsilon)$ at various photon energies varied from 2 x $10^7 s^{-1}$ to $10^{10} s^{-1}$ showing the capability of the photoacoustic technique for determining rate data over a broad range. The RRKM rate coefficient was found to reproduce the general trend in the experimental data; however, the authors note that in taking data over a broad range that rapid collisional redistribution as well as state mixing scrambles the initial state preparation by the laser. Thus the experimental method is incapable of distinguishing between dephasing, redistribution of energy, and intramolecular vibrational energy redistribution.

Chandler et al. [6.71] have investigated the unimolecular dissociation of *t*-butylhydroperoxide (*t*-BuOOH). In this case, high vibrational excitation of the *t*-BuOOH results in a dissociation of the parent compound to form OH and *t*-butoxyl radicals. The latter reacts with *t*-BuOOH to form *t*-butyl alcohol which was monitored by recording the magnitude of the photoacoustic signal at 606 nm corresponding to the 5-0 overtone of the OH stretch. In addition to quenching and unimolecular reaction, secondary reaction of both radicals from the decomposition of *t*-BuOOH was considered, as well as a termination reaction. At pressures below 40 mm Hg, however a plot of k_{app}^{-1} versus total pressure was found to be nearly linear, and could be fit by RRKM theory (with appropriate thermal averaging). Further studies by *Zare* and co-workers [6.72] where the 6-0 overtone was pumped showed that the simple Stern-Volmer kinetic model given above did not give values of $k(\epsilon)$ that agreed with RRKM theory, but that agreement could be obtained by invoking a collisional transfer mechanism.

Jasinski et al. [6.73, 74] have used the intracavity photoacoustic technique to excite the fourth and fifth CH overtone spectra of cyclobutene, which isomerizes to 1,3-butadiene. Experiments were done over the wavelength range 602 - 749 nm at pressures ranging up to 700 mm Hg. The authors define a quantum yield $\Phi = k_{app}/\rho_\nu k_a$, which, according to (6.5.4), is inversely proportional to [M]. A typical plot of the inverse of the quantum yield versus pressure is shown in Fig. 6.20. The authors find reasonable agreement between their results and RRKM theory.

Similar experiments were reported by this group on the isomerization of 1-cyclopropylcyclobutene [6.74, 75]. In particular, the authors looked for evidence of isomerization from a nonrandom energy distribution. They reported that no dependence of $k(\epsilon)$ on the kind of overtone excited was found. The explanation given was that either the excitation was not initially localized by the laser excitation or that there was rapid energy transfer before isomerization.

In further experiments with 2-methylcyclopentadiene [6.72, 76], *Moore* and co-workers formulated a model analogous to the reaction sequence (6.5.1) that included back reaction. Again the authors focused attention on finding

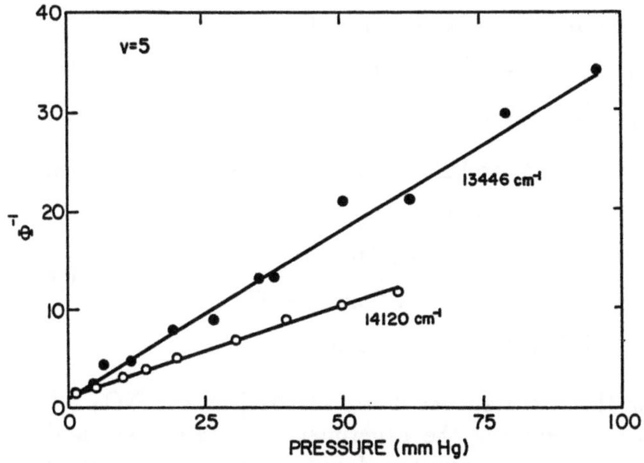

Fig. 6.20. Inverse of the quantum yield versus pressure for cyclobutane photolysis at two different excitation energies. The lines are least square fits to (6.5.4). [6.73]

specific modes that would lead to rapid reaction, but found no effect even when H atom motion along the reaction coordinate was specifically excited.

Lishan et al. [6.77] have excited several vibrational overtones of quadricyclane. The values of $k(\epsilon)$ found did not indicate any mode-specific enhancement in reaction rate; however, the authors pointed out that the opposite effect, the failure of energy deposited in some modes to produce any reaction, was equally informative from the perspective of understanding.

6.6 Direct Detection of Reactants and Products

The sensitivity of the photoacoustic effect in gases, as reviewed by *Zharov* and *Letokhov* [6.78], is quite high, on the order of $10^{-9}\mathrm{cm}^{-1}$. One of the most straightforward applications of the photoacoustic effect to gas phase chemical kinetics makes use of this high sensitivity to monitor the appearance of reaction products or the disappearance of reactants to determine mechanisms of reaction and chemical rate constants.

Colles et al. [6.79] have reported the use of the photoacoustic effect to monitor the production of CH_3NO and NO_2 in the photolysis of CH_3NO_2 by a medium pressure Hg lamp. Product formation is monitored by amplitude modulating a continuous dye laser at 1.4 kHz and detecting the resulting acoustic signal with a microphone whose output is fed to a lock-in amplifier. The buildup of CH_3NO (which was identified by scanning the dye laser and comparing band heads with a known spectrum) and NO_2 was found to take place on a time scale of a few seconds, while the decay, the same for both products, showed both a fast and a slow component. With O_2 added to the gas mixture a longer decay time for the slow component of the NO_2 decay was found. The reaction scheme that accounts for the production of CH_3NO deduced by the authors is

$$
\begin{aligned}
&(a)\quad CH_3NO_2 + h\nu \rightarrow CH_3 + NO_2\,,\\
&(b)\quad CH_3 + NO_2 \rightarrow CH_3O + NO\,,\\
&(c)\quad CH_3 + NO \rightarrow CH_3NO\,.
\end{aligned}
\qquad (6.6.1)
$$

The decay of the products was found to be consistent with the reaction sequence

$$
\begin{aligned}
&(d)\quad NO_2 + h\nu \rightarrow NO + O\\
&(e)\quad CH_3 + NO_2 \rightarrow CH_3NO_2\\
&(f)\quad CH_3NO + NO_2 \rightarrow CH_3NO_2 + NO\\
&(g)\quad CH_3O + NO_2 \rightarrow CH_3ONO_2\,.
\end{aligned}
\qquad (6.6.2)
$$

Reaction (f) accounts for the simultaneous decay of CH_3NO and NO_2. This mechanism also explains the results of the experiments where O_2 was added: with O_2 present the CH_3 radicals are scavenged and are thus prevented from forming CH_3NO through reactions (b) and (c), or the CH_3NO that is formed rapidly reacts to form CH_3ONO_2.

A simpler chemical reaction, the isotopic exchange reaction,

$$H_2S + D_2S \rightarrow 2HDS, \qquad (6.6.3)$$

has been studied by *Kadibelban* and *Hess* [6.80, 81] and *Hess* et al. [6.82] using photoacoustic detection of either D_2S or HDS. A transversely excited atmospheric CO_2 laser (TEA laser) could be tuned to the P34 line at $10.74\mu m$ to excite D_2S or the P36 line at $9.69\mu m$ to excite HDS. Data were taken by rapidly mixing equal amounts of H_2S and D_2S at 51 mm Hg and either following the decay of the D_2S concentration in time or the corresponding rise of the HDS concentration. Since TEA lasers have substantial shot-to-shot amplitude fluctuations, a reference mixture of equilibrated H_2S, D_2S, and HDS was irradiated simultaneously with the reaction cell, as shown in Fig. 6.21. The signal from

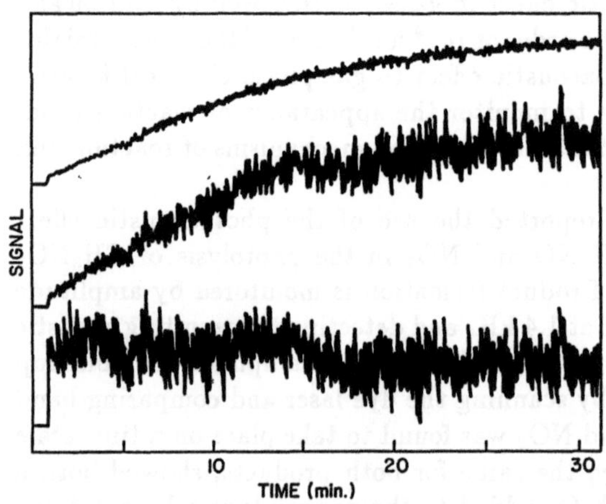

Fig. 6.21. Photoacoustic signal versus time for the isotopic exchange reaction $H_2S + D_2S \longrightarrow 2HDS$. Lower trace: signal from reference cell. Middle trace: HDS signal from reaction cell. Upper trace: divided signal from signal analyzer showing the rise in [HDS]. [6.80]

the reaction cell was divided by that from the reference cell in a signal analyzer, greatly improving the signal-to-noise ratio in the data. For pressures above 40 mm Hg the reaction rate appeared to be described as a first-order reaction (one-half for each reactant) with a rate coefficient of 1.7×10^{-3} s^{-1} at 295 K. A strong temperature dependence of the reaction rate and a marked increase in rate on substitution of a glass reaction vessel for the stainless steel one was found. These observations strongly argue for a heterogeneous reaction mechanism.

Farrow and *Richton* [6.83] used a continuous CO laser to study the reaction

$$NH_3 + HNO_3 \rightarrow NH_4NO_3 \,, \tag{6.6.4}$$

which is important from the point of view of atmospheric chemistry. Ammonia consumption was monitored with the CO laser tuned to 5.549μm and nitric acid with the laser operating at 5.929μm. At 50 mm Hg it was found that each species could be monitored without interference from the other. A photoacoustic cell constructed of stainless steel was fitted with CaF_2 windows and a gold-plated electret microphone. The reactants were rapidly mixed by opening a solenoid valve; data were taken 1.5 s after the gases were mixed. As in the pulsed experiments above, the photoacoustic signal was ratioed to the incident laser power (monitored by a pyroelectric detector) to reduce the effects of intensity fluctuations. Records of the lock-in amplifier signal versus time permitted the rate of reaction to be measured. From the rate data it was found that the reaction had a strong heterogeneous contribution to the overall rate. The homogeneous rate constant for reaction (6.6.4), however, was determined to be 2 x 10^{-18} cm^{3}s^{-1}, a figure far smaller than the literature value being used at the time the experiments were done.

The well-known reaction of nitric oxide with ozone,

$$NO + O_3 \rightarrow NO_2 + O_2 \,, \tag{6.6.5}$$

is chemiluminescent and is frequently used to determine either NO or O_3 concentrations by detection of the emitted light from excited NO_2. *Fried* and *Hodgeson* [6.84] have reported a comparison of chemiluminescent titration, ultraviolet absorption, and photoacoustic detection measurements for quantitation of the reagents in reaction (6.6.5). The photoacoustic detection was carried out by exciting NO_2 with an Ar ion laser operating at 488 nm. The NO_2 absorption is reported to be sufficiently pressure broadened at 600 mm Hg to absorb this wavelength. Using a conventional photoacoustic detection apparatus, an amplitude modulated laser, a cylindrical cell and a microphone whose output is fed to a lock-in amplifier, the authors obtained a detection limit of 5 ppb and a linear output range up to 1ppm. The results from the titration of NO by reaction (6.6.5) using all three methods showed excellent agreement with the measured NO_2 concentrations over the same range. On the other hand, the ozone titration showed a systematic error that increased with increasing ozone concentration. The source of this was suspected to be impurities such as N_2O, CH_4, CO, or CO_2, which can react to give species that subsequently undergo reaction with NO. On the whole though, excellent agreement between the NO and NO_2 concentrations produced by the titration reaction was found, indicating that NO is quantitatively converted to NO_2. The obvious advantage of the photoacoustic method for measurement of NO_2 concentrations is that a direct measurement can be made without interference from NO; previous chemiluminescence measurements required a differential measurement against a background of NO, introducing considerable uncertainties into the result, especially at high NO concentrations.

The process of mixing can be described by a chemical potential and is, broadly speaking, a chemical process. *Röper* and *Hess* [6.85] have monitored the mixing of CH_4 and N_2 by measuring the change in resonance frequency of the first radial mode of a resonator in time. As is well known, the resonance frequencies ν_{nms} of a cylindrical resonator are given by

$$\nu_{nms} = \frac{c}{2}\left[\left(\frac{\alpha_{nm}}{R}\right)^2 + \left(\frac{s}{L}\right)^2\right]^{\frac{1}{2}}, \tag{6.6.6}$$

where c is the sound speed, R and L are the radius and length of the cylinder respectively, s is an integer and α_{nm} is the nth zero of the derivative of the mth Bessel function divided by π. (See, for instance, Chap 5 or [6.86].) Since the sound speed depends on the ratio of $\bar{\gamma}$ to $\bar{M}$ (where $\bar{\gamma}$ and $\bar{M}$ are the average specific heat ratio and molecular weights of the gas, respectively), and since different gases, in general, have different values of $\bar{\gamma}/\bar{M}$, the resonance frequencies of the cylinder vary with gas composition. *Röper* and *Hess* found that the first radial resonance frequency of a 5.128 cm radius cylinder changes from 5.3 kHz to 4.2 kHz as the mole fraction of CH_4 in CH_4 - N_2 mixtures is varied between 1 and 0.

Measurements of the resonance frequency of the acoustic cell versus time were made by first filling the cell and associated gas lines with a 9.8% CH_4 in N_2 mixture and then adding pure N_2 until a pressure of 400 mm Hg was reached. A plot of the resonance frequency versus time after addition of N_2 showed that about 2 min was required for complete mixing to take place. When a 50% CH_4 in N_2 mixture was used, the mixing time increased to about 2 hours. As can be seen in Fig. 6.22, a much longer time for the same process was found when there

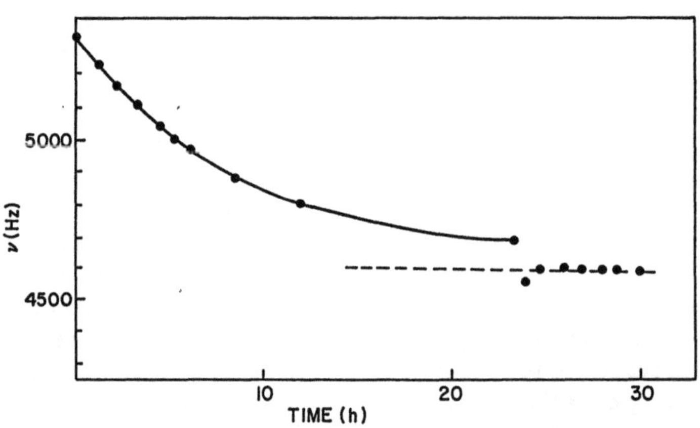

Fig. 6.22. Resonator frequency versus time in the diffusion controlled mixing of CH_4 and N_2. The dashed line is for a 50% CH_4 - 50% N_2 mixture. [6.84]

was no pressure differential between the initial gas mixture and the incoming N_2, that is, when the mixing was purely diffusive and no turbulence was created.

The application of the photoacoustic techniques to monitoring reactants and products in chemical reactions is perhaps a straightforward extension of the well-known analytical capabilities of the effect for measuring time-dependent gas concentrations. Given the difficulty of determining chemical reaction pathways in general, the photoacoustic effect with its high sensitivity and specificity may well prove to be as valuable for unravelling complex reaction mechanisms as it has been for static detection problems such as trace species monitoring.

6.7 Flames, Combustion, and Other Applications

Owing to the general commercial importance of combustion processes, the chemistry of flames continues to be an important area of research. A number of simultaneous chemical reactions take place in flames, primarily as a result of the characteristic high temperatures. In addition to the high temperature, the high pressure typical of flames and other combustion processes makes the use of a number of conventional diagnostic methods difficult. Photoacoustic and photothermal techniques have recently been employed for measurement of flow velocities, temperature, and radical concentrations. Na and Li atoms have been detected in an acetylene-air flame by tuning a flashlamp-pumped dye laser to a resonance transition and recording the acoustic signal with a fast microphone, as described by *Allen* et al. [6.87, 88]. A notable feature of the experimental apparatus was the absence of an acoustic cavity. Although the microphone was exposed to ambient noise, sub part-per-billion concentrations of Na could be detected. The temperature of the flames was found by moving the laser beam a known distance relative to the microphone and determining the sound speed. By using the expression for the sound speed $c = (\bar{\gamma}RT/\bar{M})^{\frac{1}{2}}$ along with estimates of experimental values of $\bar{\gamma}$ and $\bar{M}$, the translational temperature was determined. *Allen* et al. determined the temperature in a premixed C_2H_2-air atmospheric pressure flame to be 2280 $\pm$ 230 K; in agreement with previous results. *Howard* and *Greenhalgh* [6.89] have performed similar measurements using both a flash-lamp pumped dye laser and a Nd:YAG laser pumped dye laser. *Zapka* et al. [6.90] have reported an experimental arrangement where two probe laser beams are kept a fixed distance apart and the differential arrival time measured. They estimated that a temperature resolution of 3% is possible.

Direct detection of OH radicals produced in flames has been reported by *Smith* et al. [6.91], *Tennal* et al. [6.92] and *Rose* et al. [6.93]. The OH radical is a particularly important flame species involved in chain propagation, which has been detected using the pulsed photoacoustic effect. In the experiments reported by these authors OH was excited near 310 *nm* through the $A^2\Sigma^+ \longleftarrow X^2\Pi$ band system. *Smith* et al. recorded both the laser induced fluorescence and the photoacoustic spectrum of OH in a CH_4-air flame. From the rotational line intensities they were able to determine the rotational flame temperature. *Rose* et al. determined a spatial profile of the [OH] by positioning the laser beam at various heights above the flame.

In the case of OH, laser-induced fluorescence has a higher detection sensitivity than the photoacoustic effect; this, according to *Smith* et al. is a consequence of the unusually low quenching cross section of the A state. On the other hand, because of the relatively long radiative lifetime and high quenching rate of the excited NH_2 $\tilde{A}$ state, this molecule could not be detected by laser-induced fluorescence. The photoacoustic effect, by contrast, was found to have a detection limit of better than 10 ppm. The authors also detected CO_2 using a pulsed CO_2 laser tuned to the $P(20)$ line at 10.6 μm. This line corresponds to a hot band transition in CO_2, which cannot normally be excited at room temperature since its population is only 5×10^{-7} of the total. At 1500 K, the hot band absorption has a fractional population of 10^{-3} which is sufficient to give a photoacoustic effect.

Beam deflection techniques have proven successful for detection of several species in flames as well. *Gupta* and co-workers [6.94-97]) have used the output of a frequency-doubled dye laser pumped by a Nd:YAG laser to excite the (0,0) band of the $A^2\Sigma^+ \longleftarrow X^2\Pi$ transiton in OH produced in a C_3H_8-air flame. Relaxation of the electronic energy in the excited state produces a temperature gradient which deflects a probe HeNe laser beam. Deflection of the probe laser beam was detected with a position-sensitive photodiode whose output was fed to a differential amplifier, the output from which was filtered and displayed on an oscilloscope. The electronic filtering proved to be important in obtaining a high signal-to-noise ratio in the beam deflection signal since flames generate a significant amount of noise below 10 kHz. A detection limit for OH of 10^{14} molecules cm^{-3} for a single laser shot was reported. Approximately the same limit was found for NO_2 detection by *Rose* et al. [6.98] in a separate study. With the beam deflection technique it is possible to make the region of overlap between the probe laser and the exciting laser beams small. *Kizirnis* et al. estimate their interaction volume to be about 3 mm^3, with an interaction length of approximately 3 mm. This gives a reasonably high spatial resolution for probing the flames — much higher than is possible with the photoacoustic method which produces a cylindrical acoustic wave along the entire length of the probe laser. Typical results for a spatial profile of [OH] in a premixed natural gas – air flame are shown in Fig. 6.23. Temperature measurements in flames were carried out by the same group using a laser-heated wire as an acoustic source [6.93]. Two probe beams placed 2 *mm* apart permitted the sound speed to be measured and hence the temperature to be determined.

Soot formation in flames can be studied by the probe beam deflection technique as well, since absorption of the exciting laser beam by soot particles results in a temperature gradient. *Rose* and *Gupta* [6.99] note that it is possible, depending upon the pump and probe beam separation, to distinguish between simple heating and the laser vaporization of the particulates reported by *Dasch* [6.100, 101]. Both processes permit flow velocities to be determined. *Lawton* [6.102] has reported a study where the "Kurlbaum" temperature — a soot surface brightness temperature — was compared with the temperature determined from

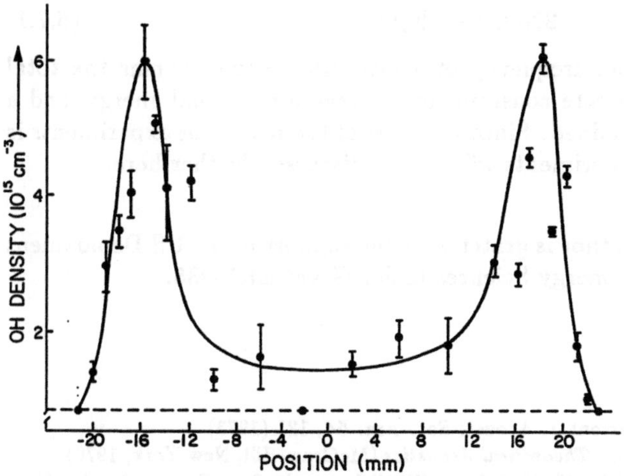

Fig. 6.23. Concentration of OH versus distance from the center of the burner determined by the beam deflection technique. [6.94]

probe beam deflection. The acoustic temperature was found to be systematically slightly higher than the Kurlbaum temperature.

Japar and *Killinger* [6.103] have recorded the photoacoustic signal of soot from a premixed C_3H_8 - O_2 flame as a function of wavelength using a continuous dye laser. They were able to determine an optical extinction coefficient and an upper limit of 0.06 μm for the particle diameter. *Roessler* and *Faxvog* [6.104] have reported measurements of absorption coefficients of soot from a C_2H_2-air burner at 514 nm and 10.6μm. The authors first used NO_2 to calibrate the spectrophone sensitivity and then recorded the photoacoustic signal as a function of mass concentration. Mie theory was used to calculate the photoacoustic signal amplitude. It was determined that in the visible spectral region 80% of the total light extinction could be attributed to absorption, with scattering accounting for the remainder.

Terhune and *Anderson* [6.105] have recorded photoacoustic signals from cigarette smoke. The particles were estimated to be 0.01 μm in diameter; ratios for the absorption and scattering relative to Rayleigh scattering were obtained.

Exhaust from diesel engines has been studied [6.106-108]. *Roessler* [6.106] studied the production of particulate clusters as a function of various combustion parameters. *Bruce* and *Richardson* [6.107, 108] used a 100 mW, millimeter wave generator to excite smoke from the combustion of diesel fuel. Since the incident radiation has such a long wavelength compared with the particulate dimensions, the Rayleigh limit in scattering theory applies. The authors were able to determine mass absorption coefficients for particles with dimensions on the order of 2 μm. Similar measurements [6.100] were made at CO_2 laser wavelengths. *Killinger* et al. [6.109] have used the photoacoustic effect to monitor soot production by automobile engines and have measured extinction coefficients at 600nm.

Fiedler and *Hess* [6.110] have recently investigated the chemical equilibrium

$$2NO_2 \longleftrightarrow N_2O_4 \tag{6.7.1}$$

by measuring the resonance frequency of a cylindrical cavity versus the total gas pressure in the cell. A rate constant for the reaction, a bond energy, and a relaxation time were determined. Since a review of the resonator experiments is given in Chap. 5, these experiments will not be discussed further here.

Acknowledgements: The author is grateful for the support of the US Department of Energy, Office of Basic Energy Sciences under Grant ER13235.

References

6.1 P. J. Westervelt, R. S. Larson: J. Acoust. Soc. Am. **54**, 121 (1973)
6.2 P. M. Morse, K. U. Ingard: *Theoretical Acoustics* (McGraw-Hill, New York, 1978)
6.3 A. S. Bebchuk, V. M. Mizin, N. Y. Salova: [English transl: Opt. Spectrosk. **44**, 158 (1978) Opt. Spectrosc. (USSR) **44**, 92 (1978)]
6.4 H. M. Lai, K. Young: J. Acoust. Soc. Am. **72**, 2000 (1982)
6.5 J. M. Heritier: Opt. Commun. **44**, 267 (1983)
6.6 S. R. Brueck, H. Kildal, L. J. Belanger: Opt. Commun. **34**, 199 (1980)
6.7 B. Sullivan, A. C. Tam: J. Acoust. Soc. Am. **75**, 437 (1984)
6.8 M. W. Sigrist, F. K. Kneubuhl: J. Acoust. Soc. Am. **64**, 1652 (1978)
6.9 P. Karcher, H. Z. Cheng, S. M. Park, G. J. Diebold: "The Photoacoustic Effect in Highly Absorbing Media", to be published
6.10 E. F. Carome, C. E. Moeller, N. A. Clark: Appl. Phys. Lett. **4**, 95 (1964); J. Acoust. Soc. Am. **40**, 1462 (1966)
6.11 L. V. Burmistrova, A. A. Karabutov, A. L. Poctnyagin, E. B. Cherepeskaya: Akust. Zh. **24**, 665 (1978) [English transl.: Sov. Phys.-Acoust. **24**, 369 (1978)]
6.12 S. Temkin: *Elements of Acoustics* (Wiley, New York 1981)
6.13 S. M. Park, G. J. Diebold: "The Photoacoustic Effect in Cl_2: Sound Wave Production Through Molecular Photodissociation and Chemical Reaction", to be published
6.14 S. Schneider, U. Moller, H. Coufal: Appl. Opt. **21**, 44 (1982)
6.15 S. E. Braslavsky, R. M. Ellul, R. G. Weiss, H. Al-Ekabi, K. Schaffner: Tetrahedron **39**, 1909 (1983)
6.16 K. S. Peters: Pure and Appl. Chem. **58**, 1263 (1986)
6.17 L. J. Rothberg: J. Phys. Chem. **91**, 3467 (1987)
6.18 H. Sontag, A. C. Tam, P. Hess: J. Chem. Phys. **86**, 3950 (1987)
6.19 W. B. Jackson, N. M. Amer, A. C. Boccara, D. Fournier: Appl. Opt. **20**, 1333 (1981)
6.20 K. Koseki, M. Koshi, H. Matsui: In *Photoacoustic and Photothermal Phenomena*, ed. by P. Hess, J. Pelzl, Springer Ser. Opt. Sci., Vol. 58 (Springer, Berlin, Heidelberg 1988)
6.21 M. W. Sigrist: J. Appl. Phys. **60**, R83 (1986)
6.22 J. R. Barker, L. Brouwer, R. Patrick, M. J. Rossi, P. L. Trevor, D. M. Golden: Int. J. Chem. Kinet. **17**, 991 (1985)
6.23 L. J. Rothberg, J. D. Simon, M. Bernstein, K. S. Peters: J. Am. Chem. Soc. **105**, 3464 (1983)
6.24 A. C. Tam, H. Sontag, P. Hess: Chem. Phys. Lett. **120**, 280 (1985)
6.25 W. B. Lewis, A. H. Zeltmann: J. Photochem. **12**, 51 (1980)
6.26 T. F. Hunter, D. Rumbles, M. G. Stock: J. Chem. Soc., Faraday Trans. 2, **70**, 1010 (1974)
6.27 T. F. Hunter, M. G. Stock: J. Chem. Soc., Faraday Trans. 2, **70**, 1022 (1974)
6.28 B. Stepanov, O. Girin: Zh. Eksp. Teor. Fiz. **20**, 947 (1950)
6.29 R. Kaiser: Can. J. Phys. **37**, 1499 (1959)
6.30 T. F. Hunter, K. S. Kristjansson: J. Chem. Soc., Faraday Trans. 2, **75**, 1284 (1975)

6.31 T. F. Hunter, K. S. Kristjansson: Chem. Phys. Lett. **58**, 291 (1978)
6.32 T. F. Hunter, K. S. Kristjansson: Chem. Phys. Lett. **90**, 35 (1982)
6.33 T. F. Hunter, C. M. Leong: Chem. Phys. **111**, 145 (1987)
6.34 T. F. Hunter, S. Lunt, K. S. Kristjansson: J. Chem. Soc., Faraday Trans. 2, **79**, 303 (1983)
6.35 T. F. Hunter, K. S. Kristjansson: J. Chem. Soc., Faraday Trans. 2, **78**, 2067 (1982)
6.36 W. R. Harshbarger, M. B. Robin: Acc. Chem. Res. **6**, 329 (1973)
6.37 W. R. Harshbarger, M. B. Robin: Chem. Phys. Lett. **21**, 462 (1973)
6.38 M. J. Colles, E. E. Marinero: In *Lasers in Chemistry*, ed. by M. West (Elsevier, New York 1977)
6.39 M. J. Colles, N. R. Geddes, E. Mehdizadeh: Contemp. Phys. **20**, 11 (1979)
6.40 M. B. Robin: In *Optoacoustic Spectroscopy and Detection*, ed. by Y.-H. Pao (Academic, New York 1977)
6.41 See M. daSilva, F. Lepoutre: In *Physical Acoustics XVII*, ed. by W. Mason, R. Thurston, (Academic, Boston 1988) for a review of energy relaxation processes
6.42 K. Kaya, W. R. Harshbarger, M. B. Robin: J. Chem. Phys. **60**, 4231 (1974)
6.43 K. Kaya, C. L. Chatelain, M. B. Robin, N. A Kuebler: J. Am. Chem. Soc. **97**, 2153 (1975)
6.44 C. M Hall, T. F. Hunter, K. S. Kristjansson: Chem. Phys. Lett. **43**, 404 (1976)
6.45 C. M Hall, K. S. Kristjansson, M. G. Stock: Chem. Phys. Lett. **44**, 145 (1976)
6.46 D. Kumar, R. Nauman, T. L. Mathers, S. P. McGlynn: J. Indian Chem. Soc. **63**, 10 (1986)
6.47 G. Louis, P. Peretti, P. Ranson: J. de Phys., Colloq. **44**, C6-541 (1983)
6.48 R. N. Dixon, D. A. Haner, C. R. Webster: Lasers Chem. **22**, 166 (1977); Chem. Phys. **22**, 199 (1977)
6.49 M. B. Robin, N. A. Kuebler: J. Am. Chem. Soc. **97**, 4822 (1975)
6.50 G. J. Diebold: J. Phys. Chem. **84**, 2213 (1980)
6.51 G. J. Diebold: In Proc. 1980 Ultrasonics Symposium, Boston, ed. by B. R. McAvay (IEEE, Piscataway, NJ 1980)
6.52 R. Bellman: *Methods of Nonlinear Analysis IV* (Academic, New York 1970) p. 24
6.53 M. T. O'Connor, G. J. Diebold: J. Chem. Phys. **81**, 812 (1984)
6.54 An extra factor of 2 appears to be in the expression for $\hat{\alpha}$ given in [5.53].
6.55 M. B. Robin, N. A. Kuebler, K. Kaya, G. J. Diebold: Chem. Phys. Lett. **70**, 93 (1980)
6.56 M. S. deGroot, C. A. Emeis, I A. M. Hesselman, E. Drent, E. Farenhorst: Chem. Phys. Lett. **17**, 332 (1974)
6.57 C. A. Emeis, E. Drent, E Farenhorst, I. A. M. Hesselman, M. S. de Groot: Chem. Phys. Lett. **27**, 17 (1974)
6.58 W. D. Chandler, L. Goodman, J. Mol. Spectrosc.**37**, 33 (1971)
6.59 S. Benson: *The Foundations of Chemical Kinetics* (McGraw-Hill, New York 1960)
6.60 K. J. Laidler: *Chemical Kinetics* (McGraw-Hill, New York 1950)
6.61 G. J. Diebold, J. S. Hayden: Chem. Phys. **49**, 429 (1980)
6.62 M. T. O'Connor, G. J. Diebold: Nature **301**, 321 (1983)
6.63 S. G. Wyers, G. J. Diebold: In *Photoacoustic and Photothermal Phenomena*, ed. by P. Hess, J. Pelzl, Springer Ser. Opt. Sci., Vol. 58 (Springer, Berlin, Heidelberg 1988)
6.64 M. T. O'Connor, R. B. Stewart, G. J. Diebold: J. Phys. Chem. **90**, 711 (1986). Eq. (10) is missing a factor of k_1 and Eq. (9) should have k_2 replaced by k_1; the errors are typographical
6.65 G. J. Diebold: In Proc. of the 1984 Conf. on Lasers (Soc. for Opt. and Quantum Elect., McLean, VA. 1985)
6.66 J. G. Choi, G. J. Diebold: Anal. Chem. **57**, 2989 (1985)
6.67 G. M. Burnett, H. W. Melvill: *Techniques of Organic Chemistry*, Vol. 8 (Peterson, New York 1983)
6.68 K. V. Reddy, M. J. Berry: Chem. Phys. Lett. **52**, 111 (1977)
6.69 K. V. Reddy, M. J. Berry: Chem. Phys. Lett. **66**, 223 (1979)
6.70 K. V. Reddy, M. J. Berry: Faraday Discuss. Chem. Soc. **67**, 188 (1979)

6.71 D. W. Chandler, W. E. Farneth, R. N. Zare: J. Chem. Phys. **77**, 4447 (1982)
6.72 M. C. Chuang, J. E. Baggott, D. W. Chandler, W. E. Farneth, R. N. Zare: Faraday Discuss. Chem. Soc. **75**, 301 (1983)
6.73 J. M. Jasinski, J. K. Frisoli, C. B. Moore: J. Chem. Phys. **79**, 1312 (1983)
6.74 J. M. Jasinski, J. K. Frisoli, C. B. Moore: Faraday Discuss. Chem. **75**, 289 (1983)
6.75 J. M. Jasinski, J. K. Frisoli, C. B. Moore: J. Chem. Phys. **87**, 3826 (1983)
6.76 J. M. Jasinski, J. K. Frisoli, C. B. Moore: J. Phys. Chem. **87**, 2209 (1983)
6.77 D. G. Lishan, K. V. Reddy, G. S. Hammond, J. E. Leonard: J. Phys. Chem. **92**, 656 (1988)
6.78 V. P. Zharov, V. S. Letokkov: *Laser Optoacoustic Spectroscopy* (Springer Ser. Opt. Sci.,Vol. **37** (Springer, Berlin, Heidelberg, 1986) Chap 6
6.79 M. J. Colles, A. M. Angus, E. E. Marinero: Nature **262**, 681 (1976)
6.80 R. Kadibelban, P. Hess: In *Photoacoustic Effect, Principles and Applications*, ed. by E Lüscher, P. Korpiun, H. J. Coufal, R. Tilgner (Vieweg, Braunschweig 1981)
6.81 R. Kadibelban, P. Hess: Appl. Opt. **21**, 61 (1982)
6.82 P. Hess, R. Kadibelban, A. Karbach, J. Röper: J. de Phys., Colloq. **44**, C6-419 (1983)
6.83 L. A. Farrow, R. E. Richton: In *Laser Spectroscopy for Sensitive Detection*, SPIE **286**, 18 (1981)
6.84 A. Fried, J. Hodgeson; Anal. Chem. **54**, 278 (1982)
6.85 J. Röper, P. Hess: Appl. Phys. Lett. **39**, 946 (1981)
6.86 L. B. Kreuzer: In *Optoacoustic Spectroscopy and Detection*, ed. by Y.-H Pao (Academic, New York 1977)
6.87 J. E. Allen, Jr., W.R. Anderson, D. R. Crosley: Opt. Lett. **1**, 118 (1977)
6.88 W. R. Anderson, J. E. Allen, Jr., T. D. Fansler, D. R. Crosley: Natl. Bur. of Stand. Special Pub. **561**, 1153 (1979)
6.89 A. G. Howard, D. A. Greenhalgh: Anal. Chim. Acta **106**, 361 (1979)
6.90 W. Zapka, P. Pokrowsky, A. C. Tam: Opt. Lett. **7**, 477 (1982)
6.91 G. P. Smith, M. J. Dyer, D. R. Crosley: Appl. Opt. **22**, 3995 (1983)
6.92 K. Tennal. G. J. Salamo, R. Gupta: Appl. Opt. **21**, 2133 (1982)
6.93 A. Rose, J. D. Pyrum, G. J. Salamo, R. Gupta: Appl. Opt. **21**, 1573 (1984)
6.94 S. W. Kizirnis, R. J. Brecha, B. N. Ganguly, L. P. Goss, R. Gupta: Appl. Opt. **23**, 3873 (1984)
6.95 A. Rose, R. Gupta: In Twentieth Symposium (International) on Combustion (The Combustion Institute, Pittsburgh 1984) p.1339
6.96 A. Rosè, G. J. Salamo, R. Gupta: Appl. Opt. **23**, 781 (1984)
6.97 R. Gupta: "Photothermal Deflection Spectoscopy", to be published
6.98 A. Rose, J. D. Pyrum, C. Muzny, G. J. Salamo, R. Gupta: Appl. Opt. **21**, 2663 (1982)
6.99 A. Rose, R. Gupta, Opt. Commun. **56**, 303 (1986)
6.100 C. J. Dasch: Appl. Opt. **23**, 2209 (1984)
6.101 C. J. Dasch: Opt. Lett. **9**, 214 (1984)
6.102 S. A. Lawton: Appl. Opt. **25**, 1263 (1986)
6.103 S. M. Japař, D. K. Killinger: Chem. Phys. Lett. **66**, 207 (1979)
6.104 D. M. Roessler, F. R. Faxvog: J. Opt. Soc. Am. **69**, 1699 (1979)
6.105 R. W. Terhune, J. E. Anderson: Opt. Lett. **1**, 70 (1977)
6.106 D. M. Roessler: Appl. Opt. **21**, 4077 (1982)
6.107 C. W. Bruce, N. M. Richardson: Appl. Opt. **23**, 13 (1984)
6.108 C. W. Bruce, N. M. Richardson, Appl. Opt. **22**, 1051 (1983)
6.109 D. K. Killinger, J. Moore, S. M. Japar: In Laser Probes of Combustion Chem., Am. Chem. Soc., Symposeum Series **134**, 457 (1980).
6.110 M. Fiedler, P. Hess: In *Photoacoustic and Photothermal Phenomena* , ed. by P. Hess, J. Pelzl, Springer Ser. Opt. Sci., Vol. 58 (Springer, Berlin, Heidelberg 1988)

7. Atmospheric and Exhaust Air Monitoring by Laser Photoacoustic Spectroscopy

M.W. Sigrist, S. Bernegger, and P.L. Meyer

With 10 Figures

The application of laser photoacoustic spectroscopy to air pollution monitoring is reviewed and compared to other spectroscopic and nonspectroscopic techniques. Emphasis is put on sensitivity, selectivity and on temporal resolution of the detection scheme. The problem of interfering absorptions by different trace gases present in multi-component mixtures is studied in detail and a new mathematical procedure for the analysis of measured photoacoustic (PA) spectra is presented. The versatility of the PA method is demonstrated by our studies performed with computer-controlled PA systems on ambient air as well as on motor-vehicle and industrial exhausts.

7.1 Introduction

7.1.1 Air Pollution

The pollution of the lower atmosphere, i.e. the troposphere, by numerous gaseous and particulate substances is of great concern today. At present the gaseous pollutants sulfur dioxide (SO_2), the nitric oxides (NO_x) and the hydrocarbons (HC) as well as their respective products like acids and oxidants are of main interest. On the basis of their sequential occurrence, SO_2, NO_x and the hydrocarbons are called primary pollutants while their products are denoted as secondary pollutants. In industrial countries with a high traffic density air pollution is mainly caused by human activities, i.e. by anthropogenic sources. In comparison, the contribution of natural sources to air pollution is insignificant [7.1].

In earlier times, the effects of air pollution were restricted to industrial areas where increased concentrations of toxic compounds occurred near their sources. More recently, large-area damage to nature, e.g. the acidification of soil and water (e.g. [7.2]) or the serious forest decline in Europe (e.g. [7.3,4]) have been traced to the increase of air pollution. However, the primary pollutants can no longer be the primary origin of these damages. Today, complex mechanisms like the formation of acid rain (e.g. [7.5]) or the generation of photo-smog (e.g. [7.6,7]) are dominant. In the former process the primary pollutants SO_2 and NO_x are converted to sulfuric and nitric acids, which are subsequently deposited on the earth as acid rain or fog. In the latter process solar UV irradiation causes the conversion of NO_x and HC to most toxic oxidants such as ozone (O_3) or peroxyacetylnitrate (PAN). In the course of these conversion processes the resultant secondary pollutants can be transported over large distances. Therefore, their concentrations may be smaller in the vicinity of emission sources, e.g. near roads or large cities, than in rural areas which were previously regarded as clean-air regions.

7.1.2 Methods for Monitoring Gaseous Pollutants

The sensitive and selective detection of the numerous trace constituents is a prerequisite for understanding the various tropospheric air-pollution processes. In the past, many detection schemes have been introduced [7.8-14]. They can be separated into spectroscopic and nonspectroscopic analytical techniques. The most widely used nonspectroscopic methods are gas chromatography (GC) [7.15-19] and various other schemes. Their main features are summarized in Table 7.1. Since these methods do not meet all the requirements with respect to sensitivity, selectivity, kind and number

Table 7.1. Nonspectroscopic techniques for trace gas detection

Detection scheme	Main features	Sensitivity	Selectivity
Gas chromatography (GC), usually combined with FID	Separation/extraction; complex, low time resolution	excellent	excellent
Flame ionization (FID)	Flame ionization in combustion, yields total HC concentraton	good	poor
Photoionization (PID)	UV ionization, linearity depends on design	excellent	partial
Electron capture (ECD)	Ionization by β-radiation, not used as direct monitoring device, limited dynamic range	Excellent for halogenated compounds	poor
Mass spectrometry (MS)	Fragmentation and ionization by electron bombardment, complex, sometimes combined with GC (GC-MS)	good	good
Thermal detection	Based on thermal conductivity or heat of combustion, limited application	----	poor
Chemiluminescence	Emission of radiation in chemical reaction, commonly applied for NO_x and O_3 detection	good	------
Flame photometry (FPD)	Emission of radiation from excited molecules in flame, small concentration ranges	Good (for sulfur compounds)	------
Electrolytic conductivity	Ionization in appropriate solution, only for compounds containing sulfur, halogens or nitrogen, simple setup	good	------ Depends on specificity of ionization process
Coulometry	Stoichiometric reaction with appropriate, electrically generated reagent, only for compounds containing sulfur, nitrogen or chloride	good	Depends on specificity of ionization process
Colorimetry	Change of color by reaction with specific reagent in solution	good	------

of substances to be detected, temporal resolution, practical applicability, etc., novel physical or chemical techniques have to be developed in addition to the conventional ones.

Today, spectroscopic techniques are of growing interest due to the recent development of powerful lasers tunable over large wavelength ranges, which permits the detection of numerous substances at low concentrations. In contrast to conventional, e.g. wet-chemical or chromatographic schemes, spectroscopic methods exhibit some unique advantages. They are in general based on absorption measurements. The minimum detectable concentration of a trace gas is thus determined by the minimum measurable absorption coefficient α_{min}. Since the absorption spectrum is characteristic for each molecule, these methods often permit the simultaneous detection of many substances, depending on the tunability and

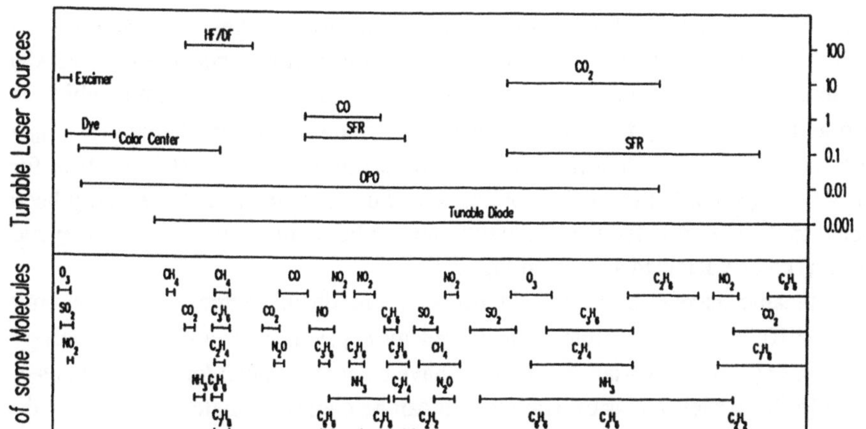

Fig. 7.1. Emission ranges of available laser sources including typical powers (top), absorption regions of various molecules of environmental concern (center), and transmission of the atmosphere for a 500 m horizontal path for wavelengths between 0.25 µm and 15 µm

bandwidth of the radiation source and on the spectral resolution of the detector. The main absorption ranges of various molecules that are of environmental concern are presented in the center part of Fig. 7.1. The list of molecules essentially includes hydrocarbons (HC), yet also nitric oxides (NO, NO_2, N_2O), ammonia (NH_3), ozone (O_3), sulfur dioxide (SO_2), carbon oxides (CO, CO_2), etc. In many cases, long absorption pathlengths ℓ are required, which often restricts the wavelength range to atmospheric windows. The bottom of Fig. 7.1 shows the atmospheric transmission for a horizontal path of 500 m at an altitude of 500 m, a total pressure of 950 hPa, a temperature of 20°C and a relative humidity of 50 %, based on a calculation with the LOWTRAN computer code [7.20]. The main atmospheric windows occur for wavelengths shorter than 2.5 μm, from 3 to 5 μm, and 8 to 14 μm. Atmospheric H_2O vapor and CO_2 drastically reduce the transmission beyond these spectral windows [7.21].

The main schemes based on absorption spectroscopy and their main features are listed in Table 7.2 [7.22-25]. Differential optical absorption spectroscopy (DOAS) [7.26] makes use of a broadband radiation source, usually a Xe lamp, which is located up to 10 km away from the receiver in the free atmosphere. Typically, $\alpha_{min}\,\ell \approx 2\cdot10^{-5}$ is achieved, which corresponds to a detection limit of e.g. 0.1 ppmV m for sulfur dioxide (SO_2). Hitherto, the application of these systems has essentially been restricted to the UV and visible wavelength range where nitric oxide (NO_2), SO_2 and ozone (O_3) can be detected as seen in Fig. 7.1.

With tunable lasers (top of Fig. 7.1) instead of broadband sources a detection limit of $\alpha_{min}\,\ell \approx 10^{-8}$ can in principle be achieved by applying derivative spectroscopy [7.23]. More recently, lead-salt diode lasers have been introduced for trace gas detection [7.27]. Since their emission wavelength can be chosen to match exactly the absorption line of a specific molecule, high sensitivities can be achieved, e.g. 0.5 ppbV for ammonia (NH_3) [7.28]. However, since the tunability of one laser is limited to a few wavenumbers or less, a different diode laser is in general required for each substance to be detected.

LIDAR (Light Detection and Ranging), and in particular DIAL (Differential Absorption LIDAR) systems permit remote sensing and three-dimensional profiling of pollutants [7.29-32]. Single-ended schemes rely on effective backscattering in the atmosphere. This approach is most useful in the UV and visible spectral regions, where Rayleigh and Mie scattering are strong, and less successful in the mid-infrared region where many more molecules could be detected (cf. Fig. 7.1). Detection limits in the sub-ppmV range have been reported for SO_2, NO_2 and O_3.

Photoacoustic spectroscopy (PAS) differs completely from the previously discussed techniques because the absorbed energy is determined directly and not via the measurement of the intensity of transmitted or backscattered radiation. Thus PAS is primarily a calorimetric technique. It offers several advantages compared to the methods mentioned above. Its basic principles for trace gas detection are presented in Sect. 7.2 while Sect. 7.3 is devoted to experimental arrangements, including laser sources, modulation techniques, cell designs and detection schemes. In Sect. 7.4 previous laser photoacoustic (PA) studies on gaseous pollutants in nonabsorbing

Table 7.2. Spectroscopic techniques for trace gas detection

Detection scheme	Main features	Sensitivity	Selectivity
Raman scattering	Inelastic interaction between photons and irradiated molecules, small cross sections, selectivity by frequency shift, only one wavelength λ needed, better for small λ	poor	excellent
Laser-induced fluorescence (LIF)	Fluorescence of irradiated species, small cross sections, only suitable for detection of atoms or radicals	poor	------
Absorption correlation spectroscopy (COSPEC)	Sky emission measured with spectrometer from ground, integrated response, limited to UV and visible, detection of SO_2 and NO_2	good	------
Differential optical absorption spectroscopy (DOAS)	Broadband source, separated from receiver by open path in the atmosphere, dual-wavelength operation, integrated response, mainly applied to UV (NO_2, SO_2, O_3)	good	good
Long-path absorption spectroscopy with lasers	Long single-pass or multipass cells, single- or double-ended schemes in open atmosphere, integrated response, UV-vis-IR	good	good
Light detection and ranging (LIDAR)/Differential absorption LIDAR (DIAL)	Atmospheric backscattering of laser pulses, 3 dim. profiling of pollutants, complex, mainly applied to UV and vis (SO_2, NO_2, O_3)	good	good
Photoacoustic spectroscopy (PAS) with lasers	Measurement of absorbed energy in cell, mainly applied to IR, selective monitoring of numerous pollutants	good	good

buffer gases and in real air samples are discussed. In Sect. 7.5 we present our stationary CO-laser PA system, which has mainly been applied to the analysis of motor-vehicle exhausts. In Sect. 7.6 we discuss some results obtained with our mobile CO_2-laser PA apparatus, which has been used for in situ measurements in ambient air as well as for pollution monitoring in exhausts.

7.2 Basic Principles of Trace Gas Detection by Laser Photoacoustic Spectroscopy

7.2.1 Generation of Photoacoustic Signal

a) Heat Production

If modulated light is absorbed by molecules a fraction of the ground state molecular population is excited into higher energy levels. These excited molecules subsequently relax through radiative or nonradiative decay. In PAS the nonradiative relaxation that generates heat is of primary concern. The local increase of thermal energy causes the generation of pressure waves, which can be detected by a suitable sensor, e.g. by a microphone.

The PA signal generation in gases has been reviewed previously [7.33,34]. In general, it concerns two aspects, namely the heat production and the generation of acoustic waves. The heat production can easily be calculated for the simple case where the absorbing gas with molecule density N can be modeled by a two-level system consisting of the vibrational ground state and the excited state with densities N-N' and N', respectively. Since the excitation rate R, given by

$$R = \Psi \cdot \sigma \tag{7.2.1}$$

where Ψ is the incident photon flux and σ the absorption cross section, is usually small, the density N' is considerably smaller than N and the stimulated emission rate from the excited state can be neglected. Consequently, N' can be expressed by the following rate equation:

$$dN'/dt = N \cdot R - N' \, 1/\tau \; . \tag{7.2.2}$$

The inverse time τ^{-1} can be expressed by the sum of reciprocal time constants of nonradiative (τ_n^{-1}) and radiative (τ_r^{-1}) relaxation:

$$(\tau)^{-1} = (\tau_n)^{-1} + (\tau_r)^{-1} \; . \tag{7.2.3}$$

For typical atmospheric conditions τ_n is on the order of 10^{-6} - 10^{-9} s while τ_r is between 10^{-1} and 10^{-3} s [7.35]. Consequently, τ is similar to τ_n.

For a complex harmonic modulation of the incident radiation described by

$$\Psi = \Psi_0 \cdot (1 + e^{i\omega t}) \tag{7.2.4}$$

only the real part has a physical meaning. For the generation of the PA signals only the time-dependent term with the circular frequency ω is of concern. The solution of (7.2.2) is thus given by

$$N' = \frac{N\Psi_0\sigma\tau}{[1 + (\omega\tau)^2]^{1/2}} \cdot e^{i(\omega t - \phi)} \tag{7.2.5}$$

with $\phi = \arctan(\omega\tau)$ (7.2.6)

representing the phase lag between the number density N' of the excited state and the photon flux Ψ. The heat production rate H is related to N' by

$$H = N' \cdot \frac{hc \cdot \Delta\tilde{\nu}}{\tau_n}, \tag{7.2.7}$$

where $hc\Delta\tilde{\nu}$ is the average thermal energy released due to nonradiative deexcitation of the excited state. In (7.2.7) h denotes Planck's constant, c the speed of light in vacuum and $\Delta\tilde{\nu}$ is the energy spacing in wavenumbers between the upper and lower state. If the deexcitation process results in conversion of the excited state to the ground state as assumed in our model, then $\Delta\tilde{\nu} \approx \tilde{\nu}_{las}$, the laser photon energy.

Since $\tau_n \approx \tau$, (7.2.7) can be written as

$$H = N' \cdot \frac{hc \cdot \tilde{\nu}_{las}}{\tau}. \tag{7.2.8}$$

By introducing (7.2.5,6) in (7.2.8) one finds

$$H = H_0 \cdot e^{i(\omega t - \theta)} \tag{7.2.9a}$$

where $$H_0 = \frac{N\sigma I_0}{[1 + (\omega\tau)^2]^{1/2}} \tag{7.2.9b}$$

and $\theta = \phi$, determined by (7.2.6). The amplitude I_0 of the beam intensity can be represented by

$$I_0 = \Psi_0 \cdot hc \cdot \tilde{\nu}_{las}. \tag{7.2.10}$$

For low modulation frequencies $\omega \ll 10^6\ s^{-1}$, i.e. for $\omega\tau \ll 1$, the amplitude H_0 of the heat rate of (7.2.9b) can be simplified to

$$H_0 = N \cdot \sigma \cdot I_0 \tag{7.2.11}$$

while the phase shift θ vanishes. Equation (7.2.11) contains the essence of most PA studies since the conditions of slowly chopped light beams in the kilohertz range or below, i.e. $\omega \ll \tau^{-1}$, and absence of optical saturation, i.e. $R \ll \tau^{-1}$, are usually fulfilled. However, it should be pointed out that care has to be taken because the condition $\omega \ll \tau^{-1}$ is not always automatically fulfilled. The most important exception concerns PA studies performed with a CO_2 laser on air samples containing CO_2 [7.36]. These molecules increase the lifetime of the excited states drastically due to the kinetic cooling effect [7.37]. In atmospheric air samples a fast resonant energy transfer takes place between the excited vibrational modes ν_3 of CO_2 and ν_1 of N_2. This results in an effective V-T relaxation time τ_{eff} of ≈ 0.8 ms for standard atmospheric conditions, depending also on the water-vapor concentration. As a

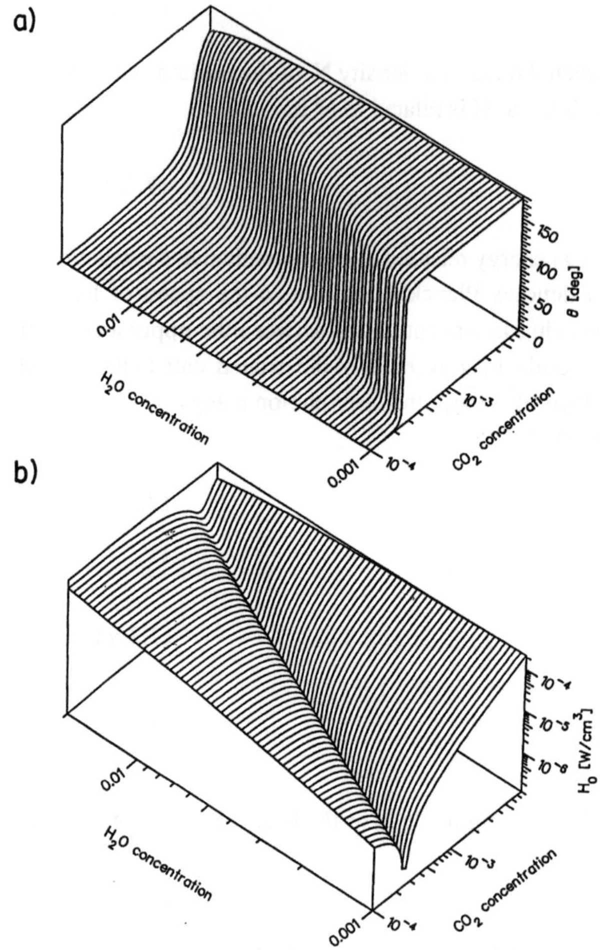

Fig. 7.2. Calculated phase θ (a) and amplitude H_0 (b) of heat production rate for a CO_2-N_2-O_2-H_2O system as functions of the concentrations c_{CO_2} and c_{H_2O} and for constant $c_{N_2} = 0.8$ and $c_{O_2} = 0.2$

consequence, a phase reversal of the heat production rate H with respect to the modulation of the incident radiation may occur. This result is illustrated in three-dimensional plots in Fig.7.2 for a CO_2-N_2-O_2-H_2O mixture as functions of the concentrations c_{CO_2} and c_{H_2O} for the phase θ (Fig. 7.2a) and the amplitude H_0 (Fig. 7.2b) of the heat production rate. The plots are based on data for the 10R(20) $^{12}C^{16}O_2$ laser transition, i.e. $I_0 = 20$ W/cm^2, $\sigma_{H_2O} = 3.5 \cdot 10^{-23}$ cm^2, $\sigma_{CO_2} = 1.0 \cdot 10^{-22}$ cm^2 and a total pressure of 1 bar. As demonstrated in Fig. 7.2a the phase reversal only occurs within narrow concentration ranges. Hence, although heat-rate phases θ different from 0° or 180° are expected rarely for low H_2O-vapor and CO_2 concentrations, the actual phase θ should always be taken into account.

b) Generation of Acoustic Waves

The second aspect concerns the generation of the acoustic waves in relation to the heat-production rate H. This problem was intensively studied in the literature (e.g. [7.38]) for nonresonant and resonant PA cells. Our discussion is restricted to resonant spectrophones, which offer certain advantages compared to nonresonant cells for trace gas detection, mainly with respect to sensitivity, continuous measurements and range of modulation frequencies. For a sinusoidal modulation of the incident light of intensity I(**r**) at an angular frequency ω the pressure amplitude p(**r**,ω) is represented in terms of a superposition of the normal acoustic modes $p_k(\mathbf{r})$ of the spectrophone

$$p(\mathbf{r},\omega) = \sum_k A_k(\omega)\, p_k(\mathbf{r}) \;, \tag{7.2.12}$$

where **r** denotes the space coordinate. The coefficients $A_k(\omega)$ are proportional to the coupling integral

$$A_k(\omega) = a(\omega) \cdot \int_V p_k(\mathbf{r}) \cdot H(\mathbf{r},\omega) \cdot dV \;, \tag{7.2.13}$$

where V denotes the cell volume and H the heat-production rate discussed in the previous section. The solutions for the pressure amplitudes $p_k(\mathbf{r},\omega)$ and hence for the PA signals are mainly determined by the cell configuration, the excitation geometry and the types of excited modes. For example, one finds for the pressure amplitude $p_1(r)$ of the first radial mode in the resonant cell used for our CO_2-laser PA setup (Sect.7.6)

$$p_1(r) = I_0 \cdot (\gamma-1) \cdot G \frac{\ell}{V} \cdot \frac{Q_1}{\omega_1} N\sigma \;. \tag{7.2.14}$$

Here $\gamma = C_p/C_v$ denotes the ratio of specific heats at constant pressure and volume, respectively, G is a geometrical factor which depends on the transverse beam profile yet not on the cell length ℓ, $Q_1 \approx \omega_1/\Delta\omega_1$ represents the quality factor of the acoustic resonance at the resonance frequency ω_1 with resonance width $\Delta\omega_1$.

The PA signal is usually recorded by a microphone whose signal S can thus be expressed by

$$S = C\, P\, N\, \sigma \;, \tag{7.2.15}$$

where P represents the average laser power and C the cell constant. The latter is in general determined by calibration measurements, where one single absorbing substance with known absorption spectrum is investigated. In most cases however, the gas sample consists of n > 1 different absorbing components, which results in a total absorption coefficient $\alpha_{tot}(\lambda)$ represented by

$$\alpha_{tot}(\lambda) = N_{tot} \cdot \sum_{j=1}^{n} c_j \sigma_j(\lambda) \quad \text{with } j = 1,\ldots,n \;. \tag{7.2.16}$$

Here c_j is the concentration of the gas component j with absorption cross section $\sigma_j(\lambda)$. On the basis of (7.2.16 and 15) the concentrations c_j of the individual components with known $\sigma_j(\lambda)$ can thus be deduced by scanning the wavelength λ of the incident radiation and simultaneously recording $S(\lambda)$. In practice, absorption interferences may cause problems which may reduce the selectivity and sensitivity of the method. These questions are addressed in Sects. 7.5 and 7.6 in connection with particular measurements. It should also be kept in mind that the phase of the PA signal is taken into account because V-T relaxation times may alter in gas mixtures, as mentioned in the previous section.

7.2.2 Main Characteristics

On the basis of the previous discussion and experience, PA spectroscopy, performed with tunable lasers as radiation sources, offers the following main characteristics relevant to in situ trace gas monitoring [7.39]:

i) A high sensitivity, which permits the measurement of absorption coefficients on the order of 10^{-8} cm^{-1}, corresponding to densities of $\mu g/m^3$ or to concentrations of ppbV (10^{-9}) for most substances.

ii) The experimental setup (Sect. 7.3) is rather simple and, for example, does not require cryogenic cooling of infrared detectors etc.

iii) The calibration with certified gases and gas mixtures is straightforward and reliable.

iv) A wide dynamic range of at least 5 orders of magnitude is offered, i.e. the same apparatus can be used for low (immission) and high (emission) concentrations.

v) Using a widely tunable IR laser numerous substances can be measured with a single apparatus.

vi) The short absorption path of typically 10-20 cm permits measurements even in wavelength regions beyond atmospheric windows.

vii) Specially designed PA cells permit continuous measurements on flowing gas mixtures, i,e. a much better temporal resolution than e.g. by gas chromatography can be achieved.

7.3 Experimental Arrangements for Laser Photoacoustic Spectroscopy

The typical setup for PA studies in the gas phase has been presented in various reviews [7.34, 40-42]. It comprises a tunable cw laser as radiation source whose amplitude-modulated beam is directed through the PA cell, which contains the gas under investigation. The generated acoustic waves are detected by a microphone whose signal is fed to a lock-in amplifier locked to the modulation frequency. In addition the laser power P is measured in order to normalize the microphone signal S according to (7.2.15). In the past, many different versions and modifications of this general scheme were introduced. The main aspects are summarized in the following.

7.3.1 Tunable Lasers

Since the PA amplitude is proportional to the laser power, the high spectral brightness offered by lasers is advantageous for the detection of low gas concentrations. Unfortunately, continuously tunable, intense laser sources throughout the mid-IR region are not available at present (Fig. 7.1, top). Many studies have been performed with CO_2 lasers, including isotopes of CO_2, and some with CO lasers. These lasers offer high power yet they are only stepwise tunable. A continuously tunable CO-SFR (spin-flip Raman) laser has been used for the sensitive detection of NO. A DF laser at a wavelength of 3.8 μm has been applied for monitoring CH_4 and NO_2. Carbon monoxide has been detected with a 96 μW Pb S_{1-x} Se_x laser tuned to 4.8 μm. In the visible and UV spectral range, PA studies have been performed with a Kr^+ laser, with dye and frequency-doubled dye lasers as well as with excimer lasers. See Tables 7.3 and 7.4.

7.3.2 Modulation Techniques

In PA spectroscopy various modulation methods are applied [7.34, 40]. It is necessary to distinguish between the modulation of the incident radiation and that of the sample absorption. The first schemes include the widely used amplitude modulation by mechanical choppers, electro-optic and acousto-optic modulators as well as the modulation of the laser emission itself by pulsed excitation, Q-switching and mode-locking. On the other hand, frequency or wavelength modulation of the incident radiation offers the advantage of eliminating the continuum background PA signal caused by a wavelength-independent absorption, e.g. of the cell windows. For example, an improvement of the detection sensitivity of nitric oxide by a factor of 10-50 resulted by using frequency modulation of a SFR laser [7.43]. This type of modulation is well suited for absorbers with narrow linewidth, such as atomic and diatomic species. With respect to laser sources, lead salt diode lasers are convenient because their wavelength can be rapidly tuned within a few wavenumbers directly by the current.

Modulation of the absorption characteristics of the sample can be achieved on the basis of the Zeeman or Stark effect, i.e. by applying modulated magnetic or electric fields to the sample. Consequently, the absorption wavelength of the sample is varied, which corresponds to a wavelength modulation method. The result is again a suppression of the continuum background. For example, a reduction of the background by a factor of 500 was achieved by Stark modulation in comparison to that in the same PA cell with conventional amplitude modulation by a chopper [7.44]. However, it should be noted that the application of the Stark modulation scheme in trace gas detection is restricted to molecules with a permanent electric dipole moment like ammonia (NH_3), nitric oxide (NO), etc. Nevertheless, a considerable increase in sensitivity and, even more important, in selectivity in multicomponent mixtures can be achieved.

7.3.3 Cell Design

A crucial point in many PA applications is an optimum design of the PA cell. In nonresonant spectrophones a high sensitivity is attained for small volumes of 1-10 cm^3 and low modulation frequencies up to several tens of hertz. The main disadvantage is the relatively large continuum background signal caused by absorption at the windows and cell walls. On the other hand, the use of an acoustically resonant cell where the modulation frequency is tuned to a resonance frequency offers several advantages:

i) continuous gas flow through the cell is possible;
ii) reduction of surface-to-volume ratio minimizes wall adsorption effect;
iii) operation at higher frequencies reduces the influence from outside noise.

In the past, numerous resonant PA cells operating on longitudinal, azimuthal, radial or Helmholtz resonances were developed [7.34,42,45,46]. Furthermore, PA cells for multi-pass [7.47] or intracavity [7.48] operation were designed. The effect of the window heating in amplitude modulation schemes has been minimized by introducing acoustic baffles [7.46], or even by the development of a "windowless" cell (Sect. 7.6.2). In many cases the window-heating signal can be markedly reduced by positioning the entrance and exit of the light beam at nodes of the mode being excited. Alternatively, a pulsed source permits the discrimination between gas and window absorption by a time gating method because the acoustic transit time between the window and the microphone differs from that between the bulk of the gas and the microphone.

Here we present a novel type of resonant PA cell, which we developed for our dual-beam setup with amplitude-modulated CO-laser radiation (Sect. 7.5). The following requirements had to be fulfilled:

i) The operational frequency should be between 400 Hz and ca. 1 kHz where the noise of the microphones in use is minimum.

ii) The resonance frequency should be tunable and the quality factor Q of the operational mode should not exceed 50 in order to operate the sample and reference cell in the dual-beam setup at identical frequencies.

iv) The cell should permit operation with continuous gas flow. Consequently, not only the cell windows but also the gas in- and outlets have to be positioned at pressure nodes.

We developed our cell design on the basis of a formalism which permits the numerical simulation of the frequency response of one-dimensional PA cells. Our general model is based on the theory by Morse and Ingard [7.49] yet instead of analog acoustic impedances it uses infinitesimal impedances and sources and a matrix formalism [7.50]. From these calculations we derived the geometry of a longitudinal resonator that best fulfills all our demands. The cell is shown in Fig. 7.3a. Its resonance

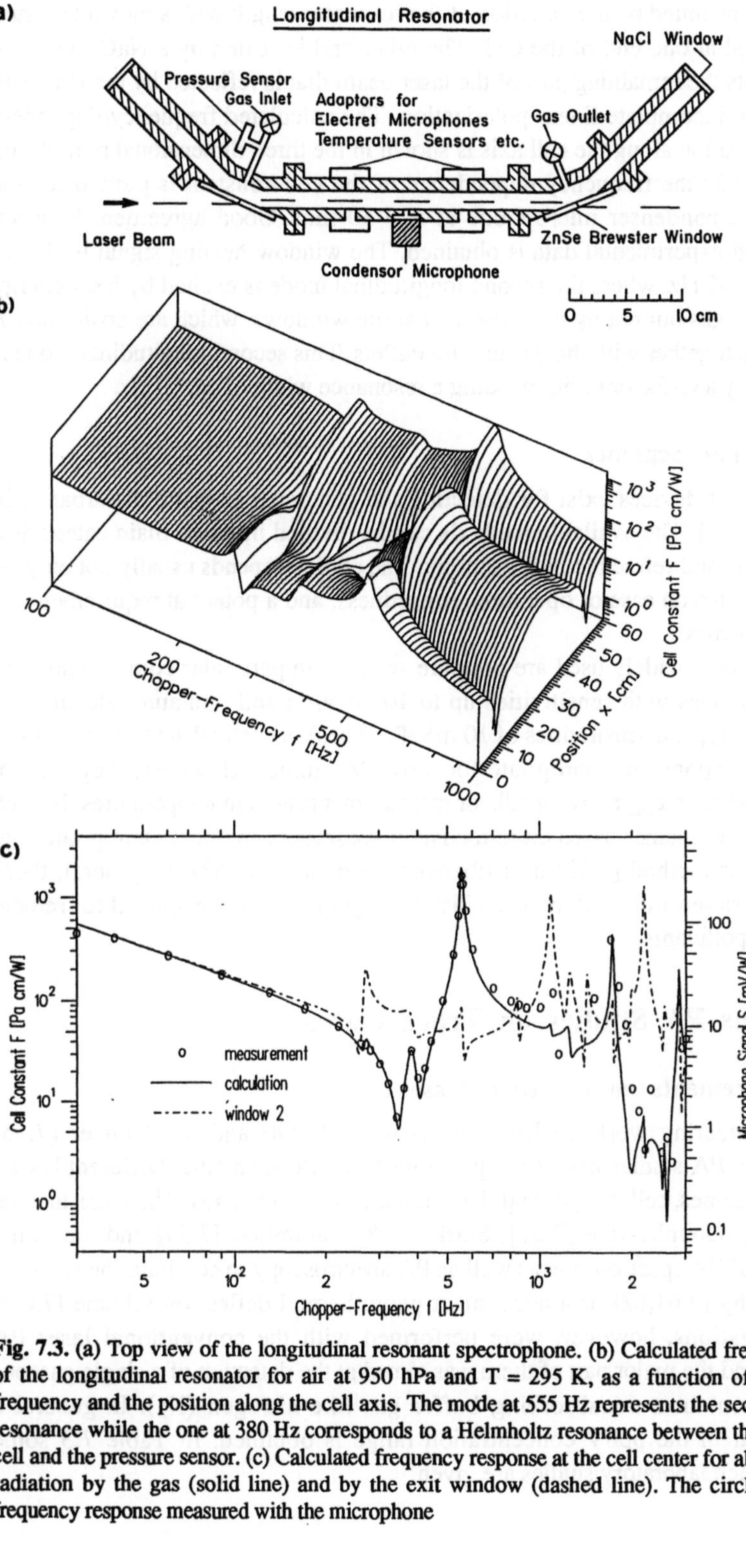

Fig. 7.3. (a) Top view of the longitudinal resonant spectrophone. (b) Calculated frequency response of the longitudinal resonator for air at 950 hPa and T = 295 K, as a function of the modulation frequency and the position along the cell axis. The mode at 555 Hz represents the second longitudinal resonance while the one at 380 Hz corresponds to a Helmholtz resonance between the volumes of the cell and the pressure sensor. (c) Calculated frequency response at the cell center for absorption of laser radiation by the gas (solid line) and by the exit window (dashed line). The circles represent the frequency response measured with the microphone

frequency can be tuned by the variation of the resonator length with a movable piston which is placed at one end of the cell. The other end is sealed by a NaCl window, which transmits the remaining part of the laser beam that is reflected by the Brewster window due to incomplete linear polarization. The calculated frequency-dependent pressure distribution along the cell axis is shown in the three-dimensional plot of Fig. 7.3b. In Fig. 7.3c the frequency dependence of the cell constant is plotted for the position of the condenser microphone at x = 30 mm. Good agreement between calculation and experimental data is obtained. The window-heating signal is clearly suppressed at 555 Hz, where the second longitudinal mode is excited by laser energy absorbed in the gas but not by that absorbed at the windows, which are positioned at pressure nodes together with the gas in- and outlets. This second longitudinal mode is characterized by a Q-factor of 50, yielding a resonance width of ca. 10 Hz.

7.3.4 Detection Schemes

Various different devices exist for detecting the generated acoustic disturbance in gases [7.40,41,51]. The available methods can be divided into two main categories: pressure sensors and refractive index sensors. The choice depends usually not only on sensitivity but also on ease of operation, ruggedness, and a potential requirement for noncontact schemes.

By far the most widely used are pressure sensors, in particular commercial condenser microphones with sensitivities up to 100 mV/Pa and miniature electret microphones with typical sensitivities of 10 mV/Pa. These acoustic detectors are easy to use and their response is appropriate for most PA studies. However, they are not suitable for studies in aggressive media or measurements at high temperatures. In such cases noncontact schemes based on refractive index sensors are more appropriate, e.g. the thermal-lens method [7.52] or a fiber-optic technique [7.53]. In general, these schemes are less sensitive and more sophisticated yet they may be applied for remote sensing of air pollutants.

7.4 Previous PA Studies on Trace Gases

7.4.1 Measurements on Certified Gases

Since the pioneering work by Kerr and Atwood [7.54] and by Kreuzer [7.55] numerous laser PA studies have been performed in gaseous media. Different lasers, modulation schemes, cell designs and detection methods were used. The latter include Raman [7.56], Doppler-free [7.57], Stark [7.58], saturation [7.59] and harmonic saturated [7.60] PA spectroscopy as well as PA spectroscopy in combination with gas chromatography [7.61,62] or a noncontact photothermal deflection scheme [7.63]. Most investigations, however, were performed with the conventional laser PA arrangement and the major part of them was aimed at the detection of a single gaseous trace component in a nonabsorbing buffer gas like nitrogen (N_2). In general a detection limit in the ppbV concentration range is obtained. In Table 7.3 some examples of such laboratory studies are given.

Table 7.3. Examples of laboratory PA studies on trace gases. Unless otherwise noted the minimum detectable concentrations c_{min} refer to N_2 as buffer gas

Laser	Spectral region	Species	c_{min} [ppbV]	Reference
HeNe	3.39 μm	CH_4	200	Kreuzer, 1971 [7.55]
CO CO_2	5.2-6 μm 9.2-10.8 μm	10 different trace gases	≤ 1	Kreuzer et al., 1972 [7.64]
HeNe	3.39 μm	CH_4, n-butane	10^3	Dewey et al., 1973 [7.65] Goldan and Goto, 1975 [7.66]
CO_2	9.2-10.8 μm	NH_3	< 3	Max and Rosengren, 1974 [7.67]
HF	2.7-2.9 μm	HF, NO, CO_2	10^3	Gomenyuk et al., 1974 [7.68]
cw dye	0.57-0.62 μm	NO_2	10	Angus et al., 1975 [7.69]
DF	3.8 μm	CH_4, N_2O	0.5-5	Deaton et al., 1975 [7.70]
CO_2	9.2-10.8 μm	C_2H_4, NH_3, freons, etc.	≤ 4	Schnell and Fischer, 1975 [7.71]
CO / $^{13}CO_2$	≈ 6 μm/≈ 11 μm	Explosives (in air)	1-10	Claspy et al., 1976 [7.72]
Pulsed dye	480-625 nm	NO_2 (in air)	4	Claspy et al., 1977 [7.73]
CO_2	9.2-10.8 μm	C_2H_4 (in air)	5	Shtrikman and Slatkine, 1977 [7.74]
CO	4.75 μm	CO	150	Gerlach and Amer, 1978 [7.75]
CO_2	9.2-10.8 μm	C_2H_4	0.3	Kritshman et al., 1978 [7.76]
Frequency-doubled cw dye	290-310 nm	SO_2	0.12	Koch and Lahmann, 1978 [7.47]
CO_2	9.2-10.8 μm	Explosives	10^3	Grane, 1978 [7.77]
CO_2	9.2-10.8 μm	SF_6	0.01	Nodov, 1978 [7.78]
Ar^+	≈ 488 nm (multiline)	NO_2 (in air)	5	Ioli et al., 1979 [7.79]
CO_2	9.2-10.8 μm	SF_6, NH_3, O_3, C_2H_4, C_6H_6, SO_2 in air	ppbV-ppmV	Adamowicz and Koo, 1979 [7.80]
CO_2	9.2-10.8 μm	Hydrazines	< 30	Loper et al., 1980 [7.81]
Pb S_{1-x} Se_x	4.8 μm	CO	$5 \cdot 10^4$	Vansteenkiste et al., 1981 [7.82]
CO_2	9.2-10.8 μm	C_2H_4, NH_3	≈ 50	Chen et al., 1982 [7.83]
CO_2	9.2-10.8 μm	30 species in dry air	50	Wang et al., 1982 [7.84]
CO_2 incl. isotopes	9-12 μm	22 species out of 250		Hubert, 1983 [7.85]
XeCl (pulsed)	308 nm	Acetaldehyde	25	Leugers and Atkinson, 1984 [7.86]
CO_2	9.2-10.8 μm	SF_6, C_2H_4	3	Fung and Lin, 1986 [7.48]
CO_2	9.2-10.8 μm	Hydrazines, toxic vapors	≤ 10	Loper et al., 1986 [7.87]
Frequency-doubled pulsed dye	300.05 nm	SO_2 in purified air	0.2	Vujkovic Cvijin et al., 1987 [7.88]
CO_2	9.2-10.8 μm	Freons (in air)	20	Zelinger et al., 1988 [7.89]
Frequency-doubled pulsed dye	303.59 nm	H_2CO in selected gas mixtures	51	Boutonnat et al., 1988 [7.90]

7.4.2 Measurements on Real Air Samples

Hitherto only a few PA studies have been performed on real air samples, either in the laboratory or in situ. Probably the first PA study on real gas samples was devoted to the detection of NO in ambient air and in automobile exhausts with a spin-flip Raman laser system [7.43]. Later on, NO was monitored in the stratosphere in real time in situ with a balloon-borne apparatus [7.91-93]. Perlmutter et al. [7.94] detected ethylene (C_2H_4) in air samples taken from a main road and from fruit storage chambers. The authors found a minimum detectable and identifiable concentration for C_2H_4 of about 5 ppbV in most cases. This limit may be increased to 50 ppbV primarily due to inaccuracy in a priori knowledge of IR spectral signatures of interfering gases. Thus, interfering absorptions limit the detection sensitivity, and not background noise. This issue is further discussed in the following with respect to our own PA studies on trace gas monitoring in ambient and exhaust air [7.95,96]. In addition, applications in biology and agriculture [7.97,98] should be mentioned. Table 7.4 summarizes examples of trace gas detection in samples of real air.

Table 7.4. Examples of PA studies on trace gases in samples of real air

Laser	Species	Source/Location	Reference
CO-SFR	NO	Ambient air, automobile exhaust	Kreuzer and Patel, 1971 [7.43]
CO-SFR	NO	Stratosphere, in situ	Patel et al., 1974 [7.91] Burkhardt et al., 1975 [7.92] Patel, 1976 [7.93]
CO_2	C_2H_4	Ambient air, fruit storage chamber	Perlmutter et al., 1979 [7.94]
CO	Various hydrocarbons	Motor-vehicle exhausts	Bernegger et al., 1988 [7.95]
CO_2	C_2H_4	Ambient air, in situ	Meyer et al., 1988 [7.96]
CO_2	NH_3	Inoculated beef	Bicanic et al., 1988 [7.97]
CO_2	C_2H_4	Exhalation of flowers	Harren et al., 1988 [7.98]

7.5 Stationary CO-Laser PA System

7.5.1 Spectral Range

Our investigations on automobile exhausts were primarily performed with a CO-laser PA apparatus because the wavelength range of a CO laser between 5 and 6.5 µm coincides with the absorption ranges of most important components present in vehicle exhausts. This is indicated in Table 7.5, which shows the CO laser emission range as well as the absorption ranges of some gases together with their absorption strengths

Table 7.5. List of selected pollutants including their absorption ranges within the CO-laser wavelength region, their mean absorption strengths, and their detection limits in N_2 and in clean air

CO Laser – PAS System

cw sealed off CO laser (Edinburgh Instr Typ PL-3)

5.0 λ [µm] 6.5

ca. 90 laser lines
1 mW – 1.5 W cw

Emission range of CO laser

Absorption strength [cm^{-1}/atm]	Absorption range of different gases and vapors	Detection Limit in N_2 [ppbv]	Detection Limit in Air [ppmv]
60	Nitrogen Dioxide NO_2	0.4	0.025
20	Acrolein (Propenal) CH_2CHCHO	1.25	0.15
11	Acetaldehyde CH_3CHO	2.2	0.1
7.4	Formaldehyde HCHO	3.4	0.25
5.4	Vinyl Chloride C_2H_3Cl	4.6	0.25
2.9	m-Xylene $C_6H_4CH_3CH_3$	8.6	0.5
2.2	Nitric Oxyde NO	11	0.15
2.2	1,3-Butadiene $CH_2CHCHCH_2$	11	0.3
1.2	Ethylene C_2H_4	21	1.5
1.2	Toluene $C_6H_5CH_3$	21	1.0
1.1	Propylene CH_3CHCH_2	23	1.5
0.75	Benzene C_6H_6	33	0.5
0.75	o-Xylene $C_6H_4CH_3CH_3$	33	1.0
0.55	p-Xylene $C_6H_4CH_3CH_3$	45	0.7
0.004	Carbon Dioxide CO_2	6250	500
10^{-4} – 2	Water Vapor H_2O	12.5	< 50 >

on the left. The minimum detectable concentrations c_{min} that we achieved with our apparatus are listed on the right-hand side for the presence of these components either in N_2 or in clean air. The data for c_{min} increase from top to bottom according to the decreasing absorption strength. Various hydrocarbons but also the nitric oxides NO and NO_2 appear in the list. It should be noticed that in contrast with the conventional chemiluminescence technique NO and NO_2 can be detected interference-free, because their absorption ranges do not overlap. The increased detection thresholds in clean air compared to N_2 are essentially caused by the interfering water vapor absorption, which is present throughout the entire CO laser emission range. However, our PA system permits a reliable detection of pollutants in this wavelength region even in the presence of water vapor, which would not be possible with long-path measurement schemes.

7.5.2 Experimental Arrangement

The CO laser in use (Edinburgh Instruments, Type PL 3) impedes an easy solution to the problem of absorption interference by water vapor because it exhibits multiple emission on adjacent laser transitions not resolved by the grating. Thus, the spectral content of the laser emission and the water-vapor absorption depend not only on the actual wavelength but also on the operating conditions of the laser. We have solved this problem by constructing a dual-beam arrangement with two PA cells. The first cell contains the gas sample under study whereas the second is filled with water vapor in a buffer gas, usually synthetic air (80% N_2, 20% O_2) at identical temperature and total pressure. In this case, the mean water-vapor absorption cross section for any chosen CO-laser line (consisting of 2-3 adjacent laser transitions) can be determined for each measurement, provided that the absorption cross sections of water vapor are identical in both cells. The latter requires identical absorption lineshapes, i.e. identical line broadening, as well as identical spectral composition of the laser beam.

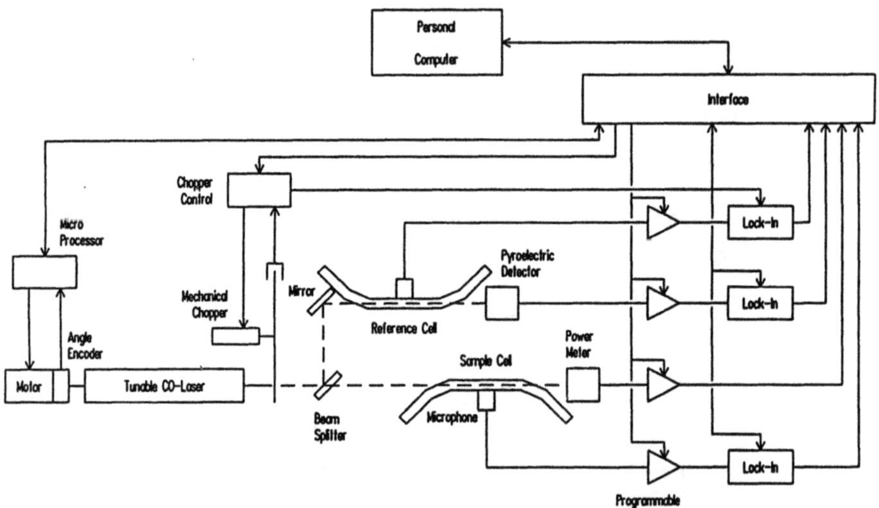

Fig. 7.4. Experimental arrangement of automated CO-laser PA system

The dual-beam setup is shown in Fig. 7.4. The CO laser is operated at -10°C with a sealed gas mixture containing CO, Xe, N_2 and He at a total pressure of 39 hPa. The laser is tuned by a computer-controlled grating with 150 lines/mm, blazed at 6.0 µm. A refilled tube emits on ca. 90 lines in the wavelength range from 5.15 µm (≈ 1940 cm^{-1}) to 6.35 µm (≈ 1575 cm^{-1}). The strongest lines exhibit powers in excess of 1 W while the power of the weakest line is only a few milliwatts. The laser beam is modulated by a mechanical chopper, which is placed in a sound isolating box. The chopper frequency can be adjusted to the resonance frequency of the spectrophone of 555 Hz with an accuracy of better than 0.05 Hz via the remote port of the chopper control. This allows an automatic compensation of changes in the resonance frequency caused by changes of the sample temperature during the measurement.

A ZnSe beamsplitter with ≈ 70 % transmission is used to divide the P-polarized CO-laser beam into two paths of the same length, in order to provide the same spectral content of the beam in the two cells. The wavelength dependence of the transmission of the beam splitter is stored in the computer and taken into account for the normalization of the PA signal.

As outlined in Sect. 7.3.3 the PA cell that best fulfills our requirements for a dual-beam setup is the longitudinally resonant cell presented in Fig. 7.3a. The Brewster windows are mounted at the pressure nodes of the standing sound wave. The PA signals in both cells are detected with condenser microphones (Brüel & Kjaer, type 4179 and 4144, respectively) which are mounted at the positions where the standing wave of the operational mode reaches the maximum amplitude. The microphone signals are processed with two identical lock-in amplifiers (Ithaco, Dynatrac 393) locked to the chopper frequency. Both the $S \cdot \sin\theta$ and the $S \cdot \cos\theta$ output channels of the lock-in amplifiers are recorded in order to numerically deduce the amplitudes S and the phases θ of both microphone signals. Furthermore, the PA cells are equipped with an absolute pressure meter and a temperature sensor.

After passing the sample cell the laser power is detected with a bolometer whose signal is used for the normalization of the PA signals. The pyroelectric detector placed behind the reference cell serves as an additional power sensor for fast feedback to the computer (IBM PC AT 03). This allows an automatic setting of the piezo voltage at the outcoupling laser mirror to the maximum power at the laser line selected by the grating.

The computer is used for the control of the entire experiment with the aid of a data acquisition and work station (Keithley Instruments, System 570) as well as for the data analysis and management. The work station permits simultaneous recording of numerous parameters. The accuracy of the measurement was considerably improved by the automation of the apparatus while the time necessary for measuring a complete spectrum was reduced from several hours to less than one hour.

7.5.3 PA Measurements on Vehicle Exhausts

a) Calibration of PA Apparatus

In principle, the cell constant C of the spectrophone introduced in (7.2.15) can be calculated by means of the formalism outlined in Sect. 7.2.1, if the microphone sensitivity at the operational frequency is accurately known. However, the calculated cell constant is not reliable because possible instrumental calibration errors are not taken into account. Therefore, C has to be determined experimentally by the measurement of the normalized PA signal of a sample with a known absorption coefficient α. For the case of our CO laser with multiple emission (Sect. 7.5.2), an accurate calibration can be performed only with heavy molecules. Since no literature data were available we performed the calibration ourselves with four certified mixtures containing $\approx$ 100 ppmV of an absorbing heavy gas (ethylene, vinyl chloride, propylene, 2,3-butadiene) buffered in nitrogen. Their absorption coefficients were derived from long-path absorption measurements at CO laser wavelengths in a commercial IR spectrometer (Perkin Elmer, model 580).

Before the analysis of unknown gas mixtures can be performed, calibration spectra of numerous well-defined gases and vapors have to be measured. We determined the absolute absorption cross sections for all CO-laser lines of 18 different components which are of interest for air pollution monitoring. Good agreement with calculated data was obtained for NO, NO_2, N_2O, CO_2 and H_2O vapor for which AFGL line parameters are available [7.99-101].

b) Detection Limits and Cross Sensitivities

The minimum detectable number density of a component in a gas mixture not only depends on its absorption strength and on the background signal, but also on the contribution of the remaining components to the total absorption as well as on the accuracy by which this contribution can be determined. Therefore, detection limits which are derived from the synchronous background signal of a spectrophone, caused by window heating, and from the microphone noise, are only valid for the ideal case where the component is diluted in a nonabsorbing buffer gas. This minimum detectable number density of a substance in a gas mixture under interference-free conditions is given by the ratio between the absorption coefficient α_{min} corresponding to the background signal and the maximum absorption cross section of the molecule within the laser wavelength range. For our CO-laser PA system the background signal measured for a filling of pure N_2 corresponds to $\alpha_{min} = 5 \cdot 10^{-8}\,cm^{-1}$. The minimum detectable concentrations obtained for these interference-free conditions range between 1 and 100 ppbV for all components investigated except CO_2. Cross sensitivities between two components can be expressed by the minimum concentration ratio for which one component exhibits the same absorption coefficient as the other one for at least one CO-laser line. The smaller these ratios, the less interferences occur for the corresponding components. In addition to the cross sensitivity, the accuracy of the measurement of the absorption coefficients is important.

Since in general water vapor is the major constituent in gas samples, it determines the detection limit of most trace components. If the water-vapor absorption is determined with an accuracy of 1%, its concentration in a gas sample has to be reduced to 100-300 ppmV if the interference-free detection limits of the trace components are to be obtained. For air at standard conditions, where the absolute water-vapor content is 1%-2%, the minimum detectable concentrations are on the order of

0.5-5 ppmV if the water-vapor contribution to the absorption is determined with an accuracy of 1%. Thus for detecting trace components at sub-ppmV concentrations, gas samples have in general to be dried.

c) Analysis of Multi-component PA Spectra

In analogy to (7.2.16) the total absorption coefficient α_i at a specific laser line i of a gas mixture containing n components is represented by

$$\alpha_i = \sum_{j=1}^{n} \alpha_{ij} = N_{tot} \sum_{j=1}^{n} \sigma_{ij}\, c_j \quad \text{for} \quad i = 1 \ldots m, \qquad j = 1 \ldots n \text{ with } n \le m\ . \tag{7.5.1}$$

Here N_{tot} is the total number density of molecules present in the gas mixture, c_j denotes the concentration of absorbing molecules of species j and σ_{ij} their absorption cross section at the laser transition i. The corresponding PA amplitudes are usually measured by a microphone, yielding the signals S_i. According to (7.5.2) the absorption coefficients $\hat{\alpha}_i$ derived from the measurements are given by

$$\hat{\alpha}_i = \frac{S_i}{C\, P_i}\ , \tag{7.5.2}$$

where P_i is the incident laser power at transition i and C the cell constant. The goal is to extract the n concentrations c_j from the set of m absorption coefficients $\hat{\alpha}_i$ derived experimentally. The measured and calculated absorption spectra can be represented as vectors $\underline{\hat{\alpha}}$ and $\underline{\alpha}$ in an m-dimensional space. Equation (7.5.1) can be written as

$$\underline{\alpha} = N_{tot}\ \underline{\Sigma}\ \underline{c} \tag{7.5.3}$$

where $\underline{\Sigma}$ is a m•n matrix which contains the elements σ_{ij} of the calibration spectra of the n components in the columns. The vector $\underline{c}$ contains the concentrations c_j. In practice, the calculated $\underline{\alpha}$ always differs from $\underline{\hat{\alpha}}$ due to the noise of the measured signals and absorption by components not taken into account for the calculation of $\underline{\alpha}$. Therefore, we usually measure the absorption at m > n laser transitions. Then the concentrations c_j are derived from the overestimated system of equations by minimizing the residuals $\rho_i = \hat{\alpha}_i - \alpha_i$ in a least squares fit, which yields

$$\underline{c} = (1/N_{tot})\ (\underline{\Sigma}^t\ \underline{\Sigma})^{-1}\ \underline{\hat{\alpha}} \tag{7.5.4}$$

where $\underline{\Sigma}^t$ is the transposed matrix of $\underline{\Sigma}$. The symmetric, positive definite matrix $(\underline{\Sigma}^t \cdot \underline{\Sigma})$ is nonsingular if $\underline{\Sigma}$ contains n linear independent column vectors $\underline{\sigma}_j$. This condition is well fulfilled for our applications since all spectra of the n components taken into account exhibit specific features. For various reasons, the residuals ρ_i are weighted by the introduction of normalized weighting factors W_i, which are inversely proportional to the uncertainty $\Delta\hat{\alpha}_i$. In addition these factors are chosen to decrease the weight for those laser transitions for which $\alpha_i > \hat{\alpha}_i$, yet to increase it for the transitions with $\alpha_i \approx \hat{\alpha}_i$. Since these factors W_i depend on the a priori unknown concentrations c_j the solution of (7.5.4) is obtained by iteration.

This new, iterative formalism permits the determination of individual concentrations even if additional components not considered for the fit contribute to the measured absorption spectrum.

d) Results of Motor-Vehicle Exhaust Analysis

As discussed above, a large water-vapor content hinders the accurate analysis of gas samples. This is particularly true for the determination of the trace concentrations of individual hydrocarbons in vehicle exhausts. This problem can be avoided by mixing the exhausts with a dry buffer gas, e.g. N_2 or synthetic air. In the course of our studies we collected samples from the exhaust pipes of several vehicles at idling operation of the motor. The exhausts were filled into an evacuated FEP bag and transported to the PA apparatus in the laboratory. The PA sample cell was filled to a pressure of ca. 100 hPa and buffered afterwards with dry N_2 to 950 hPa. In some cases the exhaust sample was not only diluted but also dried additionally with a dry-ice cooling trap before it was filled into the PA cell. The PA reference cell was always filled with ca. 2 hPa of H_2O vapor diluted in N_2 at the same total pressure of 950 hPa.

As an example, the absorption spectrum of the dried exhausts of a diesel engine (Renault Master), diluted in N_2 by a factor of 10, is shown in Fig. 7.5. The measured absorption coefficients are represented by solid lines while the bars indicate the contribution of the residual H_2O vapor to the sample absorption. This H_2O-vapor contribution was derived from the normalized PA signal of the sample cell at two strong H_2O-vapor absorption lines and the H_2O-vapor spectrum measured in the

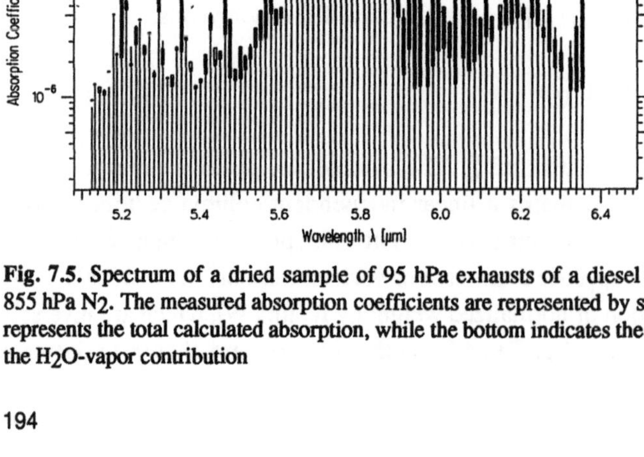

Fig. 7.5. Spectrum of a dried sample of 95 hPa exhausts of a diesel "Renault Master", diluted in 855 hPa N_2. The measured absorption coefficients are represented by solid lines. The top of the bars represents the total calculated absorption, while the bottom indicates the calculated absorption without the H_2O-vapor contribution

reference cell. This correction was performed interactively with the aid of the computer by assuming a linear temporal increase of the H_2O-vapor concentration in the sample cell caused by H_2O vapor desorbing from the cell walls during the measurement. The top of the bars represents the total absorption coefficient, which is calculated on the basis of calibration spectra of ethylene, propylene, benzene, toluene, the three isomers of xylene, the aldehydes, CO_2, NO and H_2O vapor. The bottom of the bars thus indicates the calculated absorption without H_2O vapor, caused by those components which are of special interest in the investigation of car exhausts. As demonstrated in Fig. 7.5 excellent agreement between measurement and calculation is obtained for most CO laser lines. The small deviations are mainly due to additional exhaust constituents not taken into account for the analysis.

An iterative fit on the basis of the formalism discussed under c) yields the individual concentrations and their uncertainties for the 12 components considered. The results for the analysis of the diesel exhausts together with other vehicle exhausts are summarized in Table 7.6. The concentrations in the ppmV range refer to the pure exhausts, i.e. the data were multiplied by the dilution ratio of $\approx$ 10. As is demonstrated, the concentrations of 12 of the most important components could be derived separately, including even those of the different isomers of xylene, which is difficult to obtain with conventional methods like gas chromatography. To our knowledge it is the first time that this selectivity and sensitivity has been achieved with pure PA techniques.

The low NO concentrations in Table 7.6 are explained by the well-known strong adsorption of the polar NO molecules during the sampling and drying process. The latter reduces the NO concentration by a factor of 2-3. On the other hand, however, the concentrations of the other components are almost unaffected by the drying process. The low NO concentrations of the motorcycle exhausts are a consequence of the low combustion temperature. The CO_2 concentrations are in good agreement with data expected from the air/fuel mixing ratios, which are different for the various engines investigated.

The hydrocarbon (HC) concentrations are considerably higher in the motorcycle exhausts than in all the other exhausts. The diesel exhausts contain the least HC but higher concentrations of aldehydes than the other automobile exhausts. Apart from the type of engine the operating conditions also affect the exhaust composition. This is demonstrated by the diesel exhausts. The samples of the Renault Master and the FBW truck were taken from the exhausts of cold engines, in contrast to that of the Unimog, which was taken from a hot engine. As demonstrated, the HC concentration in the exhaust of the hot engine is considerably reduced. On the other hand, the NO concentration in the truck exhaust is higher. This is a consequence of the higher combustion temperature of this engine, which is operated with direct fuel injection in contrast to the other two diesel engines. Finally, the HC concentrations derived for the VW Bus exhaust confirm the fact that a catalytic converter only reduces the HC emissions if it has reached the operational temperature, not in the 'cold' state as was the case for the measurement reported in Table 7.6. We found that under optimum conditions the converter reduces the HC and the NO emissions by a factor of ca. 10.

Table 7.6. CO-laser photoacoustic analysis of vehicle exhausts collected at idling operation of the engine

Vehicle	Dried	Concentration [ppmV] ± Uncertainty											
		CO_2 [%V]	NO	Ethylene	Propylene	Benzene	Toluene	m-Xylene	o-Xylene	p-Xylene	Formald.	Acetald.	Acrolein
Volvo (leaded)	Yes	10.3±2.5	105±9	260±20	191±9	37±5	26±14	16±10	42±18	4±19	5.5±1.9	2.7±0.8	2.5±0.4
Jeep (unleaded)	Yes	9.1±1.1	23±2	191±11	109±3	48±3	268±9	67±7	114±10	7±8	1.2±0.6	2.6±0.2	1.2±0.1
	No	11.1±1.3	40±4	206±13	107±5	78±4	196±13	188±12	103±18	67±13	0.1±1.5	4.1±0.3	2.4±0.2
VW Bus (cold cat. conv.)	Yes	10.8±1.1	11.8±1.5	172±10	109±4	34±3	149±7	31±7	59±10	2.8±8	0.1±0.8	3.6±0.3	1.7±0.2
	No	11.1±1.5	27±3	178±14	98±13	45±5	87±15	39±14	33±19	11±11	1.2±2.6	3.4±0.9	1.9±0.4
Renault Master (diesel)	Yes	2.8±0.3	10.4±0.9	32±3	18.2±1.2	2.7±0.7	18.5±1.9	4.0±1.7	7.7±2.6	0.5±2.5	8.1±0.8	7.6±0.4	3.2±0.2
	No	2.7±0.6	24.5±2.2	33±6	14±8	0.3±2.8	16±7	6.4±6	5.6±10	1.2±6	7.1±2.5	8.7±1.0	4.4±0.5
Unimog (diesel)	Yes	2.6±0.4	15.7±2.2	5±3	5±2	2.0±0.8	6±2	2.5±2	1±3	5±3	2.6±0.7	2.5±0.3	1.3±0.1
	No	2.2±1.0	43±5	7±10	-	-	-	3.5±4	-	8±8	3±7	0.9±2	1.5±1.2
FBW Truck (diesel)	Yes	1.9±0.4	37±3	16±3	17.5±1.3	-	12±2	1.1±1.5	0.3±3	-	4.2±0.6	7.4±0.2	3.4±0.1
	No	2.7±0.9	78±6	23±10	23±9	2±5	20±10	6±7	11±12	3±10	3±2	8.4±0.5	6.9±0.3
Puch Motorcycle (2-stroke)	Yes	7.9±3	0.3±3	220±10	329±7	203±7	700±20	47±18	135±15	15±10	15.9±1.8	4.1±0.7	6.2±0.3
	No	8.5±3	0.1±2	234±10	284±7	217±16	405±16	95±13	148±15	20±9	12.6±2.2	5.2±0.8	6.5±0.3

7.6 Mobile CO_2-Laser PA System

7.6.1 Spectral Range

For our PA studies on urban air and on industrial exhausts with a mobile system a CO_2 laser is most appropriate. It offers simple operation, high power and an emission range that covers the absorption range of numerous species of interest. Furthermore, the interference with water vapor does not impose the problems faced with the CO laser due to the weak absorption [7.102]. However, the interference with CO_2 present in most samples has to be taken into account.

Table 7.7 summarizes the absorption ranges and strengths of some gases and vapors of interest for environmental studies. At the top the characteristic emission

Table 7.7. List of selected pollutants including their absorption ranges within the CO_2-laser wavelength region, their mean absorption strengths, typical emission sources and immission concentrations

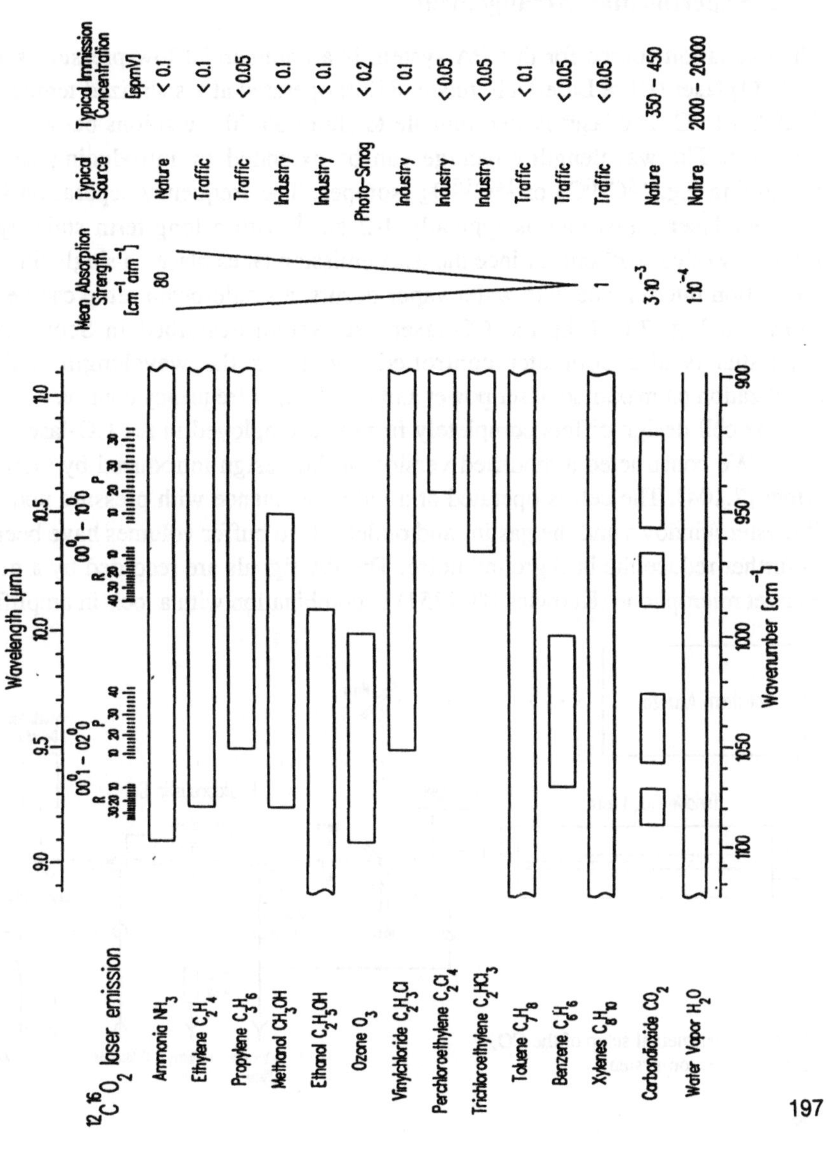

spectrum of a $^{12}C^{16}O_2$ laser is presented. The list contains various hydrocarbons like ethylene (C_2H_4), which presumably plays an important role for the serious forest decline [7.103]. On the other hand, ammonia (NH_3) is important for the production of acid rain [7.5] whereas ozone (O_3) is the well-known photo-oxidant present in summer smog. The mean absorption strength of the components listed decreases from top to bottom, but only substances j with an expected absorption coefficient $\alpha_j \geq \alpha_{min} = 5 \cdot 10^{-8}$ cm^{-1} are mentioned. With respect to air pollution monitoring, Table 7.6 also contains typical sources and immission concentrations of the molecules listed. These sub-ppmV concentrations restrict the applications in the field of environmental monitoring to certain trace gases and often to urban areas even with the high sensitivity achieved with PA methods.

7.6.2 Experimental Arrangement

The excitation source for this PA system is a commercial low-pressure sealed-off $^{12}C^{16}O_2$ laser (Ultra Lasertech, model 8122) operated at a stabilized temperature of 20.5 ± 0.1 °C. The laser is step-tunable to about 65-70 transitions between 9.2 and 10.8 µm. The wavelength coverage can be extended by introducing laser tubes containing e.g. $^{13}C^{16}O_2$ or $^{14}C^{16}O_2$ isotopes. The frequency separation between adjacent laser transitions is typically 1-2 cm^{-1} with a long-term stability of the frequency of ca. 10^{-3} cm^{-1}. Since the laser emission takes place on single lines and no absorption interference with water vapor occurs, a single-beam setup can be used as shown in Fig. 7.6. Like the CO-laser PA system described in Sect. 7.5.2 this apparatus is also computer controlled, including the wavelength tuning, the stabilization on maximum laser power and the chopper frequency control.

The cell design differs completely from that employed in the CO-laser arrangement. We constructed a modified version of the design introduced by Gerlach and Amer [7.104]. The cell is operated at a radial resonance with pressure nodes at the Brewster windows and the gas in- and outlets. Two buffer volumes have been added to further reduce the background noise. The PA signals are recorded by a miniature electret microphone (Knowles BT 1751) in combination with a lock-in amplifier. The

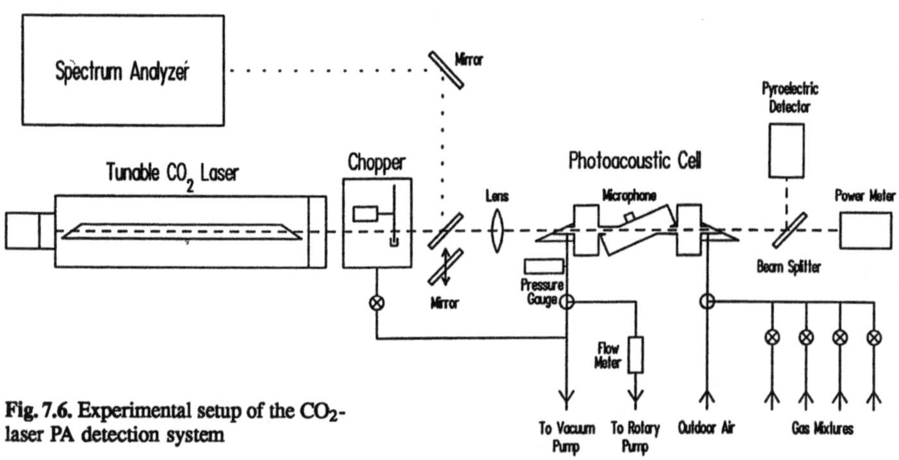

Fig. 7.6. Experimental setup of the CO_2-laser PA detection system

first radial mode of our cell exhibits a resonance frequency of ca. 2650 Hz, which is high enough to allow measurements to be performed also in a noisy environment. Furthermore, this cell design allows continuous recording with a gas flow rate up to ca. 1 l/min. Gas mixtures used for calibration and air samples are admitted to the PA cell by a greaseless filling system as shown in Fig. 7.6. The cell and the whole filling system can be evacuated by a turbomolecular pump to a pressure of less than 10^{-8} bar, which corresponds to a contamination concentration of 10 ppbV at atmospheric conditions.

In order to permit in situ monitoring of air pollutants, the entire PA apparatus is installed in a mobile, air-conditioned trailer of 4 m length, shown in Fig. 7.7. In the course of our studies carried out at various locations in Switzerland the performance of this mobile system was never impaired by transportation.

Fig. 7.7. Photograph of the mobile CO_2-laser PA system

7.6.3 Trace Gas Measurements
a) Calibration of PA Apparatus

We calibrated our spectrophone with the widely used reference gas ethylene (C_2H_4) whose absorption cross sections including their pressure and temperature dependences are accurately known at CO_2-laser wavelengths [7.105]. Furthermore, C_2H_4 is known to interact only weakly with common cell surface materials [7.106] and it exhibits a highly structured spectrum in the 9-11 μm range. Our calibrations, which were performed periodically, yielded a mean cell constant, averaged over eight laser transitions, $C \approx 3.5$ Vcm/W over an investigation period of 4 years. A long-term increase of the microphone responsivity resulted in a continuous growth of C by 0.5 Vcm/W. For shorter time intervals the variation of C was below 2%.

Our arrangement yields a minimum detectable absorption coefficient $\alpha_{min} \approx 1 \cdot 10^{-8} - 3 \cdot 10^{-8}\ cm^{-1}$ depending on wavelength for a laser power of 1 W. This ultimate sensitivity which is somewhat better than that of the CO-laser setup, corresponds to a minimum detectable concentration of 0.3-1.0 ppbV C_2H_4 in pure synthetic air, i.e. under interference-free conditions. In a multi-component analysis of the real atmosphere this theoretical detection limit is usually impaired due to interferences with other absorbing substances.

In the course of our studies we recorded the absorption spectra for $^{12}C^{16}O_2$-laser transitions of 18 high-purity hydrocarbons (gases and vapors) in addition to H_2O vapor and CO_2. These spectra represent the basis for the analysis of multi-component gas mixtures. Furthermore, we performed sealed-off calibration measurements on two-component certified gas mixtures as well as dynamic PA measurements on flowing mixtures with more than two components, which were analyzed by gas chromatography for comparison.

b) Partial Selectivity Γ_j and Cross Sensitivity Q_{jk}

Detection limits in multi-component mixtures can be characterized by partial selectivities [7.107] and cross sensitivities [7.96]. The partial selectivity Γ_j with $0 \geq \Gamma_j \geq 1$ represents a normalized measure of the influence of the remaining n-1 absorbing components on the detection of substance j. The case $\Gamma_j = 1$ implies full selectivity, i.e. the presence of the remaining n-1 other substances does not alter the variance of the calculated concentration c_j. If $\Gamma_j = 0$ a quantitative determination of c_j is impossible since $(\underline{\Sigma}^t\, \underline{\Sigma})$ defined by (7.5.4) is singular. In this case the spectrum of substance j is very similar to the spectrum of another substance or to a linear combination of the spectra of other substances. The concept of cross sensitivity describes interference effects more quantitatively. For the determination of the concentration c_j of substance j at the appropriate laser transition j the influence of a potential interference by substance k is taken into account by the mutual cross sensitivities

$$Q_{jk} = \Sigma_{jk} / \Sigma_{jj}\ , \qquad (7.6.1)$$

where Σ_{jk} and Σ_{jj} represent the absorption at laser transition j of substances j and k, respectively. An uncertainty Δc_k of the concentration c_k of substance k raises the uncertainty Δc_k of the measured concentration c_j of substance j by

$$\Delta c_j \geq Q_{jk}\ \Delta c_k\ . \qquad (7.6.2)$$

Equation (7.6.1) assumes the existence of one appropriate laser transition for the detection of each substance j. However, this condition is fulfilled only in those cases where the partial selectivities Γ_j are sufficiently large, in practice for $\Gamma_j \geq 0.3$.

c) In Situ Air Pollution Monitoring

Today, considerable amounts of ethylene (C_2H_4) are emitted into the atmosphere by road traffic and various industries. The resulting C_2H_4 concentrations are a continual source of stress for plant growth in urban areas [7.108]. Discernible effects have been reported at concentrations around 50 ppbV [7.109]. The 10P(14) CO_2 laser transition at 949.49 cm^{-1} is ideally suited for detecting this important pollutant due to the proximity of the Q-branch of its vibrational band ν_7 located at 948.77 cm^{-1} [7.110]. The resulting large absorption cross section $\sigma = 1.3 \cdot 10^{-18}$ cm^2 yields $c_{min} < 1$ ppbV under interference-free conditions. In real air samples the detection limit is raised to ≤ 5 ppbV. In contrast to Perlmutter et al. [7.94], who introduced a NaOH scrubber to eliminate the interfering CO_2, we achieved a similar or even better sensitivity with direct monitoring without any scrubber by measuring the PA amplitude and phase simultaneously as discussed in Sect. 7.2 (Fig. 7.2). In the course of our studies on urban air we found that up to 95 % of its total absorption within the whole 10 μm branch of the CO_2-laser emission is caused by C_2H_4, CO_2 and H_2O vapor. As a consequence, the multi-component detection of urban-air samples by direct $^{12}C^{16}O_2$-laser PA spectroscopy is essentially limited to these molecules, in particular if the analysis is restricted to the 10 μm branch. It should be emphasized, however, that this statement may not be valid near local emission sources or if other CO_2 isotopes or PA Stark cells etc. are used.

Since C_2H_4 and H_2O vapor exhibit well-structured absorption spectra within the CO_2-laser wavelength range, the spectral interferences between C_2H_4, CO_2 and H_2O vapor are relatively small. This small spectral overlap is also reflected by the high partial selectivities Γ_j mentioned under b), which are 0.9 for C_2H_4, 0.75 for CO_2 and 0.8 for H_2O-vapor, when including the entire 10 μm branch. A restriction to the three dominant laser transitions 10P(14), 10P(20) and 10R(20) reduces the Γ_j only slightly to 0.65, 0.73 and 0.53, respectively. As a consequence, it is possible to perform continuous determinations of the concentrations of C_2H_4, CO_2 and H_2O by measuring the absorption coefficients at the attributed laser transitions 10P(14), 10P(20 and 10R(20). The effects of the small H_2O and CO_2 fluctuations on the determination of the C_2H_4 concentration can be quantitatively determined by the cross sensitivities Q_{jk} defined by (7.6.1). Consequently, possible changes of the CO_2 and H_2O concentrations of 10 ppmV and 0.1%, respectively, cause uncertainties of only 1.6 and 2.2 ppbV for the C_2H_4 concentration. The total estimated uncertainty of the C_2H_4 concentration is therefore approximately ± 2ppbV.

Figure 7.8 shows an example of multi-component monitoring of urban air in Biel, Switzerland. Diurnal concentration profiles for H_2O vapor, CO_2, C_2H_4 and ozone (O_3) are presented together with wind characteristics. The top three measurements were performed in a city park at a distance of about 10 m from a main road on a sunny summer workday with considerable photosmog production. The H_2O-vapor, CO_2 and C_2H_4 profiles were taken with our mobile station by measuring the absorption at the 10R(20), 10P(20) and 10P(14) laser transitions, respectively. The O_3 and wind data were recorded by commercial equipment at a distance of ca. 1000 m from the park.

As demonstrated at the top of Fig. 7.8, good agreement between our PA data for H_2O vapor and independent humidity measurements is obtained. Whilst the H_2O and CO_2 concentration show only small daily fluctuations, the C_2H_4 concentration exhibits a pronounced temporal profile with peak values of up to 35 ppbV,

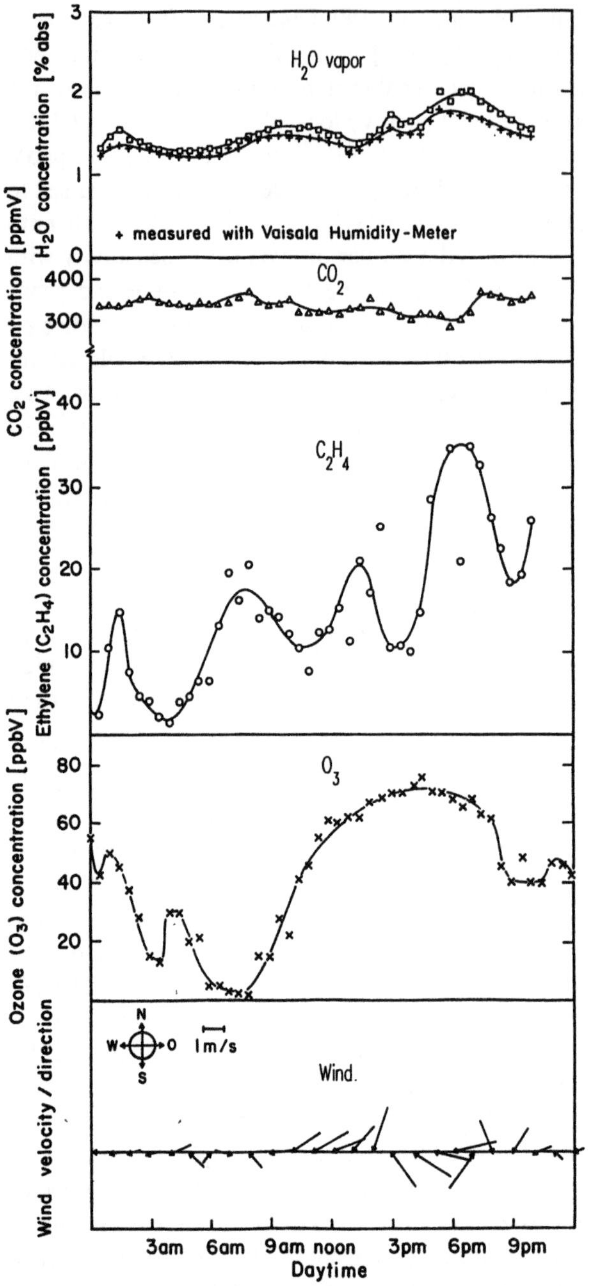

Fig. 7.8. Temporal H_2O-vapor, CO_2 and C_2H_4 concentration profiles, derived from PA measurements, and an O_3 concentration profile and wind data obtained by commercial equipment. The H_2O-vapor concentration profile is compared to independent measurements with a humidity meter. Location: City park in Biel, Switzerland; Date: July 2, 1986

corresponding to 44 μg/m^3. Previous studies on ambient air yielded good agreement between C_2H_4 concentrations derived from PA measurements and those obtained from independent gas-chromatographic data [7.111]. The three C_2H_4-concentration peaks on the right of Fig. 7.8 coincide with traffic rush hours whereas the first peak occurring around 1:30 am was presumably caused by local truck movements in a nearby parking lot. The photosmog production, represented by the increasing O_3 concentration in the afternoon, clearly reduces the morning and noon peak concentrations of C_2H_4, which is a precursor component for the generation of photosmog. Furthermore, wind effects may influence the various concentration profiles. This was in particular observed on the following day of the measurement campaign [7.96].

d) In Situ Monitoring of Industrial Exhausts

Recent studies on industrial exhaust gases have further demonstrated the versatility of our system. Exhaust air represents a mixture of numerous components with concentrations varying in time. Thus continuous measurements of the main constituents are of great importance. On the basis of the multi-component analysis discussed in Sect. 7.5.3c, a computer program was developed for the evaluation of the best laser transitions for the detection of specific components yielding a minimum of interference with other components at optimum sensitivity, i.e. with a maximum relative absorption.

In collaboration with a chemical company in Basel, Switzerland, we performed in situ emission measurements at a pharmaceutical production site. During our measurement campaign we investigated the exhausts from two different sources A and B. For emission source A simultaneous gas-chromatographic (GC) data were available for the same exhaust stream. Therefore, we restricted our continuous measurements to that location, although interesting results on additional substances were expected from source B.

A theoretical evaluation with respect to CO_2-laser PAS monitoring of single components is presented in Table 7.8 for the exhausts of both sources A and B. The maximum of the relative averaged absorption $(\Delta\hat{\alpha}/\hat{\alpha})_{max}$ and the corresponding laser transitions are listed for each substance on the basis of the fixed concentrations indicated. These expected average concentrations were derived from previous GC measurements, from data given by the production manager and from our own PA analyses. It should be noticed that the actual concentrations may deviate markedly from these mean concentrations. Furthermore, 400 ppmV CO_2 and 0.7% absolute H_2O vapor have been included in the theoretical analysis as interfering substances. The assumed H_2O-vapor concentration corresponds to a relative humidity of ≈ 100% at a temperature of 0°C and a pressure of 950 hPa according to the conditions of our continuous measurements. The CO_2 concentration is not expected to be significantly larger since neither emission source involves combustion processes.

With respect to source A the vapors of methanol (CH_3OH) and ethanol (C_2H_5OH) are expected to be easily detectable due to their large relative absorptions $(\Delta\hat{\alpha}/\hat{\alpha})_{max} >$ 80% whereas the detection of the remaining constituents with $(\Delta\hat{\alpha}/\hat{\alpha})_{max} <$ 10%

Table 7.8. Average concentrations[a,b] of gaseous components in exhaust samples from two different pharmaceutical production sites. Maximum relative contributions[c] to the total absorption at the corresponding CO_2-laser transitions

Substance	Concentration [ppmV]	$(\Delta\hat{\alpha}/\hat{\alpha})_{max.}$ [%]	CO_2-laser transition
Source A[a]			
Acetone (CH_3COCH_3)	10	6.4	10 P(42)
Dichloromethane (CH_2Cl_2)	10	4.2	10 P(42)
Ethanol (C_2H_5OH)	20	84.5	9 R(26)
Ethylbenzene (C_8H_{10})	2	2.8	10 R(12)
Methanol (CH_3OH)	20	85.6	9 P(34)
Toluene (C_7H_8)	5	7.7	10 P(34)
o-Xylene (C_8H_{10})	2	6.3	10 R(28)
m-Xylene (C_8H_{10})	2	2.3	10 P(10)
p-Xylene (C_8H_{10})	2	2.5	10 P(14)
Source B[b]			
Chlorobenzene (C_6H_5Cl)	10	40.5	9 R(30)
o-Dichlorobenzene ($C_6H_4Cl_2$)	2	23.8	9 P(28)
m-Dichlorobenzene ($C_6H_4Cl_2$)	15	48.6	9 R(20)
Ethanol (C_2H_5OH)	5	53.4	9 P(12)
Ethylene (C_2H_4)	2	94.3	10 P(14)
Methanol (CH_3OH)	10	29.3	9 P(34)
Toluene (C_7H_8)	10	21.8	9 P(36)

[a] According to previous GC measurements and data given by the production manager.

[b] The composition of exhausts from source B was not completely known in advance, but has been deduced by our own PA analysis. This concerns in particular the chlorinated benzenes.

[c] In the presence of 0.7% abs. H_2O vapor (≈ 100% rel. humidity at T = 0°C) and 400 ppmV CO_2.

appears to be critical. As shown in Table 7.8 the selection of the best laser transition for the selective detection of a single substance in a multi-component mixture depends on the actual composition of the sample. For example, the 9R(26) laser transition was the best choice for the detection of C_2H_5OH vapor from source A, whereas for source B this is the 9P(12) transition. On the other hand, the 9P(34) transition is best suited for CH_3OH-monitoring from both sources.

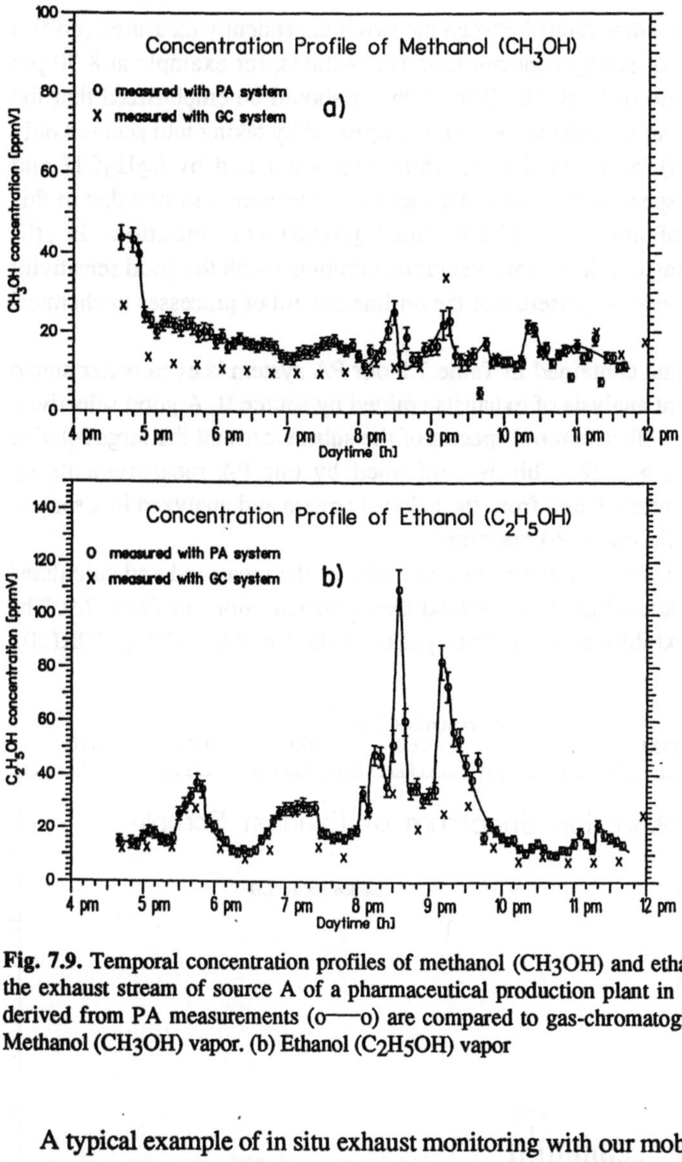

Fig. 7.9. Temporal concentration profiles of methanol (CH_3OH) and ethanol (C_2H_5OH) vapor in the exhaust stream of source A of a pharmaceutical production plant in Basel, Switzerland. Data derived from PA measurements (o—o) are compared to gas-chromatographic (GC) data (x). (a) Methanol (CH_3OH) vapor. (b) Ethanol (C_2H_5OH) vapor

A typical example of in situ exhaust monitoring with our mobile, automated station is presented in Fig. 7.9 for CH_3OH (Fig. 7.9a) and C_2H_5OH vapor (Fig. 7.9b) emitted by source A. Both components can be monitored continuously and simultaneously without difficulty. The plotted concentration profiles were obtained by periodical measurement of the PA signals at the 9R(26), 9P(30), 9P(34), 9P(36), 9P(14), 10P(38) and 10P(40) laser transitions and subsequent two-component analysis of the data. Both PA concentration profiles are compared to data deduced from GC measurements, indicated by the symbols x in Figs. 7.9a and b. In general, good agreement is obtained. However, since the temporal resolution of our system (5 min) is much better than that of the GC system (20 min), concentration peaks that are monitored with our apparatus are detected by the GC system only by accident.

Consequently, the largest deviations between the two independently measured profiles are observed for times of strong concentration fluctuations, for example at 8:30 pm and 9:10 pm for the case of C_2H_5OH (Fig. 7.9b). It should be emphasized that the temporal resolution of our system can be further improved by taking into account only the 9R(26) and 9P(34) laser transitions, which are dominated by C_2H_5OH and CH_3OH absorption, respectively. Additional measurements demonstrated that in this case a proper determination of CH_3OH and C_2H_5OH concentrations is still guaranteed. The resulting excellent time resolution combined with the good sensitivity and selectivity could prove very useful for the on-line control of processes in chemical production.

According to the data contained in Table 7.8 our PA system is even better suited for the multi-component analysis of exhausts emitted by source B. A good selectivity is expected due to the well-structured spectra of the substances and the large relative absorptions $(\Delta\hat{\alpha}/\hat{\alpha})_{max} > 20\%$. This is confirmed by our PA measurements on samples which were directly taken from the exhaust stream and analyzed in a sealed-off cell on the basis of the entire PA spectrum.

As a typical result for source B we present a plot of the measured and calculated spectrum in Fig. 7.10 and a list of the derived trace concentrations in Table 7.9. The measured spectrum exhibits characteristic peaks at the 9R(28), 9P(36), 10R(20),

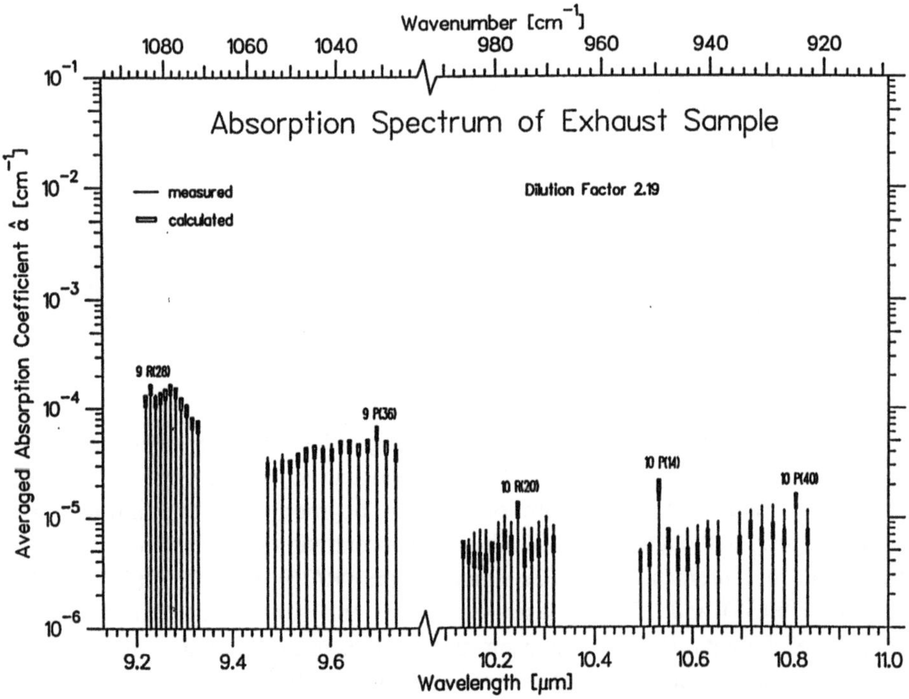

Fig. 7.10. Measured and calculated PA spectrum of an exhaust sample from source B of a pharmaceutical production site in Basel, Switzerland. The data refer to exhausts buffered in synthetic air in atmospheric conditions

Table 7.9. Concentrations of selected components in the exhausts of source B measured with the PA system

Substance	Concentration [ppmV]
Water (H_2O)	17000 ± 3000 [a]
Carbon dioxide (CO_2)	466 ± 355
Chlorobenzene (C_6H_5Cl)	27.8 ± 3.1
o-Dichlorobenzene ($C_6H_4Cl_2$)	4.4 ± 1.2
m-Dichlorobenzene ($C_6H_4Cl_2$)	80.8 ± 3.2
Ethanol (C_2H_5OH)	6.4 ± 1.8
Ethylene (C_2H_4)	1.3 ± 0.4
Methanol CH_3OH)	0.2 ± 0.4
Toluene (C_7H_8)	47.5 ± 13.6

[a] For comparison, the capacitive humidity meter yielded 15000 ppmV for the H_2O-vapor concentration.

10P(14) and 10P(40) laser transitions. These peaks are due to high concentrations of m-dichlorobenzene, toluene, water vapor and ethylene. The agreement between measured and calculated spectrum is excellent in the 9 μm branch but somewhat poorer in the 10 μm branch. In this branch the calculated absorption is generally below that measured. This indicates that even more substances have to be included in the analysis. The selective quantitative determination of the concentrations of the different isomers of a molecule, namely of dichlorobenzene, could again be demonstrated, as in the case of the xylenes in automobile exhausts (Sect. 7.5.3d). It should be noticed that no reliable data on the composition of the exhausts from source B were available, particularly no GC data. The unpredicted presence of m-dichlorobenzene in the exhausts was first found by our PA analysis on the basis of the enhanced absorption at the 9R(28) laser transition that could not be explained by the other components, notably not by o-dichlorobenzene or chlorobenzene. This confirms that PA spectroscopy is not only suited for the individual monitoring of known trace components in mixtures but can also be a valuable tool for the analysis of mixtures or for the search for additional constituents.

7.7 Conclusion

In comparison to conventional methods, PA spectroscopy offers some unique advantages in the field of air pollution monitoring. Previous studies as well as the present one demonstrate that in situ PA monitoring of numerous pollutants with high sensitivity and selectivity is feasible. The experimental arrangement is rather simple and can thus be installed in a mobile system. We put emphasis on investigations on multi-component gas mixtures and present a new, iterative formalism for analyzing PA spectra on the basis of calibration spectra. An excellent selectivity is achieved, which is demonstrated, for example, by the individual detection of the isomers of xylene in vehicle exhausts with our CO-laser PA apparatus or of the isomers of dichlorobenzene in industrial exhausts with our CO_2-laser PA system.

The excellent time resolution of the method permits an almost simultaneous and continuous monitoring of several components of interest by sequential measurements at laser wavelengths that are appropriately chosen to yield maximum sensitivity at minimum interference for the selective detection of the individual components. The PA scheme could thus present a valuable tool for the control of certain processes in the chemical industry by continuously monitoring specific substances in a 'flow-through' mode.

Finally, the high sensitivity and large dynamic range should be pointed out. For example, we were able to detect ethylene (C_2H_4) concentrations in ambient air at the ppbV level with our mobile system. The dynamic range covers at least 5 orders of magnitude and thus permits immission measurements distant from the source of pollution as well as emission monitoring at the source, e.g. in exhaust air.

With regard to the great potential of laser PA spectroscopy, we expect more applications of in situ trace gas monitoring in real atmospheres to evolve in the near future, such as those activities performed in the Netherlands (Chap.8 and [7.112]) and at Moscow [7.113]. Comprehensive reports on our own PA studies will be published elsewhere [7.36,114].

The CO- and CO_2-laser PAS projects were funded by the Swiss National Science Foundation (National Research Program NFP 14).

References

7.1 A.P. Altshuller: Atmos. Environ. **17**, 2131 (1983)

7.2 J.O. Reuss, D.W. Johnson: *Acid Deposition and the Acidification of Soils and Waters*, Ecological Studies, Vol. 59 (Springer, Berlin, Heidelberg 1986)

7.3 Rat von Sachverständigen für Umweltfragen (ed.): *Waldschäden und Luftverunreinigungen*, Sondergutachten (Kohlhammer Verlag, Stuttgart 1983)

7.4 B. Stimm: *Waldsterben* (Ecomed, München 1984)

7.5 A. Wellburn: *Air Pollution and Acid Rain* (Longman, Harlow, UK 1988)

7.6 P. Fabian: *Atmosphäre und Umwelt* , 2.Aufl. (Springer, Berlin, Heidelberg 1987)

7.7 OECD, Organization for Economic Co-operation and Development (ed.): *Photochemical Smog* (OECD, Paris 1982)

7.8 A.C. Stern (ed.): *Air Pollution*, Vol. 3, 3rd ed. (Academic, New York 1976)

7.9 W. Strauss (ed.): *Air Pollution Control*, Part III (Wiley, New York 1978)

7.10 W. Thain: *Monitoring Toxic Gases in the Atmosphere for Hygiene and Pollution Control* (Pergamon, Oxford 1980)

7.11 L.H. Keith (ed.): *Identification and Analysis of Organic Pollutants in Air* (Butterworth, Boston 1984)

7.12 R.W. Frei, J. Albaiges (eds.): *Air and Water Analysis*, Curr. Topics in Environ. and Toxicol. Chem., Vol. 9 (Gordon and Breach, New York 1986)

7.13 R.M. Harrison, R. Perry (eds.): *Handbook of Air Pollution Analysis*, 2nd ed. (Chapman and Hall, London 1986)

7.14 D.L. Fox: Air pollution. Anal. Chem. **59**, 280R (1987)

7.15 G.M. Message: *Practical Aspects of Gas Chromatography / Mass Spectrometry* (Wiley, New York 1984)
7.16 W. Jennings: *Analytical Gas Chromatography* (Academic, Orlando 1987)
7.17 W. Roedel, G. Woelm: *A Guide to Gas Chromatography* (Huethig, Heidelberg 1987)
7.18 J. Willet: *Gas Chromatography* (Wiley, Chichester, UK 1987)
7.19 F.W. Karasek, R.E. Clement: *Basic Gas Chromatography - Mass Spectrometry* (Elsevier, Amsterdam 1988)
7.20 F.X. Kneizys, E.P. Shettle, W.D. Gallery, J.H. Chetwynd Jr., L.W. Abreu, J.E.A. Selby, S.A. Clough, R.W. Fenn: *Atmospheric Transmittance/Radiance;* Computer Code LOWTRAN-6, Environ. Res. Paper No. 846, AFGL-TR-83-0187 (1983)
7.21 G. Finger, F.K. Kneubühl: "Spectral Thermal Infrared Emission of the Terrestrial Atmosphere" in *Infrared and Millimeter Waves*, ed. by K.J. Button, Vol. 12, Part II (Academic, New York 1984) pp. 145-193
7.22 E.D. Hinkley (ed.): *Laser Monitoring of the Atmosphere*, Topics Appl. Phys., Vol. 14 (Springer, Berlin, Heidelberg 1976)
7.23 C.K.N. Patel: Science **202**, 157 (1978)
7.24 I. Turn, D.N. Waters: Int. J. Environ. Stud. **21**, 165 (1983)
7.25 W.B. Grant, R.T. Menzies: J. Air Pollut. Control Assoc. **33**, 187 (1983)
7.26 H. Edner, A. Sunesson, S. Svanberg, L. Unéus, S. Wallin: Appl. Opt. **25**, 403 (1986)
7.27 R. Grisar, H. Preier, G. Schmidtke, G. Restelli (eds.): *Monitoring of Gaseous Pollutants by Tunable Diode Lasers* (D. Reidel, Dordrecht 1987)
7.28 F. Cappellani, G. Melandrone, G. Restelli: Environ. Monit. Assessment **5**, 271 (1985)
7.29 D.K. Killinger, A. Mooradian: *Optical and Laser Remote Sensing*, Springer Ser. Opt. Sci., Vol. 39 (Springer, Berlin, Heidelberg 1983)
7.30 A.I. Carswell: Can. J. Phys. **61**, 378 (1983)
7.31 R.M. Measures: *Laser Remote Sensing, Fundamentals and Applications* (Wiley, New York 1984)
7.32 R.N. Dubinsky: Lasers Optron. **7**, 93 (April 1988)
7.33 T.F. Hunter, P.C. Turtle: "Optoacoustic and Thermo-Opto Detection in Spectroscopy and in the Study of Relaxation Processes" in *Advances in Infrared and Raman Spectroscopy*, Vol. 7, ed. by R.J.H. Clark, R.E. Hester (Heyden, London 1980) pp. 283-352
7.34 A.C. Tam: "Photoacoustics: Spectroscopy and other Applications", in *Ultrasensitive Laser Spectroscopy*, ed. by D.S. Kliger (Academic, New York 1983) Chap. 1
7.35 P. Hess: Top. Curr. Chem. **111**, 1 (1983)
7.36 P.L. Meyer: "Air Pollution Monitoring with a Mobile CO_2-Laser Photoacoustic System"; PhD Thesis, ETH Zurich (1988), to be published
7.37 A.D. Wood, M. Camac, E.T. Gerry: Appl. Opt. **10**, 1877 (1971)
7.38 L.B. Kreuzer: "The Physics of Signal Generation and Detection" in *Optoacoustic Spectroscopy and Detection*, ed. by Y.-H. Pao (Academic, New York 1977) Chap. 1
7.39 M.W. Sigrist, S. Bernegger, P.L. Meyer: Infrared Phys. **29**, No. 2 / 3 (1989), in press
7.40 G.A. West, J.J. Barrett, D.R. Siebert, K.V. Reddy: Rev. Sci. Instrum. **54**, 797 (1983)
7.41 M.W. Sigrist: J. Appl. Phys. **60**, R83 (1986)
7.42 V.P. Zharov, V.S. Letokhov: *Laser Optoacoustic Spectroscopy*, Springer Ser. Opt. Sci., Vol. 37 (Springer, Berlin, Heidelberg 1986)
7.43 L.B. Kreuzer, C.K.N. Patel: Science **173**, 45 (1971)
7.44 M.J. Kavaya, J.S. Margolis, M.S. Shumate: Appl. Opt. **18**, 2602 (1979)
7.45 L.G. Rosengren: Appl. Opt. **14**, 1960 (1975)
7.46 C.F. Dewey: "Design of Optoacoustic Systems" in *Optoacoustic Spectroscopy and Detection*, ed. by Y.-H. Pao (Academic, New York 1977) Chap. 3
7.47 K.P. Koch, W. Lahmann: Appl. Phys. Lett. **32**, 289 (1978)
7.48 K.H. Fung, H.-B. Lin: Appl. Opt. **25**, 749 (1986)
7.49 P.M. Morse, K.U. Ingard: *Theoretical Acoustics* (McGraw-Hill, New York 1968) pp. 270-300, 322-325, 467-492
7.50 S. Bernegger, M.W. Sigrist: Appl. Phys. B **44**, 125 (1987)
7.51 A.C. Tam: Rev. Mod. Phys. **58**, 381 (1986)

7.52 C.K.N. Patel, R.J. Kerl: Appl. Phys. Lett. **30**, 578 (1977)
7.53 D.H. Leslie, G.L. Trusty, A. Dandrige, T.G. Giallorenzi: Electron. Lett. **17**, 581 (1981)
7.54 E.L. Kerr, J.G. Atwood: Appl. Opt. **7**, 915 (1968)
7.55 L.B. Kreuzer: J. Appl. Phys. **42**, 2934 (1971)
7.56 G.A. West, D.R. Siebert, J.J. Barrett: J. Appl. Phys. **51**, 2823 (1980)
7.57 E.E. Marinero, M. Stuke: Opt. Commun. **30**, 349 (1979)
7.58 A. Di Lieto, P. Minguzzi, M. Tonelli: Appl. Phys. B **27**, 1 (1982)
7.59 A. Di Lieto, P. Minguzzi, M. Tonelli: Opt. Commun. **31**, 25 (1979)
7.60 C.M. Klimcak, J.A. Gelbwachs: Appl. Opt. **24**, 247 (1985)
7.61 S.L. Nickolaisen, S.E. Bialkowski: J. Chromatogr. **366**, 127 (1986)
7.62 V.P. Zharov: "Laser Opto-acoustic Spectroscopy in Chromatography", in *Laser Analytical Spectrochemistry*, ed. by V.S. Letokhov (Adam Hilger, Bristol 1985) Chap. 5
7.63 D. Fournier, A.C. Boccara, N.M. Amer, R. Gerlach: Appl. Phys. Lett. **37**, 519 (1972)
7.64 L.B. Kreuzer, N.P. Kenyan, C.K.N. Patel: Science **177**, 349 (1972)
7.65 C.F. Dewey, Jr., R.D. Kamm, C.E. Hackett: Appl. Phys. Lett. **23**, 633 (1973)
7.66 P.D. Goldan, K. Goto: J. Appl. Phys. **45**, 2058 (1975)
7.67 E. Max, L.-G. Rosengren: Opt. Commun. **11**, 4350 (1974)
7.68 A.S. Gomenyuk, V.P. Zharov, D.D. Ogurok et al.: Kvantovaya Elektron. **1**, 1805 (1974)
7.69 A.M. Angus, E.E. Marinero, M.J. Colles: Opt. Commun. **14**, 223 (1975)
7.70 T.F. Deaton, D.A. Depatie, T.W. Walker: Appl. Phys. Lett. **26**, 300 (1975)
7.71 W. Schnell, G. Fischer: Appl. Opt. **14**, 2058 (1975)
7.72 P.C. Claspy, Y.-H. Pao, S. Kwong, E. Nodov: Appl. Opt. **15**, 1506 (1976)
7.73 P.C. Claspy, C. Ha, Y.-H. Pao: Appl. Opt. **16**, 2972 (1977)
7.74 S. Shtrikman, M. Slatkine: Appl. Phys. Lett. **31**, 830 (1977)
7.75 R. Gerlach, N.M. Amer: Appl. Phys. Lett. **32**, 228 (1978)
7.76 E. Kritshman, S. Shtrikman, M. Slatkine: J. Opt. Soc. Am. **68**, 1257 (1978)
7.77 R.A. Grane: Appl. Opt. **17**, 2097 (1978)
7.78 E. Nodov: Appl. Opt. **17**, 1110 (1978)
7.79 N. Ioli, P. Violino, M. Meucci: J. Phys. E **12**, 168 (1979)
7.80 R.F. Adamowicz, K.P. Koo: Appl. Opt. **18**, 2938 (1979)
7.81 G.L. Loper, A.R. Calloway, M.A. Stamps, J.A. Gelbwachs: Appl. Opt. **19**, 2726 (1980)
7.82 T.H. Vansteenkiste, F.R. Faxvog, D.M. Roessler: Appl. Spectrosc. **35**, 194 (1981)
7.83 C. Chen, Y. Liu, C. Ming, C. Wang, W. Wang, J. Xu, Z. Li, L. Wang: Chin. Phys. **2**, 451 (1982)
7.84 W. Wang, C. Chen, Y. Liu: Kexue Tongbao **27**, 400 (1982)
7.85 M.H. Hubert: Laser Optoacoustic Detector, Report No. 83-715-1 (Natl. Res. Council, Canada 1983)
7.86 M.A. Leugers, G.H. Atkinson: Anal. Chem. **56**, 925 (1984)
7.87 G.L. Loper, J.A. Gelbwachs, S.M. Beck: Can. J. Phys. **64**, 1124 (1986)
7.88 P. Vujkovic Cvijin, D.A. Gilmore, M.A. Leugers, G.H. Atkinson: Anal. Chem. **59**, 300 (1987)
7.89 Z. Zelinger, Z. Papouskova, M. Jakoubkova, P. Engst: Collect. Czech. Chem. Commun. **53**, 749 (1988)
7.90 M. Boutonnat, K.A. Keilbach, D.A. Gilmore, N. Oliphant, P. Vujkovic Cvijin, G.H. Atkinson: In *Photoacoustic and Photothermal Phenomena*, ed. by P. Hess, J. Pelzl, Springer Ser. Opt. Sci., Vol. 58 (Springer, Berlin, Heidelberg 1988) pp. 137-138
7.91 C.K.N. Patel, E.G. Burkhardt, C.A. Lambert: Science **184**, 1173 (1974)
7.92 E.G. Burkhardt, C.A. Lambert, C.K.N. Patel: Science **188**, 1111 (1975)
7.93 C.K.N. Patel: Opt. Quantum Electron. **8**, 145 (1976)
7.94 P. Perlmutter, S. Shtrikman, M. Slatkine: Appl. Opt. **18**, 2267 (1979)
7.95 S. Bernegger, P.L. Meyer, C. Widmer, M.W. Sigrist: In *Photoacoustic and Photothermal Phenomena*, ed. by P. Hess, J. Pelzl, Springer Ser. Opt. Sci., Vol. 58 (Springer, Berlin, Heidelberg 1988) pp. 122-126
7.96 P.L. Meyer, S. Bernegger, M.W. Sigrist: In *Photoacoustic and Photothermal Phenomena*, ed. by P. Hess, J. Pelzl, Springer Ser. Opt. Sci., Vol. 58 (Springer, Berlin, Heidelberg 1988) pp. 127-130

7.97 D.D. Bicanic, B.F.J. Zuidberg, R. Friedhoff, F. Harren, D.P. v.d. Akker, A. Bizzarri: In *Photoacoustic and Photothermal Phenomena*, ed. by P. Hess, J. Pelzl, Springer Ser. Opt. Sci., Vol. 58 (Springer, Berlin, Heidelberg 1988) pp. 145-147
7.98 F. Harren, J. Reuss, D.D. Bicanic, E. Woltering: In *Photoacoustic and Photothermal Phenomena*, ed. by P. Hess, J. Pelzl, Springer Ser. Opt. Sci., Vol. 58 (Springer, Berlin, Heidelberg 1988) pp. 148-149
7.99 L.S. Rothman, A. Goldman, J.R. Gillis, R.R. Gamache, H.M. Pickett, R.L. Poynter, N. Husson, A. Chedin: Appl. Opt. **22**, 1616 (1983)
7.100 L.S. Rothman, R.R. Gamache, A. Barbe, A. Goldman, J.R. Gillis, L.R. Brown, R.A. Toth, J.-M. Flaud, C. Camry-Peyret: Appl. Opt. **22**, 2247 (1983)
7.101 L.S. Rothman, R.R. Gamache, A. Goldman, L.R. Brown, R.A. Toth, H.M. Picket, R.L. Poynter, J.-M. Flaud, C. Camry-Peyret, A. Barbe, N. Husson, C.P. Rinsland, M.A.H. Smith: Appl. Opt. **26**, 4058 (1987)
7.102 J. Hinderling, M.W. Sigrist, F.K. Kneubühl: Infrared Phys. **27**, 63 (1987)
7.103 H. Mehlhorn, A.R. Wellburn: Nature **327**, 417 (1987)
7.104 R. Gerlach, N.M. Amer: Appl. Phys. **23**, 319 (1980)
7.105 U. Persson, B. Marthinsson, J. Johansson, S.T. Eng: Appl. Opt. **19**, 1711 (1980)
7.106 S.M. Beck: Appl. Opt. **24**, 1761 (1985)
7.107 G. Bergmann, B. von Oepen, P. Zinn: Anal. Chem. **59**, 2522 (1987)
7.108 I.M. Campbell: *Energy and the Atmosphere* (Wiley, London 1977)
7.109 F.B. Abeles, H.E. Heggestad: J. Air Pollut. Control Assoc. **23**, 517 (1973)
7.110 F. Herlemont, M. Lyszyk, J. Lemaire: J. Mol. Spectrosc. **74**, 400 (1979)
7.111 M.W. Sigrist: In *Photoacoustic and Photothermal Phenomena*, ed. by P. Hess, J. Pelzl, Springer Ser. Opt. Sci., Vol. 58 (Springer, Berlin, Heidelberg 1988) pp. 114-121
7.112 R.A. Rooth, A.J.L. Verhage, L.W. Wouters, L. van de Beld: In Proc. 4th Int. Conf. on IR Physics (CIRP 4), Zurich, ed. by R. Kesselring, F.K. Kneubühl (ETH Zurich 1988) pp. 590-592
7.113 V.P. Zharov, Moscow High Technical School (USSR), private communication
7.114 S. Bernegger: "CO-Laser Photoacoustic Spectroscopy of Gases and Vapors for Trace Gas Analysis"; PhD Thesis, ETH Zurich (1988), to be published

8. Trace Detection in Agriculture and Biology

D.Bicanic, F. Harren, J. Reuss, E. Woltering, J. Snel, L.A.C.J. Voesenek,
B. Zuidberg, H. Jalink, F. Bijnen, C.W.P.M. Blom, H. Sauren
M. Kooijman, L. van Hove, and W. Tonk

With 31 Figures

The use of the photoacoustic effect for new applications in agricultural biology is the subject of this chapter. Section 8.1 deals with the photoacoustic detection of ethylene that is produced by specific plant species during senescence and inundation. It was jointly written by the groups of Molecular and Laser Physics (F. Harren, J. Reuss, F. Bijnen) and Experimental Botany (L. Voesenek, C. Blom) from the Catholic University, Nijmegen, and E. Woltering, associated with the Sprenger Institute in Wageningen.

Section 8.2 discusses the fluorescence and photoacoustic transients in spinach leaves and is the work of J. Snel and M. Kooijman of the Department of Plant Physiological Research, Agricultural University Wageningen. D. Bicanic, H. Jalink and H. Sauren (Department of Physics and Meteorology, Agricultural University Wageningen) and L. van Hove and W. Tonk (Department of Plant Physiological Research, Agricultural University Wageningen) provided Sect. 8.3, dealing with the potentiality of photoacoustic sensing in greenhouses, stables and storage compartments, and studies of adsorption, aerobic meat spoilage and soilless growth, respectively.

The need for experiments in the field of olfactory psyhophysics is suggested as well. Finally, the photoacoustic cell that operates on the principle of the heat pipe at elevated temperatures is described and its initial performance results evaluated.

8.1 Photoacoustic Detection of Ethylene Production in Plants

8.1.1 The Choice of Ethylene

Photoacoustic detection is especially useful for the detection of gaseous traces of C_2H_4 in atmospheric conditions. The clear fingerprint absorption spectrum of C_2H_4 at CO_2 laser wavelengths and its enhanced absorption at the $10P14$ CO_2 laser transition (30.4 atm^{-1} cm^{-1} at 949.48 cm^{-1}) [8.1] allow sensitive detection of this gas. Another advantage of the method is the instantaneous and continuous measurement of flowing gas mixtures. In addition, C_2H_4 causes no wall adsorption problems, which produce a loss of information about the gas mixture.

Ethylene is a very interesting gas from a biological point of view. In 1901, *Neljubov* [8.2] discovered that growth alterations in plants, caused by exposure to illumination gas, were due to the presence of ethylene in the gas. Thereafter, plant physiologists became interested in this gaseous plant growth regulator, and in 1934 it was discovered that plants themselves produce ethylene [8.3]. Until the 1960s, research in this area was greatly limited by the poor sensitivity of the equipment used for analysis of ethylene. After the introduction of a sensitive gas-chromatographic method, however, a true upsurge in research activities was observed [8.4].

Over the years, ethylene has been found to play a role in many developmental processes in plants (e.g. ripening of fruits; abscission of leaves, flowers and fruits; wilting of flowers; release of dormancy in certain bulbs and seeds; growth of petioles and stems [8.5–7] and its biosynthesis has been intensively studied. It was found that ethylene is produced from methionine via the intermediates *S*-adenosyl methionine (SAM) and 1-aminocyclopropane-1-carboxylic acid (ACC). The conversion of SAM to ACC is catalyzed by a pyridoxal phosphate dependent enzyme called ACC-synthase. This enzyme is believed to be the rate-limiting step in ethylene synthesis [8.5].

Many environmental stresses, such as excessive water loss, mechanical damage, lack of sufficient light and exposure to ethylene or other pollutants, have been shown to stimulate the activity of ACC-synthase, resulting in an increased production of ethylene by the plant. As ethylene is generally associated with ageing processes, its role in postharvest deterioration of agricultural and horticultural corps is of special interest. Furthermore, it has recently been found that plant-produced ethylene may increase sensitivity to ozone injury, implying that ethylene may play a part in the acid-rain controversy [8.8].

The traditional way of determining ethylene evolution in plants is by enclosing them in containers of appropriate volume for a certain period of time. Thereafter, the ethylene accumulated in the headspace is analyzed by gas chromatography. Besides ethylene, accumulation of carbon dioxide, water vapor and possibly other volatiles (e.g. odors) will also appear. Therefore, this method has a number of intrinsic disadvantages:

1. Accumulation of ethylene has been reported to affect its biosynthesis in a stimulatory or inhibitory way [8.5].
2. Accumulated carbon dioxide may have a stimulating effect on the rate of ethylene production [8.9].
3. Accumulation does not allow detailed kinetic studies.

These disadvantages can be overcome by placing the plants in a flow-through system, thereby continuously flushing away the produced gases.

The ethylene production of intact plants and ornamental products such as flowers, however, is generally very low and cannot be measured directly in the outlet air. We therefore employed a laser photoacoustic detection system in line with a flow-through system to study the change in ethylene production during senescence of carnations and some selected orchid flowers, and during flooding-induced growth of different *Rumex* species.

8.1.2 Experimental Setup

A CO_2 waveguide laser in combination with a resonant photoacoustic cell is used for sensitive detection of ethylene [8.10, 11]. The infrared CO_2 laser emits radiation at wavelengths in the 9–11 μm region at 90 laser lines and has an output power of 9 W.

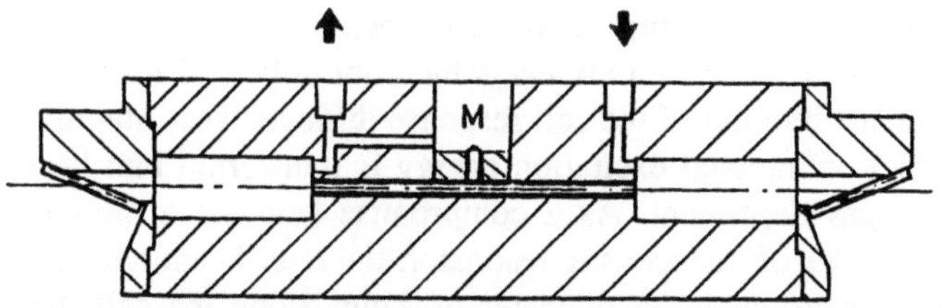

Fig. 8.1. Resonant photoacoustic cell used in the intracavity setup of Fig. 8.2. Resonator: length 100 mm, diameter 6 mm. Buffer: length 50 mm, diameter 20 mm

The photoacoustic cell with a reduced length contains a 100 mm long brass tube (6 mm diameter), which acts as a resonator excited in its first longitudinal mode [8.12] (Fig. 8.1). The ends of this resonator tube are open and coupled to buffer volumes (20 mm diameter, 50 mm length each). The length of these buffer volumes was chosen so as to minimize the acoustic background signal originating from absorption of laser power in the cell windows. A miniature electret microphone (Knowles BT 1754) is coupled to the resonator volume at the center of the tube. The first longitudinal resonance occurs at 1652 Hz with a quality factor $Q = 31.8$ and has a cell constant $F = (3.7 \pm 0.2) \times 10^3$ Pa cm/W. The cell constant F is the proportionality constant between the generated pressure amplitude p [Pa] and the absorption coefficient of the gas α [$\mathrm{atm}^{-1}\,\mathrm{cm}^{-1}$] and the laser power I_L : $p = F\alpha I_\mathrm{L}$. In this way the sensitivity of the photoacoustic cell reaches a value of $1.4 \times 10^{-8}\,\mathrm{cm}^{-1}$ with 1 W laser power and a band width of 1 Hz. Thus, a small photoacoustic cell has been designed with a volume of $35\,\mathrm{cm}^3$ to obtain a fast response of the cell for changing gas samples, allowing low flow velocities and low gas consumption.

Inserting the photoacoustic cell in the laser cavity, a detection system results which becomes more sensitive due to the high intracavity laserpower (~100 W). For the construction of the intracavity laser we used the experience gained with the construction of conventional CO_2 waveguide lasers. Figure 8.2 shows the intracavity arrangement. To prevent light scattered from the edges of the chopper blades entering the photoacoustic cell, the chopper is placed at the opposite side of the waveguide tube.

To avoid high background signal, the radius of the laser beam should be small compared to the diameter of the acoustic resonator tube. Therefore, the intracavity

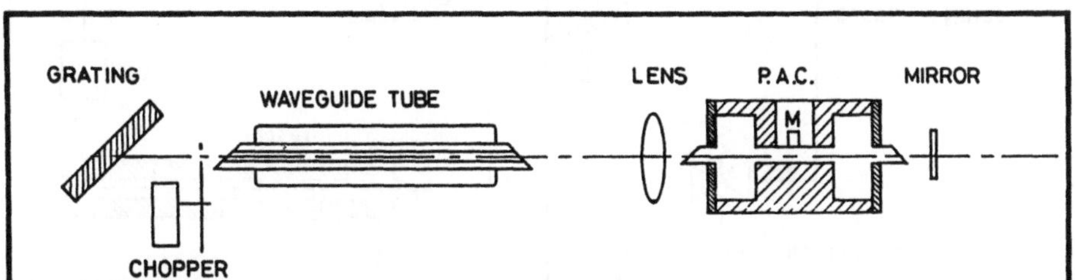

Fig 8.2. General scheme of the photoacoustic intracavity measurement with the CO_2 waveguide laser (PAC: photoacoustic cell, M: microphone)

setup is equipped with a ZnSe lens to compress the laser beam onto the reflecting mirror. Care is taken to minimize saturation effects, which not only lead to loss of sensitivity but also make the calibration of the cell response difficult. The saturation depends on the rate at which vibrational excitation energy is transferred into heat; the degree of saturation is gas dependent. As a compromise between a minimal saturation effect and the diameter of the acoustic resonator, the laser beam has been made parallel with a nearly constant waist diameter of 2 mm along the cell. For C_2H_4 on the 10*P*14 CO_2-laser transition, the microphone signal for full laser power deviates by less than 25% from the linear response due to the saturation effects.

As mentioned above, the plants were placed in a small cuvette and flushed with purified air at a flow rate of 0.9 l/h. Before this flow was admitted to the plants, the hydrocarbons present in the air were removed by interaction with a catalyst. (Platinized aluminum oxide grains at 350°C; Fig. 8.3.) After passage through the sampling cell the CO_2 and H_2O concentrations were reduced by using a KOH based scrubber. In this way interference effects of other gases with the C_2H_4 signal were negligible. During the experiment a practical detection limit of 6 ppt (parts per trillion; $1 : 10^{12}$) C_2H_4 in air is reached, comparable to a sensitivity of $1.5 \times 10^{-10}\,cm^{-1}$.

8.1.3 Ethylene Production During Senescence of Carnation and Orchid Flowers

Senescence in a number of different cut flowers (e.g. roses, carnations, *Petunia*, various orchids) is accompanied by a sudden increase in their endogenous ethylene production [8.13–15]. This increase, which is also observed during ripening

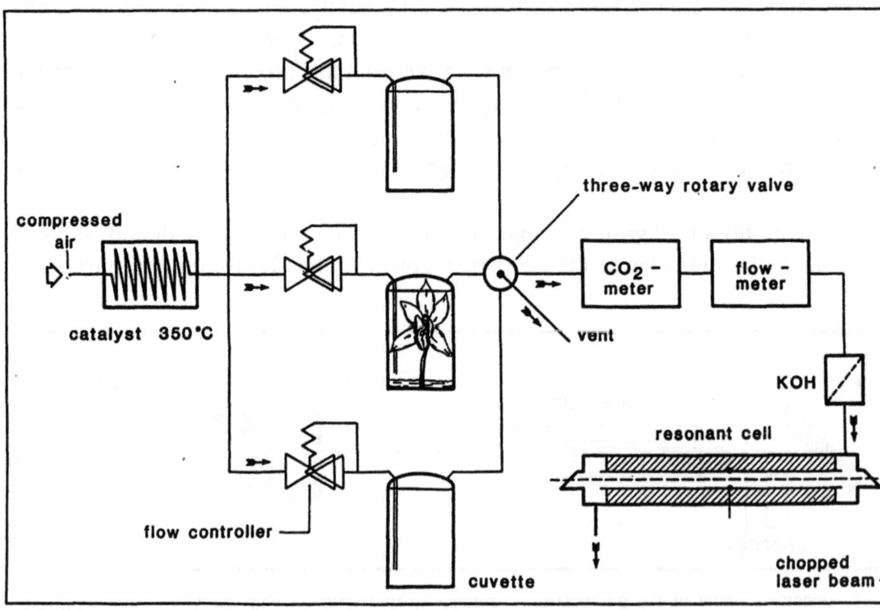

Fig. 8.3. Schematic experimental setup for measurement of ethylene production

of climacteric fruits, is thought to be the result of an ethylene-stimulated increase of the activity of enzymes involved in ethylene biosynthesis (autostimulation) [8.5]. Although presumably necessary for integration of the ripening process(es), this upsurge observed in fruits may not be the primary event. Changes in ethylene sensitivity rather than ethylene production may initiate the autostimulated ethylene production [8.16].

The pattern of ethylene production in cut carnation flowers has been intensively studied by many researchers. These flowers show a low production rate during the first 4 or 5 days of vase life, while an enormous upsurge of up to 500–1000 times the basal level is observed just before the flower shows visible signs of wilting [8.13]. Treatment of fresh flowers with low concentrations of ethylene evokes the same symptoms as observed during natural wilting [8.17], which indicates that ethylene may be the cause of wilting.

However, recently it has been postulated that the upsurge in ethylene production during wilting may not be the cause but merely the result of the wilting process, as senescence-associated changes in the cell membranes were already observed when ethylene production was still at a very low or undetectable level [8.18, 19].

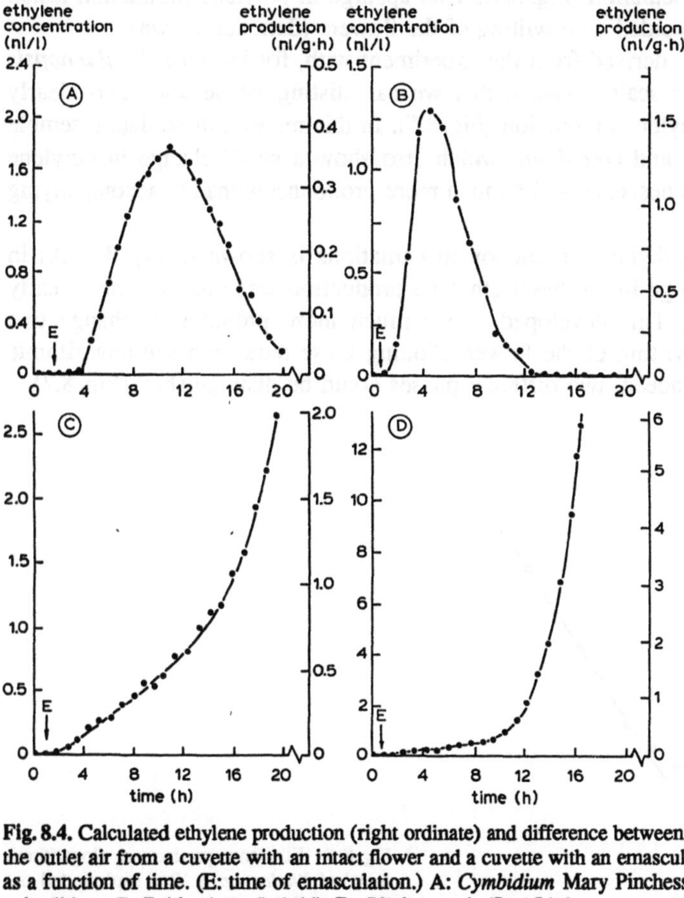

Fig. 8.4. Calculated ethylene production (right ordinate) and difference between ethylene concentration in the outlet air from a cuvette with an intact flower and a cuvette with an emasculated flower (left ordinate) as a function of time. (E: time of emasculation.) A: *Cymbidium* Mary Pinchess 'Del Rey', B: *Oncidium splendidum*, C: *Epidendrum* (hybrid), D: *Phalaenopsis* 'Red Lip'

In orchids, senescence is also accompanied by a pronounced upsurge in ethylene production. This process is often dramatically advanced by emasculation (removal of the pollinia and the anther cap). The reason for this phenomenon is not yet clear [8.15, 20].

The above-mentioned processes were studied using the laser photoacoustic set up.

The changes in ethylene production following emasculation in different orchid species were calculated from the ethylene concentration measured in the outlet air. Ethylene production of emasculated flowers was expressed as the difference between emasculated and intact flowers. The latter produced ethylene at a steady, very low rate (0.01 – 0.02nl/g h) that did not show significant fluctuations during the course of the experiment.

In *Cymbidium* and *Oncidium* flowers, a relatively small ethylene peak appeared shortly after emasculation (Fig. 8.4), followed some days later by a more pronounced increase accompanying senescence (data not shown). In *Phalaenopsis* and *Epidendrum* there was a gradual increase in ethylene production starting almost immediately after emasculation, which developed into a much more pronounced increase starting about 12 h after emasculation (Fig. 8.4). This upsurge in ethylene production lasted several days and coincided with wilting of the flowers (data not shown).

Plotting the data derived from the experiment with, for instance, *Phalaenopsis* on a semilogarithmic scale revealed that we can distinguish at least two clearly distinct phases in ethylene production (Fig. 8.5). In this respect, these data resemble those of *Cymbidium* and *Oncidium*, which also show a small change in ethylene production prior to senescence and a much more pronounced change accompanying senescence.

The pattern of ethylene production in carnations is shown in Fig. 8.6. As in orchids, a small change in the basal ethylene production was observed at an early stage of senescence. This developed into a much more pronounced change that accompanied rapid wilting of the flower. Plotting these data on a semilogarithmic scale revealed that, indeed, two different phases could be distinguished (Fig. 8.7).

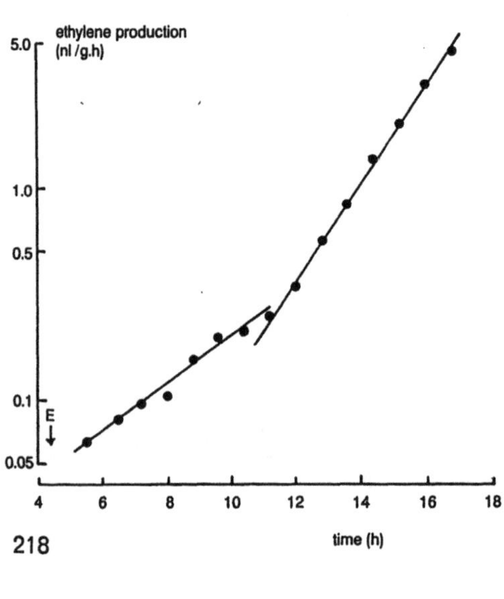

Fig. 8.5. Ethylene production (logarithmic scale) of *Phalaenopsis* 'Red Lip' plotted as a function of time (E: time of emasculation)

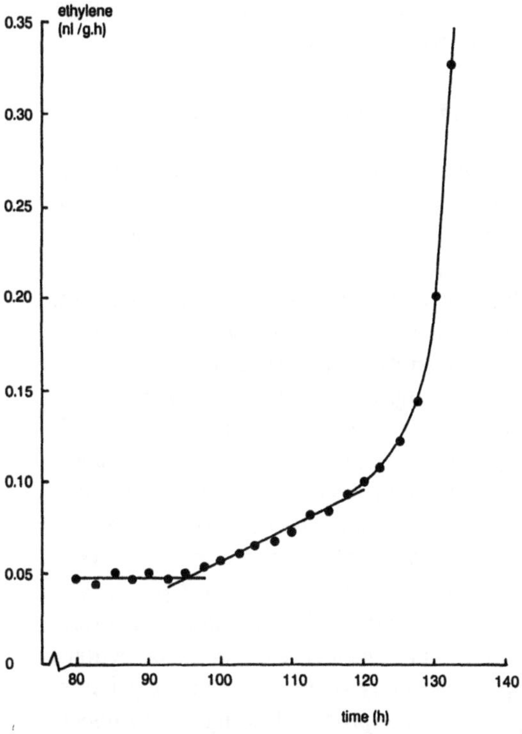

Fig. 8.6. Ethylene production of carnation 'Bagatel' plotted as a function of time after cutting

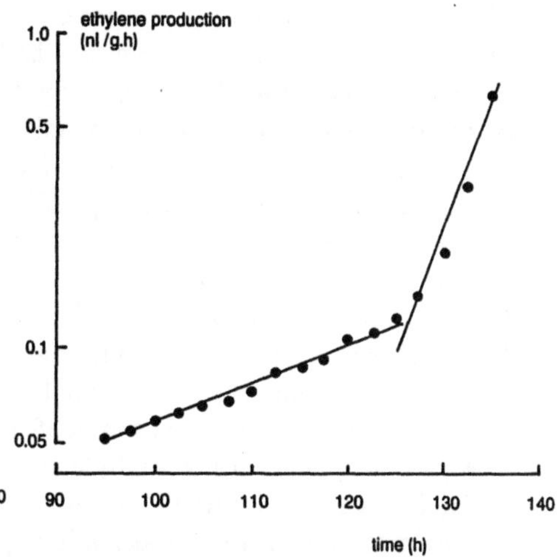

Fig. 8.7. Ethylene production (logarithmic scale) of carnation 'Bagatel' plotted as a function of time

The experiments collectively indicated that the sudden upsurge in ethylene production that accompanies wilting in carnation and orchid flowers is preceeded by a much smaller change in production. This two-phase behavior thus seems to be a general phenomenon in flower senescence and we may speculate about the physiological role.

When carnations are pretreated with amino-oxyacetic acid (AOA) (an inhibitor of ethylene production) the rise in ethylene production is consideably delayed [8.21, 22]. It has also been shown that AOA alleviates the effect of drought stress, which in untreated flowers hastens the onset of the sudden upsurge in ethylene production [8.22, 23]. This implies a role for ethylene in the initiation of senescence. Furthermore, treatment of carnations with exogenous ethylene induces a rise in ethylene production and increase in membrane permeability [8.19].

These observations all indicate a role for endogenously produced ethylene in the senescence of flowers. The suggestion that ethylene is not involved in the initiation of senescence in carnations [8.19, 24] may be invalid because this conclusion was based on the observations that membrane properties started to change before a detectable rise in endogenous ethylene production. Our data show that the correlation between ethylene production and senescence may be closer than can be derived from these results. We therefore suggest that the ethylene produced during the early phase of senescence is responsible both for the changes in cell membrane properties and for the initiation of the process(es) that lead(s) to the much more pronounced upsurge in ethylene production that accompanies rapid wilting of the flower.

8.1.4 Growth of Docks (Rumex Species) Under Flooded Conditions

Docks are common in river areas in the Netherlands. These areas are characterized by strongly changing water levels caused by irregular runoff of melting snow and rainwater. Summertime floodings greatly influence the distribution, population biology and physiological processes of plants [8.25]. It is hypothesized that resistance to flooding mainly determines the location of *Rumex* species in the river area. *Rumex crispus* is found on frequently flooded parts of the river floodplain, whereas *R. acetosa* mainly occurs on high, seldom flooded dykes and river levels.

Flooding causes chemical and biological changes in the soil, leading to a sharp decline of oxygen concentration [8.26]. Oxygen is necessary to maintain root respiration, a process essential to provide energy for growth, ion uptake and maintenance [8.27]. An important adaptation to aeration stress in flooded roots is oxygen diffusion from the atmosphere via air channels in petioles and roots to the root tips [8.28]. This process, although important for plant survival in flooded areas, can only take place when the leaves are not submerged. Flooding-resistant *Rumex* plants have developed a mechanism to ensure leaf–atmosphere contact during a flooding period. The gaseous plant hormone ethylene mediates an enhanced elongation of petioles under flooded conditions [8.29]. A comparative study of the ethylene physiology of two contrasting *Rumex* species is important in the process leading to full understanding of the ethylene–flooding interaction. We therefore studied the changes in ethylene production of *R. crispus* and *R. acetosa* under partially inundated conditions with the laser photoacoustic detection system in line with the flow-through system. After flooding, two phases in the ethylene production of *R. crispus* can be distinguished (Fig. 8.8). In the first phase, a few hours after onset of the flooding treatment

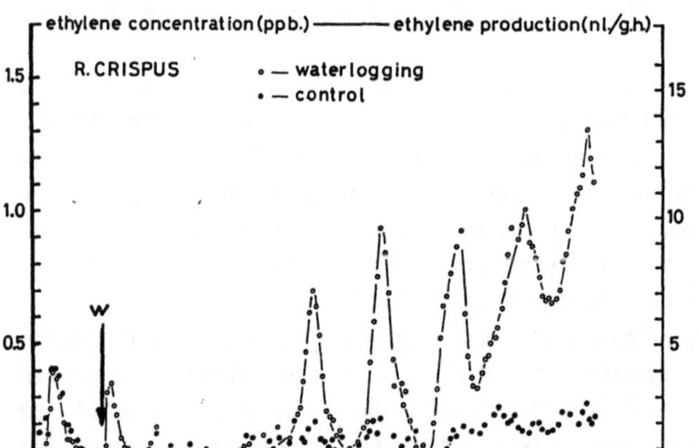

Fig. 8.8. Ethylene production of *R. crispus* after flooding (marker *w*). Two different phases can be distinguished: direct stress-induced production and a gradual increase with a circadian rhythm correlated with an elongation of the youngest perioles on a larger time scale (flow 0.9 l/h; 1 g fresh weight)

the ethylene production increases up to 4 nanoliter/(gram dry weight) · (hour). This stress-induced increase was followed by a sharp decrease of the production after 10–15 h. The second phase was characterized by a gradual increase of the ethylene production during the flooding period. In this phase a circadian rhythm in the ethylene production was observed, although temperature and light conditions were constant. At the end of the experimental period the ethylene production was 20 times the control production.

In *R. crispus* this increase in ethylene production strongly correlates with elongation of mainly the youngest petioles. It is not yet known whether the observed circadian rhythm is directly related to circadian fluctuations in the ethylene production or indirectly to fluctuations in the internal CO_2 levels of the plant, which stimulate ethylene production [8.30].

Rumex acetosa showed a different response to the onset of the flooding treatment (data not shown). Only a very restricted stress peak in ethylene production was observed. Following the treatment in time it could be concluded that in *R. acetosa* flooding causes a small, but significant, increased production of ethylene ($\approx$ 2 times control production). A circadian rhythm like that in *R. crispus* was not observed. The increased ethylene production in *R. acetosa* is correlated with a very small growth inhibition of the petioles. Exogenously applied high ethylene concentration (5–10 ppmv) caused a strong reduction of petiole growth in this species [8.29].

A flooding-resistant *Rumex* species like *R. crispus* shows a strong increase in ethylene production in response to flooding. Increased ethylene concentrations in the tissue of this species are related to elongation of the petioles, an important adaptation to overcome aeration stress in the flooded soil. A *Rumex* species not resistant to flooding (*R. acetosa*) is able to set bounds to ethylene production. We suggest that this is a mode of *R. acetosa* to avoid a strong growth inhibition caused by high ethylene concentrations. This species is not able to enhance its petiole growth to avoid aeration stress in the root region.

8.2 Comparison of Chlorophyll Fluorescence and Photoacoustic Transients in Spinach Leaves

8.2.1 Photosynthetic Energy Conversion, Chlorophyll Fluorescence and Photoacoustic Transients in Spinach Leaves

In plant photosynthesis the energy of light is used to transfer hydrogen from water to carbon dioxide, resulting in uptake of CO_2, evolution of O_2 and accumulation of sugar phosphates. The primary reactions take place in the photosystems 1 and 2, where excitation energy is transferred from excited chlorophyll pigments to the reaction center. Here this energy is used to drive vectorial electron flow. These processes are accompanied by radiative (mainly chlorophyll *a* fluorescence) and nonradiative (thermal) losses, which can be measured by respectively chlorophyll fluorescence and photoacoustic spectroscopy. In addition, photoacoustic spectroscopy can detect photosynthetic oxygen evolution. Therefore, a combination of these techniques could provide a fast and nondestructive method of tracking energy conversion in photosynthesis.

Chlorophyll fluorescence in intact plants originates mainly from chlorophyll *a* associated with photosystem 2. Several mechanisms of fluorescence quenching have been described. The major mechanisms are photochemical quenching (q_Q), related to deexcitation by photochemistry in photosystem 2, and nonphotochemical quenching (q_E), which is mainly caused by enhanced nonradiative deexcitation related to energization of the thylakoid membrane [8.31, 32]. Recent developments in instrumentation and measuring protocol have made chlorophyll fluorescence a valuable parameter in photosynthesis research. *Schreiber* et al. [8.33] were able to measure photochemical and nonphotochemical quenching of chlorophyll fluorescence in intact leaves with a pulse-amplitude-modulated fluorometer by removing photochemical quenching with a short saturating light pulse (saturating pulse method). At low oxygen pressure a good correlation has been demonstrated between electron flow estimated from carbon dioxide uptake and electron flow calculated on the basis of q_Q and/or q_E [8.34, 35].

Photoacoustic spectroscopy (PAS) has been successfully applied in the field of photosynthesis research [8.36]. In a PAS experiment the sample is placed in a sealed chamber and illuminated with modulated light. Absorption of light will lead to the generation of modulated light. Absorption of light will lead to the generation of modulated heat in the sample. The heat diffuses from the site of generation to the surface where the temperature changes produce pressure changes in the gas phase in the PA cell (photothermal component). In plants, additional pressure changes are caused by the production of molecular oxygen during photosynthesis (photobaric component). Due to the different mechanisms for the generation of the photobaric and the photothermal components, these two components have a different delay (phase angle) with respect to the modulated light. The photobaric component is damped out strongly at modulation frequencies above 120 Hz [8.37]. Therefore the PA signal at high frequency mainly consists of the photothermal component. The quantum yield of photochemistry can effectively be reduced to zero by illumination of the leaf with nonmodulated saturating light [8.37]. By measuring the photothermal signal in the presence and in the absence of nonmodulated saturating light, the photochemical loss (PL), which is a useful parameter for describing the efficiency of photosynthetic energy convesion [8.38], can be determined. The photobaric component, which is suppressed by saturating light, can then be estimated by subtracting the photothermal contribution from the total signal.

Both the separation of photochemical and nonphotochemical quenching of chlorophyll fluorescence and the separation of the photothermal and the photobaric components of the PA signal require saturating light. The short, usually 500 ms, light pulse introduced with the saturating pulse method [8.33] has the advantage that (1) only photochemical quenching is removed and (2) the pulse can be given repetitively with a rate up to 0.3 Hz. This allows the study of the kinetics of fluorescence quenching components, thermal losses and oxygen evolution. Here we describe experiments in which chlorophyll fluorescence and photoacoustic transients were measured simultaneously using the saturating pulse method.

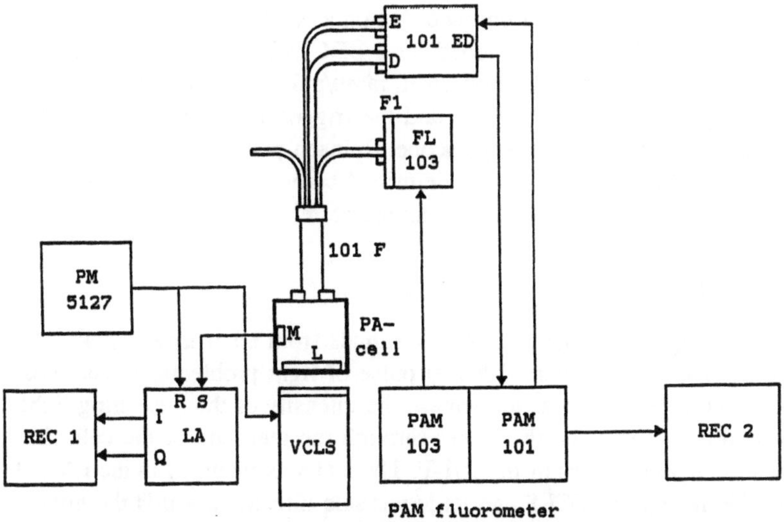

Fig. 8.9. Schematic of the experimental setup. Components: PAM 101: pulse-amplitude-modulation fluorometer (basic unit); 101 ED: emitter/detector unit; 101 F: 4-armed flexible lightguide; FL 103: high-intensity halogen light source; PAM 103: timer module for FL 103; PM-5127: function generator; VCLS: voltage controlled light source; M: microphone; L: leaf; F1: optical low-pass filter (λ_{cutoff} = 700 nm); LA: 2-phase lock-in amplifier; REC 1,2: digital oscilloscope or pen recorder

8.2.2 Experimental Setup

The experimental setup is schematically depicted in Fig. 8.9. It consists basically of the following components: a photoacoustic cell plus a two-phase lock-in amplifier, a pulsed chlorophyll fluorometer, a voltage-controlled actinic light source plus function generator, a four-armed fiber-optics lightguide for the measuring pulse, the fluorescence signal and the saturating light. The output of the integrated preamplifier in the microphone of the photoacoustic cell (Yeda Inc., Rehovot, Israel) was connected to a two-phase lock-in amplifier (EG&G/PAR model 5204). Chlorophyll *a* fluorescence yield was measured using a PAM fluorometer (Heinz Walz, Mess- und Regeltechnik, Effeltrich, FRG). The configuration consisted of the basic module PAM-101, the emitter/detector unit ED-101, the high-intensity light source FL-103 with its control module PAM-10, and the optical fiber assembly 101-F. The apparatus and the measuring principles have been described by *Schreiber* [8.39] and *Schreiber* et al. [8.33].

Modulated actinic light for the measurement of the PA signal was generated by a voltage controlled light source (VCLS), a previous version of which has been described [8.40]. This version uses twelve high-intensity LEDs (Stanley Electric Co. Ltd. type H-3k), driven by a single transconductance amplifier and coupled to a twelve-armed optical fiber (Schott, custom made). The emitted red light is centered around 660 nm with a spectral half-bandwith of approximately 25 nm. The sinusoidal input voltage for the VCLS was provided by a function generator (Philips, type PM-5127). The light intensity of the VCLS was measured at the leaf surface using a United Detector Technology 80X Optometer. The output signals of the lock-in amplifier and the PAM fluorometer were recorded on a Nicolet 2090 digital storage scope in the dual channel mode and plotted on a BBC *XY* recorder or otherwise recorded on a pen recorder. A Tektronix 2601 ramp/pulse generator (not shown) delivered the gating voltage for the VCLS and the triggering pulses for the PAM-103 module and the Nicolet 2090 storage scope. Spinach plants (Spinacia oleracea cv. Amsterdams breedblad) were grown as described before [8.41]. All experiments were carried out on detached leaves at room temperature.

8.2.3 Results and Discussion

The relatively poor signal-to-noise ratio of PA signals from plant leaves makes the detection of the response to a short (700 ms) pulse of light problematic. The most convenient solution in our case was to increase the intensity of the measuring light. Therefore we chose to illuminate the leaves through the rear side of the cell, thus avoiding losses in the fiber optics of the 101-F. The leaf was mounted in the PA cell with its upper side facing the VCLS and its lower side directed towards the microphone and the 101-F fiber optics for the measurement of chlorophyll fluorescence.

Figure 8.10 shows recordings of chlorophyll fluorescence (A) and photoacoustic signals at 35 Hz (B,C) and 370 Hz (D) of a spinach leaf after 45 min dark adaptation. The signals in Fig. 8.10A, B and C were record simultaneously, while the signal in D

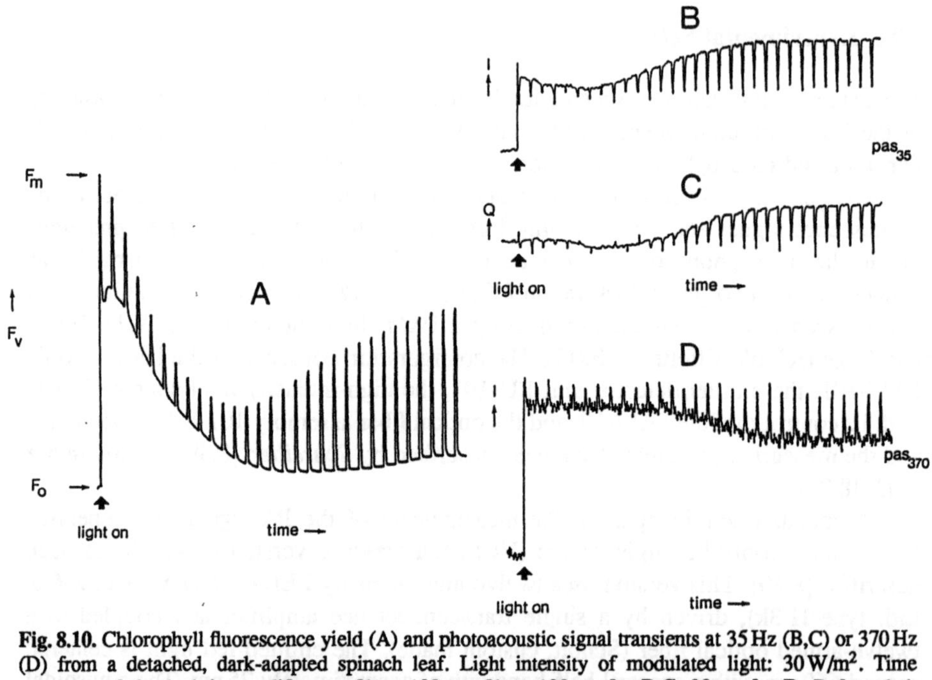

Fig. 8.10. Chlorophyll fluorescence yield (A) and photoacoustic signal transients at 35 Hz (B,C) or 370 Hz (D) from a detached, dark-adapted spinach leaf. Light intensity of modulated light: 30 W/m^2. Time constants of the lock-in amplifier: prefilter: 100 ms; filter: 100 ms for B,C, 30 ms for D. Before dark adaptation the *Q*-channel (C) was adjusted to zero during the saturating pulse by adjustment of the phase of the lock-in amplifier, while in D the phase was adjusted to maximal signal amplitude. Saturating light pulses of 700 ms duration were given every 6 s, starting at the beginning of illumination with modulated PA light. F_o: ground-state fluorescence; F_m: maximal fluorescence; F_v: variable fluorescence

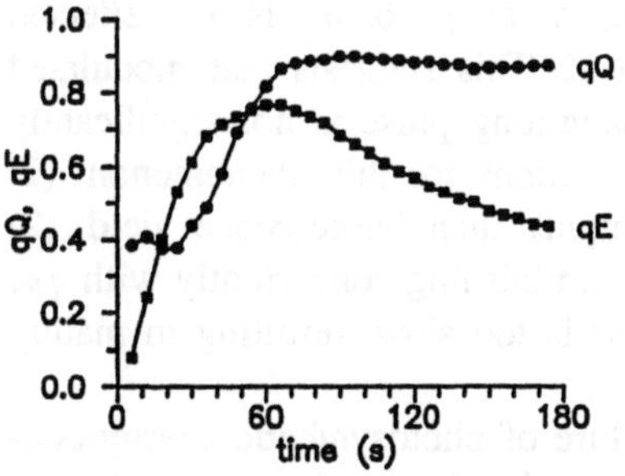

Fig. 8.11. Time course of the photochemical (q_Q) and the nonphotochemical (q_E) quenching parameters during the experimental depicted in Fig. 8.10. The parameters were calculated after [8.35]

was recorded in a separate experiment with the same leaf after a second 45 min dark adaptation period. The Q-channel contains a part of the photobaric component and the I-channel contains a mixture of the photothermal and the photobaric components. This means that Fig. 8.10C depicts the kinetics of modulated oxygen evolution with a time resolution of 200 ms, which is much faster than a conventional Clark-type oxygen electrode. The PA signals do contain some switching transients as a result of the saturating light pulses. These artefacts, which are discernible by their faster kinetics, were excluded for determination of the signal amplitude during the pulse.

Several photosynthetically relevant parameters were calculated from the recordings shown in Fig. 8.10. The fluorescence quenching parameters q_Q and q_E are given in Fig. 8.11, and Fig. 8.12 shows the calculated time courses of the photochemical loss parameter PL (A) and the total oxygen signal (B). The data show that first there is a fast decrease of q_Q, followed by a slower increase. Subsequently q_E, the photochemical loss PL and the total oxygen signal A_{ox} begin to increase, and finally q_E starts to decrease.

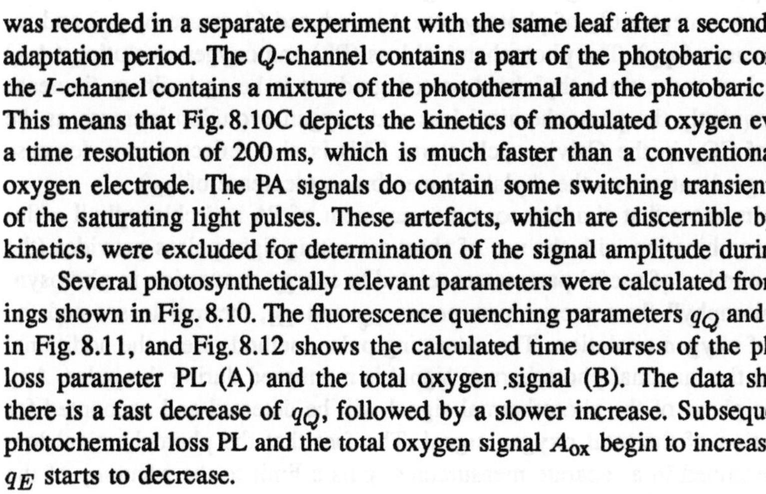

Fig. 8.12. Time course of the photochemical loss PL (A) and the total oxygen signal (B) during the experiment depicted in Fig. 8.10; calculated after [8.42]

Figure 8.10C, which contains only a photobaric component, indicates that the oxygen evolution rate is zero, or even negative, during a certain period. This is however not the case in the total oxygen signal depicted in Fig. 8.12B. This is due to the contribution of the in-phase channel (Fig. 8.10B), which still contains a significant photobaric component.

The observation that the amplitude of the PA signal at 370 Hz is more or less constant during the saturating pulse (Fig. 8.10A) while q_E increases during the induction phase (Fig. 8.11), demonstrates that a change of q_E is not reflected in a change in overall heat production in the leaf. This suggests that modulated heat production by photosystem 2 during the saturating pulse is not significantly affected by q_E. There are several possible explanations for this phenomenon: (1) the yield of heat production is already much higher than fluorescence yield, (2) the absorption cross section of photosystem 2 is diminishing concurrently with q_E, or (3) the reaction rate of the dissipating process is too slow, resulting in mainly nonmodulated heat production.

Figures 8.11 and 12 give the following picture of photosynthetic energy conversion in a spinach leaf during a dark-to-light transition from a dark-adapted state. First the primary acceptor Q_a of photosystem 2 is reduced (decrease of q_Q) and re-oxidized (increase of q_Q). The photochemical loss (PL) and oxygen evolution (A_{ox}) are relatively low and a trans-thylakoid proton gradient is being built up (increase of q_E). Subsequently the photochemical loss and oxygen evolution increase as the assimilation of CO_2 in the Calvin cycle starts. This is also reflected in a decrease of the proton gradient across the thylakoid membrane (decrease of q_E).

The results show that simultaneous measurement of PA and chlorophyll *a* fluoroscence in combination with the use of short saturating light pulses provides (the kinetics of) a number of useful parameters related to energy conversion in photosynthesis: the chlorophyll fluorescence parameters q_Q and q_E, thermal losses and the relative rate of oxygen evolution. The saturating pulse method offers the additional advantage that the maximal photothermal signal is monitored during the pulse. Any changes in the phase of the photothermal signal will be detected and accounted for in the calculation of the total oxygen signal. The fact that the photochemical loss has to be determined in a separate measurement puts a limit to the accuracy of the separation of the photothermal and photobaric components. True simultaneous measurement at high and low frequency could be achieved by using a second lock-in amplifier and modulation of the VCLS simultaneously with two frequencies.

8.3 Potentialities of Photoacoustic Sensing

8.3.1 Potential Use of Photoacoustic Sensing in Greenhouses, Stables, Fumigation Chambers, Storage Compartments and in Meteorological Studies

The need for trace detection, in general, in different fields of the agricultural sciences, is self-explanatory, and hence the range of potential applications is obviously quite large. It is practically an impossible task to cover all of them in a single chapter of this size. Because a great diversity of other reliable and sensitive trace detection techniques are presently available, until recently, the use of photoacoustic and photothermal methods in agriculture was incidental.

In this section experimental trials leading to new applications in the gaseous and liquid phases are described. Some of these studies have already been completed, while the remaining ones being carried out jointly with other institutions are either currently being performed or scheduled for the immediate future.

The survey begins with some facts pertaining to gas concentration measurements in the atmosphere and a variety of agricultural buildings. It proceeds with a description of a series of experiments aimed at proving the feasibility of photothermal techniques in studying food samples and in the field of olfactory psychophysics. Finally, trace detection of phosphate in aqueous solution in relation to the soilless growth of plants is discussed.

Nowadays, the general public is frequently subjected to nonoccupational pesticides through indoor and outdoor exposure to air, drinking water and the consumption of food. There is steadily growing concern about the long-term effects of environmental pesticides on human health, hence a need to establish models by monitoring the level and evolution of these pesticides in time. However, current methods are elaborate and intrinsically slow. The initial application of photoacoustic spectroscopy to the study of herbicides was the work of *Dorofeeva* et al. on picloram and 2,4 DB [8.43]. Recently CO_2-laser photoacoustic spectroscopy has been applied to determine the absorption cross sections of 1,3 dicloropropene (DCP) (subsoil fumigant) [8.44], diazinon horticultural pesticide and dichlorvos (public fumigant) at room temperature using the resonant longitudinal cell [8.45]. Commercially available 1,3 DCP is a 1:1 mixture of cis and trans isomers. Although the abolute cross section values are not very favorable (Fig. 8.13), the experimental data reveal significantly different fingerprints, which allows detection [at well below the MAC (maximum allowed concentration) value] of these isomers at CO_2-laser wavelengths

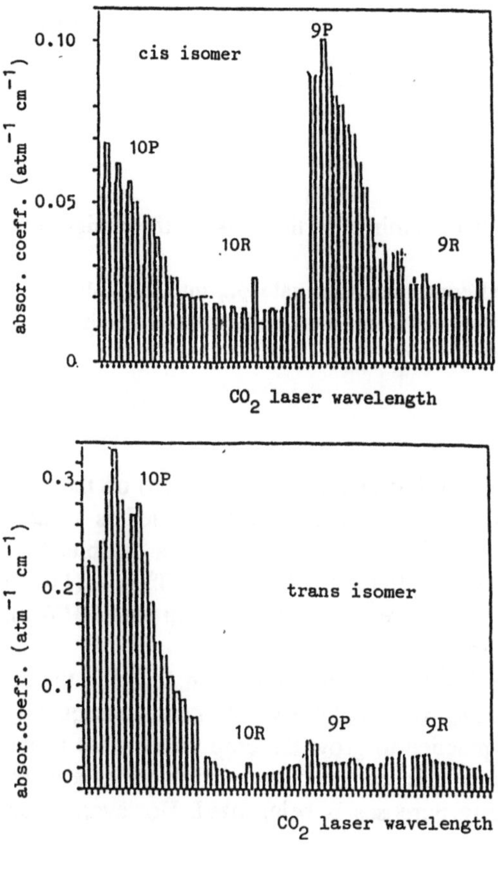

Fig. 8.13. Absorption spectra of trans isomers of 1,3 dichloropropene recorded at room temperature and a pressure of 1000 mbar in the spectral region of CO_2-laser emission [8.44]

when they are present in the mixture. In another study, malathion pesticide with large molecular weight and even lower vapor pressure (at standard temperature) has been investigated [8.46] using the heat pipe photoacoustic cell [8.47].

The indoor/outdoor exchange rate R (defined as the ratio of volume flow ϕ_v to the given volume V) of air by natural ventilation is a parameter of one of the most important physical processes taking place in greenhouses and livestock buildings because the ventilation affects the microclimate. This is of vital horticultural and economic importance during crop tillage and for optimization of yield, in particular at northern latitudes. The quantification of these processes into a dynamic model is therefore needed [8.48, 49]. The natural ventilation is difficult to determine as it shows a complicated dependence on several parameters, such as wind speed and direction, window geometry and how much it is opened, and the difference in air temperature inside and outside the greenhouse [8.48]. In practice two methods are commonly used to determine R: a decay method and the constant flow method. The decay method involves injecting a tracer gas (properties similar to that of air, concentration of tracer gas in outside air c_a) of concentration c_0 into the test area (volume V) at the time instant t_0 and subsequent monitoring of the concentration c as a function of time. Of course, monitoring takes place for selected window opening r and wind speed u. In the absence of the production term ϕ_m within V, the differential equation

$$V\left(\frac{dc}{dt}\right) = -\phi_v(c_a - c) + \phi_m \qquad (8.3.1)$$

has a solution

$$\ln\left(\frac{c_a - c_0}{c_a - c}\right) = \frac{\phi_v}{V}t \quad , \qquad (8.3.2)$$

from which the ventilation rate R can be determined as the slope of the straight line in the semilogarithmic plot.

Alternatively, tracer gas is injected continuously at rate ϕ_m into the volume V. The concentration measurement is performed when the steady state is reached, and from the above equation it follows that

$$\phi_v = \frac{\phi_m}{c - c_a} \quad . \qquad (8.3.3)$$

A typical dependence of the ventilation rate (normalized to windspeed) on the window aperture is illustrated in Fig. 8.14. Nitrous oxide is frequently used as a tracer gas. At CO_2-laser wavelengths the absorption spectrum of this gas exhibits four distinguishing lines $P(16)$, $P(7)$, $R(10)$ and $R(12)$ of the N_2O $(00^\circ 1) - (10^\circ 0)$ vibrational combination band due to a good degree of spectral overlap with $10P(42)$, $10P(32)$, $10P(16)$ and $10P(14)$ laser lines [8.50].

Carbon dioxide concentration at a level 6–7 times above that found in the atmosphere was proven to have a stimulating role on photosynthesis and consequently on the yield. It has become common practice to grow the crop under glass in the presence of CO_2-enriched air. To achieve this, carbon dioxide produced as a result of methane combustion used for heating purposes is being used. However, under

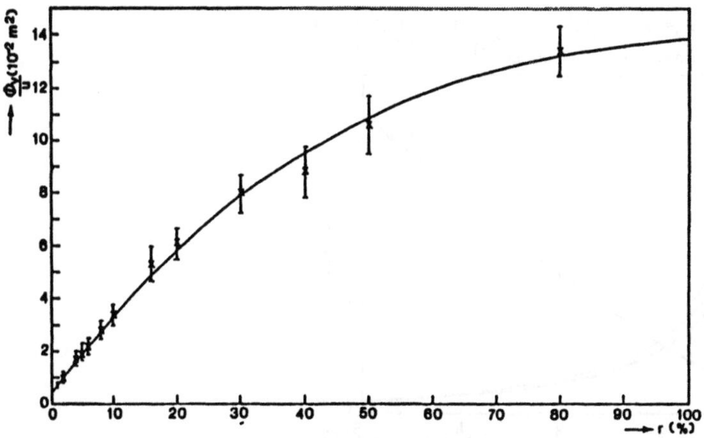

Fig. 8.14. Ventilation rate, normalized to the wind velocity, plotted as a function of the window opening for a given greenhouse [8.48]

some conditions hydrocarbons and nitrogen oxides, which are both harmful to the crop, might develop, which is of geat concern, especially to owners of well-insulated greenhouses with a low ventilation rate [8.51]. This problem is aggravated by the fact that the environmental air, in particular in highly industrial areas, contains NO_x and hydrocarbons as well. In addition, the NO originating from the soil was shown to have a negative impact on the crop growth and quality, and it can even cancel the effect obtained with CO_2-enriched gas [8.52].

A technique capable of rapid, low-concentration measurements of gases is desirable for the control of climate conditions in livestock buildings and stables. The ammonia exhaust emitted from animal slurry and manure in countries with very intensive breeding and advanced agriculture is significant because of the increased concentration level of this gas. An experiment aimed at studying the dependence of natual ventilation in livestock buildings on a variety of parameters is being prepared [8.53]. In principle, if the ventilation rate is known (see above), the production rate of ammonia ϕ_m within the stable can be determined when its concentration is measured versus time. Likewise, the effect on ammonia concentration within the stable [8.54] when the slurrly is treated with natural products could possibly be studied. Photoacoustic measurements of the ammonia emission rates resulting from fertilizers applied at various soil depths have been reported [8.55, 56], and the concentration-distance dependence established. The time profiles of ammonia concentration were used to obtain the ammonia flux emitted from the soil surface treated with the slurry applied by farm-scale spreading machinery [8.57]. When dealing with mixtures of nitrogen, water vapor and ammonia, great care has to be exercised when interpreting the data obtained during the photoacoustic measurements [8.58] due to the phase changes associated with the combined effect of kinetic cooling (CO_2 absoption) and heating (caused by deexcitation of the coupled CO_2 (001)-N_2^* level by water molecules). As an example, the calculated amplitude A and the phase of the photoacoustic signal θ obtained at $9R(28)$ and $9R(30)$ CO_2-laser transitions for a mixture containing 340 ppm cabon dioxide, 0 ppb of ammonia and 0%–5% of water

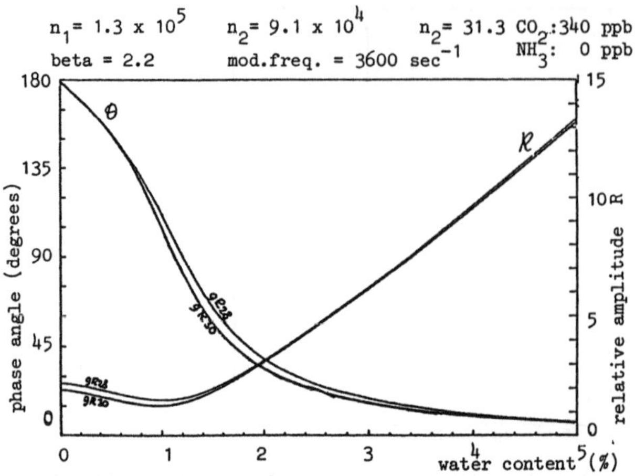

Fig. 8.15. The phase and amplitude of the photoacoustic signal for a mixture of 340 ppm carbon dioxide, 0 ppb ammonia and water vapor concentrations ranging from 0 to 5% at angular modulation frequency $3600\,s^{-1}$ and rate constants n_1, n_2 and n_3 [8.59]. (Courtesy of A. Verhage and R. Rooth, Kema N.V. Physics Laboratory, Arnhem, The Netherlands)

vapor are shown in Fig. 8.15 [8.59] at the modulated frequency $2\pi f = 3600\,s^{-1}$. The authors developed an interesting methodology by which the ammonia concentration can be computed from the measured signals and phases obtained at $9R(28)$ and $9R(30)$ transitions and knowledge of the absorption coefficients.

In order to investigate various effects (e.g. dry matter production of a specific gas, or several gases, a plant species is fumigated in closed or open top chambes, or under field conditions. Each of these has its advantages and drawbacks. For example, a closed top chamber fumigation system provides the highest degree of reproducibility, while exposing the plants under field conditions presents the most natural situation. Regardless of the fumigation type employed, accurate measurement of the concentration in the low range is the aim. The effect of sulfur dioxide concentration higher than ambient ($150\,\mu g/m^3$) on the vegetation has been studied with a fully automated field fumigation system (Fig. 8.16) [8.60]. It allows the universal insertion of a photoacoustic-based instrument as a detector.

A quantitative investigation of the uptake of ozone by a growing maize crop has been performed [8.61] by measuring diurnal and seasonal variation of these fluxes towards the maize and the soil. As standard micrometeorological and eddy correlation methods rely on a knowledge of the height-dependent profiles and accurate concentration measurements (in particular when changes are small, comparable to the accuracy of the detector) the sensitivity and the speed offered by photoacoustic monitoring might improve the reliability.

The role of ethylene during the storage of harvested agricultural products is well recognized. In harsh climate conditions, forced ventilation is economically unacceptable due to large temperature differences – hence the need for an efficient and inexpensive ethylene-selective scrubber. The oxidation of ethylene and other gaseous hydrocarbons by alkene-utilizing bacteria has been studied extensively [8.62]. In principle, monitoring the process of ethylene elimination photoacoustically, even

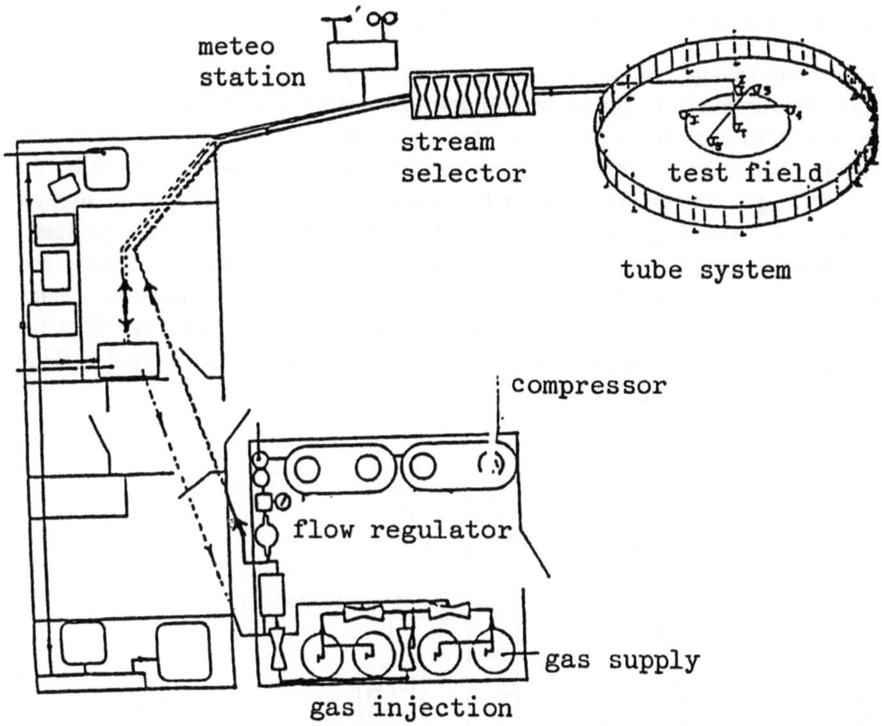

Fig. 8.16. A fully automated system for studying pollutant concentration effects on vegetation under field fumigation conditions (Courtesy of J. Mooi, Research Institute for Plant Protection, Wageningen, The Netherlands) [8.60]

at ultralow concentration levels, should not present any difficulties [8.63, 45]. The moderate absorption cross sections of vinylchloride, styrene, benzene and toluene at CO_2-laser wavelengths, suggest the use of the photoacoustic method in bacterial removal of these carcinogenic substances from industrial waste gases [8.64, 65] at low concentration levels.

Harvested potato is a living organism characterized by various intrinsic, species-dependent processes (sprouting is one of these) and the storage conditions. In general, sprouting in the post-harvest period is prevented by storage at 2° – 4°C. However, prolonged storage under such conditions leads to an increase of reducing hexose. At a later stage (e.g. frying), interaction of these sugars with aminoacids causes product quality degradation. In order to avoid this, potato treated with a mixture of propham and chloropropham sprouting inhibitors is stored at an elevated temperature (7° – 10°C). Although the toxicity of these substances to human health has not been demonstrated, a search for volatile sprouting-surpressing chemicals of plant origin that would be effective at low concentations has been carried out [8.66]. Carvone, found in caraway seed and dill seed oil, and pulegone from the *Labitae* family plant oil were shown to have such a potential. The absorpiton cross sections of these supressants in the region covered by the CO_2-laser emission are, as far as their magnitude is concerned, not comparable to those of simpler pollutant molecules and the spectrum is not vey specific either [8.67]. Nevertheless, detections sensitivity of several tens of ppbb, adequate for practical application, might readily be anticipated.

8.3.2 Leaf Chamber for Adsorption Studies

An interesting example of trace detection is the study of uptake of atmospheric ammonia by leaves and its effect on the photosynthesis and stomatal behavior [8.68, 69] as an attempt to understand better the dry deposition on vegetation. The uptake is, in analogy with Ohm's electricity law, defined as the flux (current) F driven by the difference in gas concentration Δc and restricted by the total diffusion resistance r_t of the leaf's boundary layer. Since the plant's response to the specific pollutant varies greatly with the concentration, exposure and the soil, as well as climatic and biological factors, a sophisticated design of leaf chamber (Fig. 8.17) that allows the performance of in vivo dynamic measurement under simulated and controlled but perfectly reproducible conditions (wind velocity, temperature, relative humidity) has been constructed and tested [8.70].

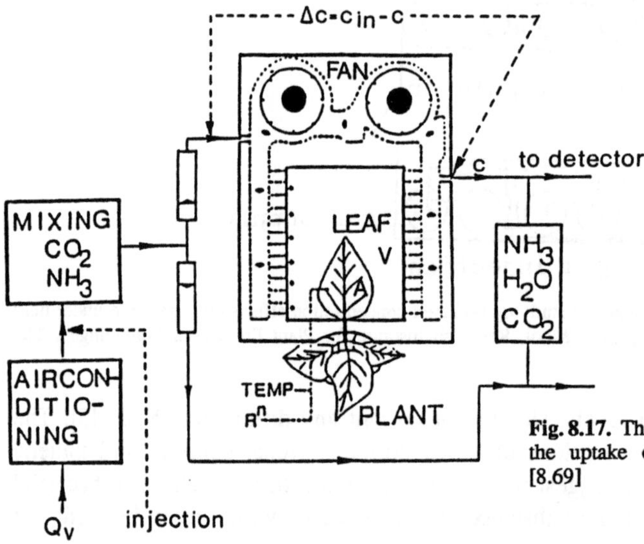

Fig. 8.17. The leaf chamber for measuring the uptake of pollutant gases by plants [8.69]

When the preconditioned gaseous mixture is caused to flow at a constant rate ϕ_V through the chamber containing the leaf (area A), flux F can be calculated from the measured difference Δc between concentrations at the inlet and outlet, since $F = \phi_V \Delta c / A$ in the steady state.

Illuminated leaves of bean and poplar plants were shown [8.68] to have considerable affinity for ammonia, as no saturation was observed even at large ammonia input concentrations (Fig. 8.18). The amount of ammonia adsorbed on the leaf surface can be determined with the same apparatus by treating the cut-off leaf with absisic acid followed by exposure to darkness, which causes the closing of stomatal openings. The concentration c_{out} measured at the outlet port of the leaf chamber is related to the c_{in} feed concentration in the preconditioned air through $c_{out} = c_{in}[1 - \exp(-kt)]$, with k being the overall time constant including the adsorption on the inner walls of the leaf chamber. By repeating the measurements, this time using the empty chamber, and subtracting the two curves, one is able to obtain information on ammonia adsorption. Proportionality between the amount of

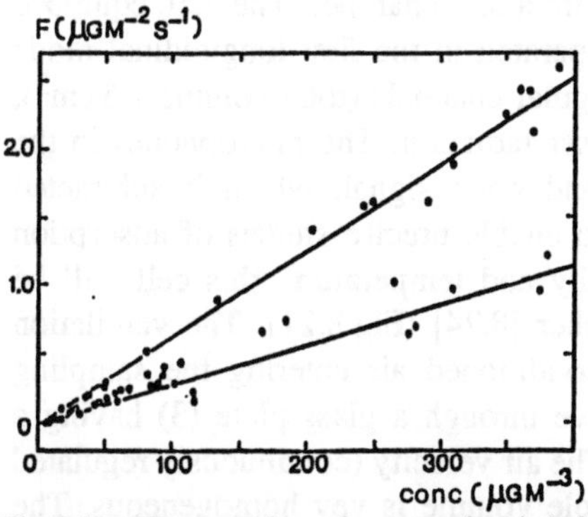

Fig. 8.18. The uptake of ammonia (relative humidity 60%, temperature 23.5° C) by leaves of *Phaseolus vulgaris L.* exposed to light of various intensities [8.68, 69]

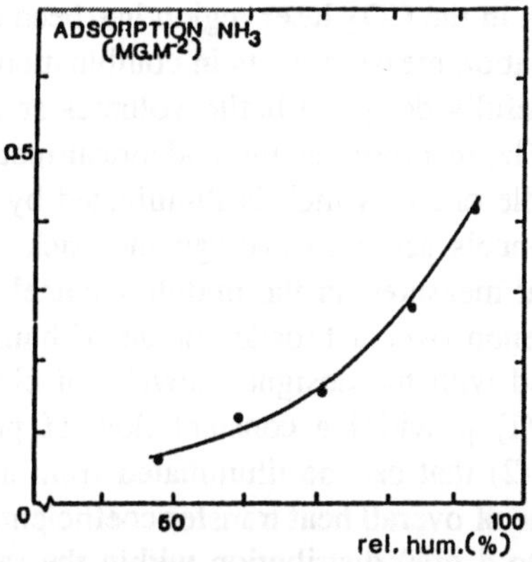

Fig. 8.19. The amount of ammonia adsorbed per unit surface of *Phaseolus vulgaris L.* leaf plotted versus the relative humidity [8.70]

adsorbed ammonia (which depends on the relative humidity) and the feed ammonia concentration (Fig. 8.19) has been observed [8.69, 70], suggesting the existence of a thin water film that contains dissolved ammonia.

The leaf chamber appears to be suitable for studying the adsorption of ammonia on different materials [8.71]. The taking up of gas by the solid surface of the photoacoustic cell resulting in bound surface layers of molecules is essential for low concentration measurements, in particular of polar gases. In general, the adsorption proper increases with decreasing surface temperature. At a given temperature, adsorption is greater for more readily condensable gases. Initial data collected with ammonia in air (15% relative humidity) indicate the lowest adsorption affinity for Teflon, followed by anodized aluminum, which agrees with the results obtained in the photoacoustic adsorption measurement [8.72]. In order to further reduce the adsorption, an additional coating treatment resulting in "closing" of the surface pores is required. A flow-through photoacoustic cell (Fig. 8.20) based on Stark tuning of

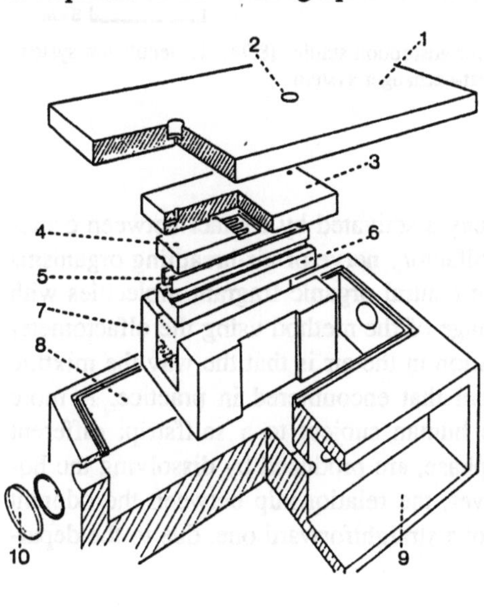

Fig. 8.20. Three-channel Stark tuned photoacoustic cell for on-stream monitoring of the ammonia concentration in ambient air [8.73]. 1: Main cover plate; 2: gas inlet; 3: Teflon plate; 4: grounded Stark plate with a microphone; 5: Teflon spacer; 6: high-voltage Stark plate; 7: Teflon block; 8: baffle volume; 9: amplifying electronics; 10: window

ammonia in the CO_2-laser region has been designed [8.73] for selective and reliable concentration measurements in combination with a leaf chamber. The cell, equipped with carefully designed baffle volumes and operated in the first longitudinal mode at 1600 Hz, incorporates three identical rectangular channels (total volume 7.5 cm^3), the middle one of which is illuminated by laser radiation. The microphones in the side channels serve to average the background noise signal, which is subtracted from that measured in the middle channel. To enable precise studies of adsorption phenomenon over a broader range of humidity and temperature, this cell will be combined with the designed novel leaf chamber [8.74] (Fig.8.21). The ventilation system (1) provides a constant flow of preconditioned air entering the sampling volume (2) that can be illuminated from above through a glass plate (3) having a low value of overall heat transfer coefficient. The air velocity (continuously regulated from 0 to 4 m/s) distribution within the sample volume is vey homogeneous. The specimen to be studied is admitted through the opening 4. The precise control of the selected conditions (temperature, relative humidity, wind speed) in the sampling volume is achieved by means of the heat-exchanger system (5).

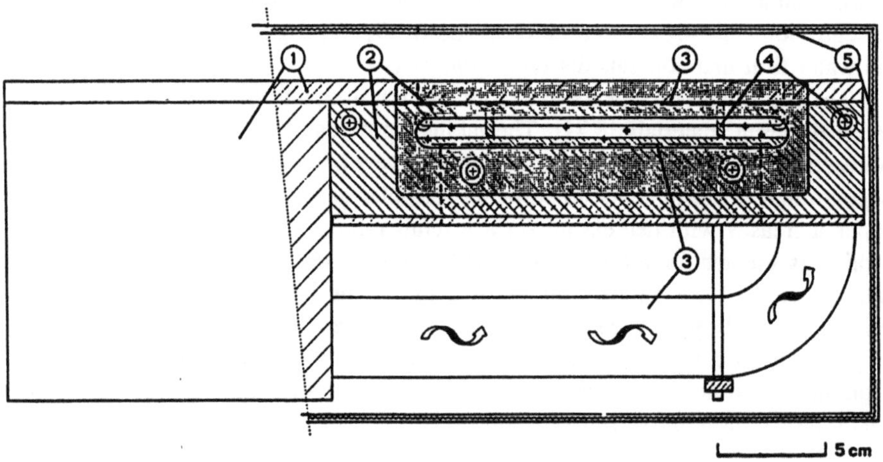

Fig. 8.21. The novel design of the chamber suitable for adsorption studies [8.74]. 1: Ventilation system; 2: sampling volume; 3: glass plate; 4: opening; 5: heat-exchanger system

8.3.3 Olfactory Psychophysics

The sensations of and taste are chemical senses activated by contact between osmogenic molecules and geminal nerves. The olfactory nerve of air-breathing organisms can detect the stimuli generated by volatile neutral organic fragrant molecules with molecular weight up to 400. The disadvantage of the method using the olfactometer for precise measurement of odor concentration in the air is that the way the mixture is delivered to the nose cone deviates from that encountered in practice. A more commonly used method is to expose the human subject to a sniffstrip; different odorant concentrations, now in the liquid phase, are produced by dissolving the liquid odorant in an odorless solvent. However, the relationship between the odorant concentration in liquid and gas phases is not a straightforward one, due to the depar-

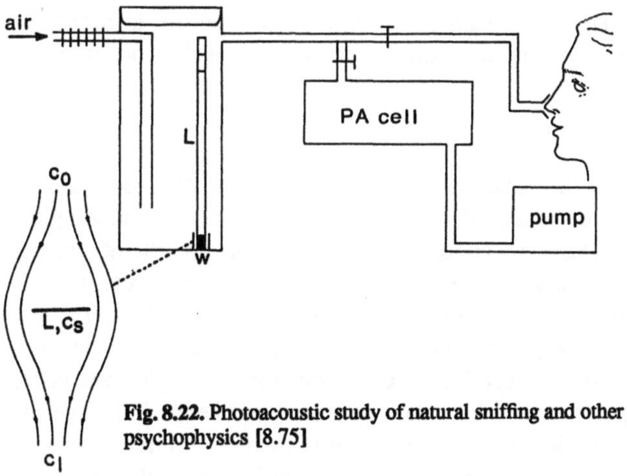

Fig. 8.22. Photoacoustic study of natural sniffing and other parameters of interest in olfactory psychophysics [8.75]

ture from Raoult's law. A psychophysical experiment that enables the measurement of odorant evaporating (rate ϕ_m) from the sniffstrip (length L, width W) in the air flow ϕ_v, thereby simulating sniffing under natural conditions (ϕ_v = 0.5 l/s for 0.4 s) (Fig 8.22) has been suggested [8.75] since

$$\frac{\phi_m}{\phi_v} = (c_l - c_0) = LW\frac{c_s k}{\phi_v} \qquad (8.3.4)$$

with c_s, c_l and c_0 indicating the saturated vapour concentration at the sniffstrip surface, the concentration in the air and the initial concentration of the odorant in the air, respectively. The parameter k is the representative heat transfer coefficient, the value of which can be found from the relationship between three dimensionless numbers: Sherwood Sh = kL/D, Reynolds Re = vL/ν and Schmitt Sc = ν/D, i.e. Sh = f(Re, Sc) = $\gamma \mathrm{Re}^n \mathrm{Sc}^m$, ($v$ and ν are the velocity and viscosity of the air, D is the odorant's diffusion coefficient) while ϕ, n and m are numerical constants. The biophysics of odor perception, in particular of the intensity of odor and malodor, is another attractive field of research where photothermal techniques could prove useful due to their high sensitivity and ease of sampling. Determination of Weber's ratio $W = dc/c$ [the ratio of minimum detectable concentration difference dc of one and the same fragrant (initial concentration c) presented successively to the test person] and the psychophysical olfactory response curve (perceived odor intensity I versus the concentration of the odorant c) are examples of applications [8.76]. The latter data can be used to determine the odorant constants ξ and η that appear in the general *Steven*'s law [8.77]

$$\log I = \xi \log c + \log \eta \quad . \qquad (8.3.5)$$

Once the psychophysical power function is known it can be used to practical advantage in measuring the abatement of odor pollution, such as malodors found in spoiled fish, air pollution and in manure. Other applications involve the measurement of human olfactory thresholds (the lowest concentration that can be discriminated with certainty from pure air), studies of odor fatigue or olfactory adaptations (resulting from long-term exposure to moderate, or short-time exposure to excessive concentration) and the study of the simultaneous presence of more than one odorant.

8.3.4 Aerobic Meat Spoilage

As for almost any fresh food, the appearance, odor and taste of meat deteriorates due to spoilage when it is stored in air. This is due to the complex microbiological activity of the mixed flora. Different conditions stimulate the growth of various microorganisms [8.78]. When meat is stored at refrigerating temperatures, bacteria grow (only on the surface) on account of a primary substrate such as glucose. Upon depletion of glucose, bacteria begin to metabolize secondary substrates etc. The volatile metabolites released during spoilage are responsible for unpleasant odors and the organoleptic sensation perceived by the consumer, and hence influence the acceptability of the product. Beside the sensoric criteria, there is a time-consuming microbiological method based on determination of the bacterial count (the logarithm of the total bacterial count per unit surface) by which the degree of spoilage can be estimated. A typical time variation of bacterial count for a beef sample inoculated with *Pseudomonas* and *Brochotrix* bacteria and stored at 5° C is shown in Fig. 8.23 [8.79]. The off odors become perceivable when bacterial counts reach values close to 7. At bacterial counts of 8 and 9, slime production and proteolysis develop, causing the emission of ammonia and sulfur compounds.

The time course of volatile compound formation during the aerobic, refrigerated storage of beef and pork contaminated naturally and by pure cultures has been investigated [8.80–82] using gas chromatography and mass spectroscopy. The main object in these experiments was to find out whether or not various volatile com-

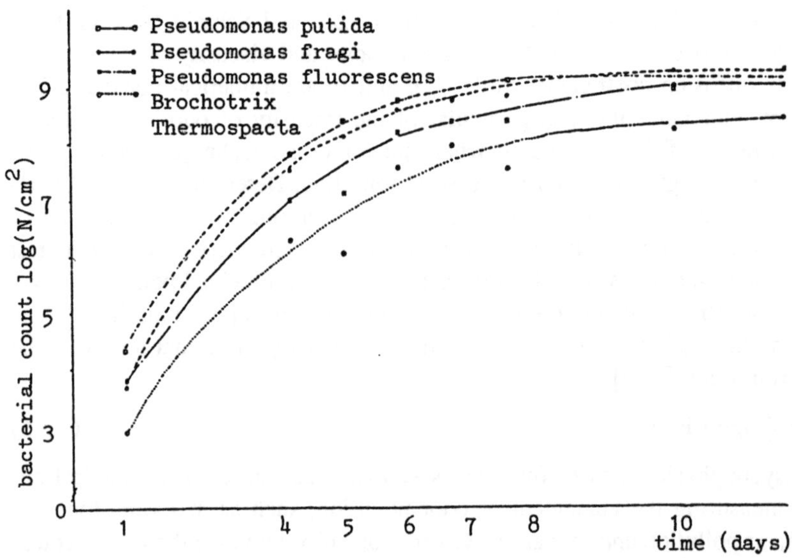

Fig. 8.23. Variation of microbial count with time for beef samples inoculated with *Pseudomonas* and *Brochotrix* and stored at 5.5° C [8.79]

pounds associated with the chemical changes due to the microbiological activity could be used as indicators of product acceptability and shelf-life [8.80]. Pertaining to the odors, alterations in sensory perception ranging from "fresh", "normal" and "meaty" (in the early stage) to "cheesy", "fatty", "sweet" and "putrid" (in this order as spoilage continues) have been found. Although more than 1000 compounds were found in the head space, detailed analysis indicated (depending on the meat sort and the bacterial flora) the presence of a well-defined time sequence in the appearance of some volatiles. For example toluene, 1-undecene, two dimethylbenzene isomers and acetoin (an aliphatic ketone) were found in the very early stages of the aerobic spoilage (under refrigerating conditions) of naturally contaminated beef [8.80], followed by ethyl-esters (major component ethylacetate) of short chain (C_2–C_6) fatty acids in the period thereafter. If the relationship between the microbial count and the concentration of a specific volatile could be established, one could, at least in principle (knowing the dependence of microbial count on time), provide a faster method to establish the spoilage stage.

To test the feasibility of a CO_2-laser photoacoustic method for early spoilage detection, the absorption cross sections of acetoin and ethylacetate (Fig. 8.24) were

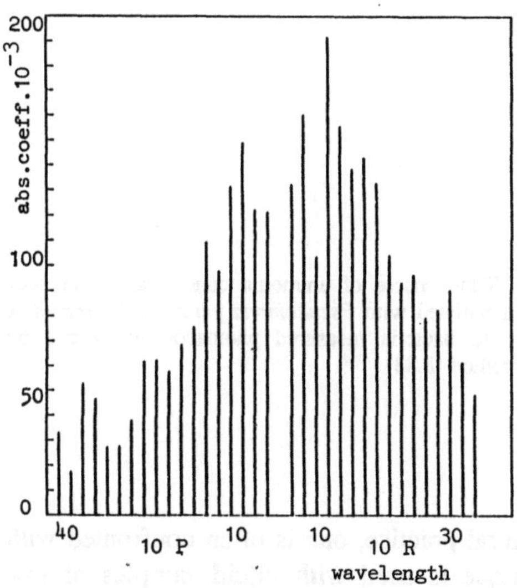

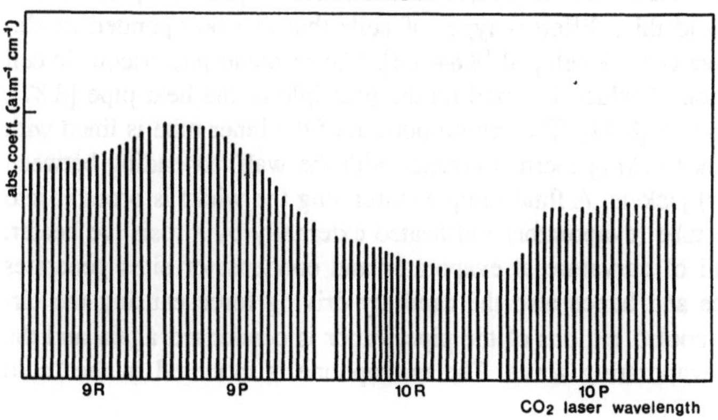

Fig. 8.24. The CO_2-laser absorption coefficient of (a) acetoin in the 10 μm band [8.83] and (b) ethylacetate [8.10], both at 291 K and atmospheric pressure

measured [8.83, 10]. Both exhibit a congested, "grasslike" spectrum appearance – the maximum value (0.19 atm^{-1} cm^{-1}) for acetoin in the $10R$ band is 50 times lower than the peak value (approximately 10 atm^{-1}) of ethylacetate in the $9P$ band. Since the absorption cross section of toluene is not significant either, interpretation of data obtained with realistic samples is not easy. However, during the study on beef samples inoculated with *Pseudomonas* and *Brochotrix* bacteria, possible evidence for the early presence of acetoin at well below ppmv level in both samples has been collected. The time formation of ammonia (Fig. 8.25) showing a strong gradient (probably associated with depletion of glucose and the onset of amino acid metabolism) could readily be monitored with the CO_2-laser system [8.83].

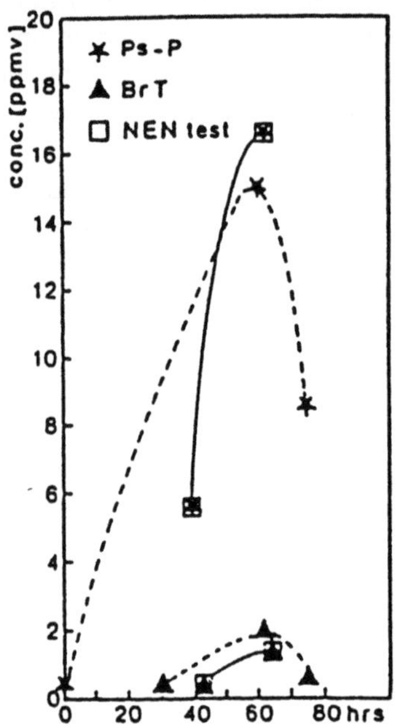

Fig. 8.25. Time course of ammonia concentration for beef samples inoculated with *Pseudomona putida* and *Brochotrix thermospacta* bacteria measured photoacoustically and by another method [8.83]

8.3.5 Heat Pipe Cell

In various situations met with in agricultural practice, one is often confronted with problems when the photoacoustic technique is used with liquid samples of low vapor pressure. To avoid this, different types of cells that can be operated at elevated temperatures have been developed [8.84–86]. The resonant photoacoustic cell (Fig. 8.26), the operation of which is based on the principle of the heat pipe [8.87], has recently been proposed [8.47]. The central portion of the inner tube is lined with a fine metal mesh wick (making thermal contact with the wall) extending between the two coolant (water) jackets. A fluid sample (saturating the wick) is brought into the central part of the tube (evaporator) and heated externally by a resistive heater. The difference in radii of curvature at evaporator and condensator sites generates the pressure difference and hence also the capillary driving force causing the circulation. In the water-cooled regions of the pipe, vapor is condensed again and the liquid returns to the evaporating zones. The microphone is mounted in the "cold

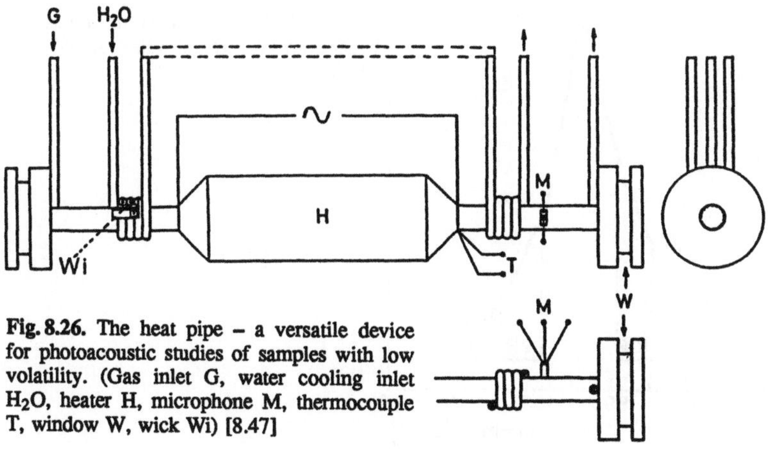

Fig. 8.26. The heat pipe – a versatile device for photoacoustic studies of samples with low volatility. (Gas inlet G, water cooling inlet H_2O, heater H, microphone M, thermocouple T, window W, wick Wi) [8.47]

section" close to the tube and terminated by a window (the maximum pressure amplitude for a closed pipe is at the tube's end). The column containing the sample vapor (at any preselected elevated temperature, provided, of course, the choice of cell material permits it) of uniform density [8.88], but separated from the baffle volume, is produced in this way. The cell windows remain clean and the cell can be easily disassembled. The suitability of the photoacoustic heat pipe for CO_2-laser spectroscopic studies has been demonstrated (Fig. 8.27) by initial studies on geraniol, a fragrant substance, used as a test sample. The characteristic spectral features obtained by Fourier transform infrared spectroscopy (Fig. 8.28) are found in the photoacoustic spectrum as well [8.89]. Additional improvement in the cell performance might be possible by optimizing the effects of various physical properties on the free convection heat transfer [8.90, 91].

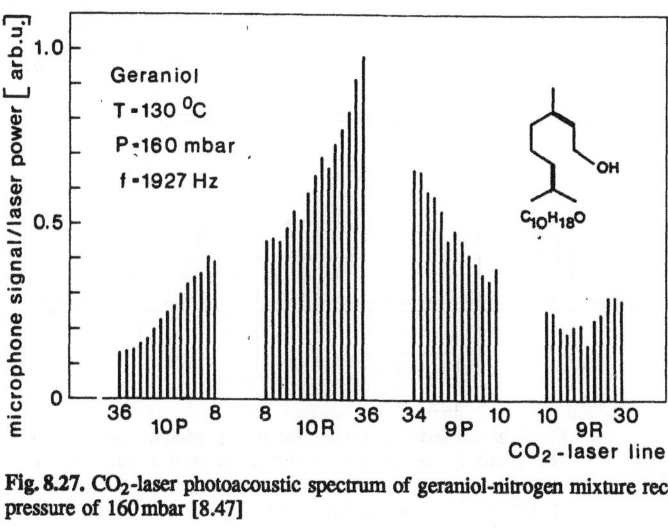

Fig. 8.27. CO_2-laser photoacoustic spectrum of geraniol-nitrogen mixture recorded at 130° C and a total pressure of 160 mbar [8.47]

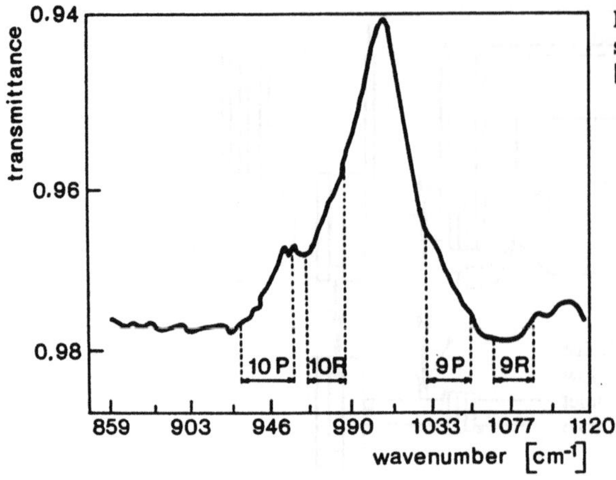

Fig. 8.28. Fourier transform infrared spectrum of geraniol vapor at 60° C [8.89]

8.3.6 Soilless Growth

A steadily increasing number of glasshouse growers in horticulture prefer soilless growth to that in conventional soil. The main advantages are the elimination of the need for expensive soil sterilization and the possibility of improved control of nutrients in water [8.92], providing better yield. Essentially, growing on a substrate requires the transport of the mixture of nutrients (well-controlled composition) in the irrigation water from the central reservoir to the plant. In order to avoid additional pollution of the environmental surface water, the early open drain design has recently been replaced by a closed-loop system (Fig. 8.29) that involves the recirculation and

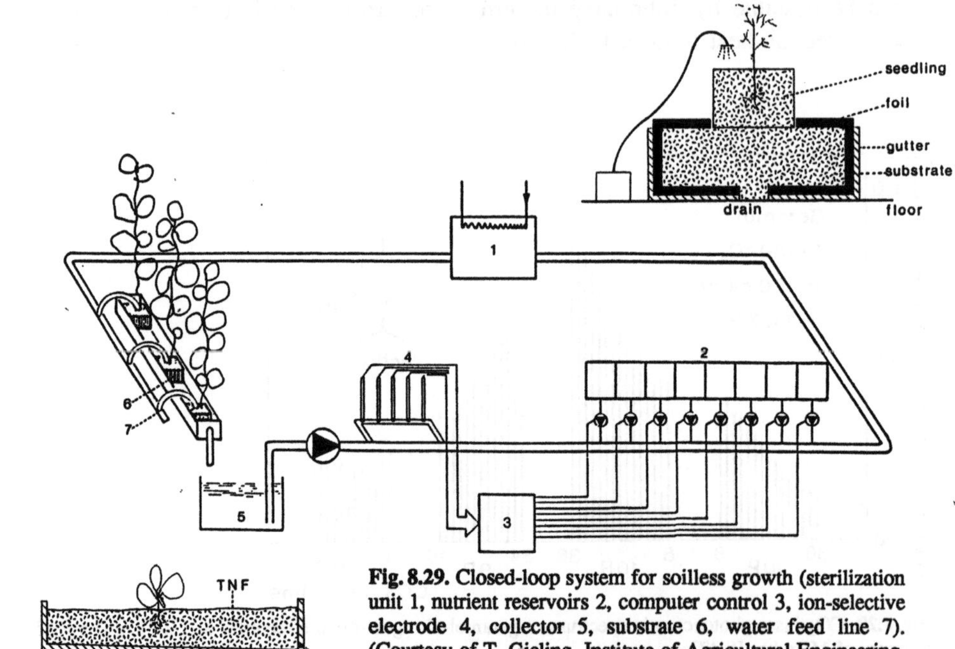

Fig. 8.29. Closed-loop system for soilless growth (sterilization unit 1, nutrient reservoirs 2, computer control 3, ion-selective electrode 4, collector 5, substrate 6, water feed line 7). (Courtesy of T. Gieling, Institute of Agricultural Engineering, IMAG, Wageningen, The Netherlands)

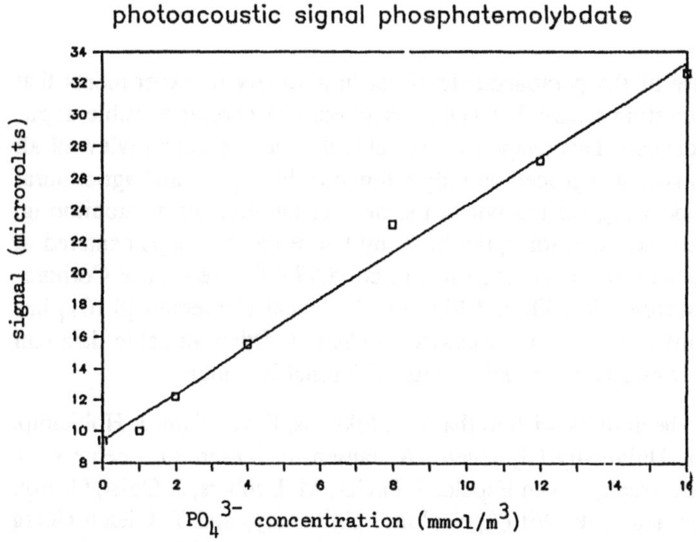

Fig. 8.30. The lock-in signal versus the concentration of orthophosphate in water obtained with 8 mW power of a HeNe laser modulated at 30 Hz [8.93]

sterilization of the surplus water. Due to the selective nutrient uptake by the plant, computer control based on pH factor and electrical conductivity measurements is not sufficient, and hence additional, selective concentration data for each ion are needed. The prospects of photoacoustic spectroscopy as a sensor for detection of orthophosphate in water has recently been evaluated [8.93] in the laboratory by the combined use of a low power HeNe laser, PZT (lead zirconate titanate) and the specific colorimetric reaction [8.94], yielding 96 ppbv detection sensitivity (Fig. 8.30) with a relatively simple cell design as shown in Fig. 8.31. Quite recently, a dual-beam thermal lensing spectrometer [with inexpensive HeNe and (0.8 μm) diode lasers] was shown capable of detecting 0.1 mmol/m^3 ($\sim$ 10 ppbv) of orthophosphate [8.95] and 0.5 mmol/m^3 of ammonia [8.96] dissolved in water.

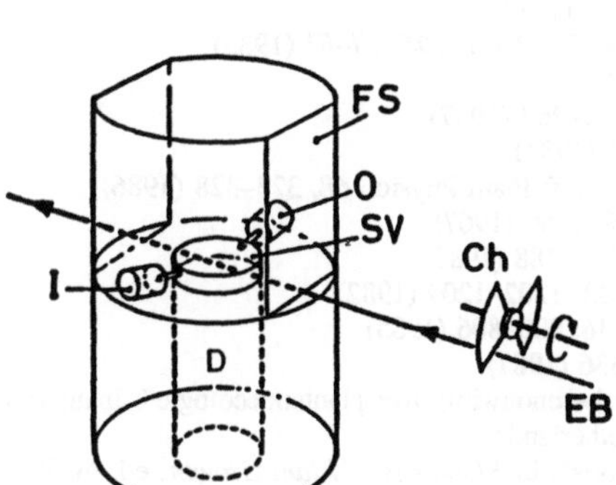

Fig. 8.31. The photoacoustic cell for detection of orthophosphate in aqueous solution (excitation beam EB, chopper Ch, sample volume SV, liquid inlet I, liquid outlet O, detector D, polished front face FS). (Courtesy of E. Strauss, University of Oldenburg, FRG)

8.4 Conclusion

In this chapter the use of the photoacoustic effect in a variety of experiments that produced several interesting results has been described, and several possible applications have been indicated. This chapter is undoubtedly not a detailed review of all research work – emphasis was placed on only a few new biological and agricultural applications. The work being carried out in the field of environmental monitoring, which is of great relevance also for agriculture, by the research groups centered at the Swiss Federal Institute of Technology in Zurich [8.97–99], Aerospace Corporation, El Segundo, California [8.100] and Moscow Technical University [8.101] has been deliberately omitted, since it is discussed in Chap. 7. Other valuable data can be found in books, reviews and a widely scattered journal literature.

Acknowledgements: The authors wish to thank C. Sikkens, F. van Rijn, J. Holtkamp, L. Hendriks (Catholic University Nijmegen), A. Jansen, P. Jansen, G. Leenters, D. van der Akker, K. van Asselt, P. van Espelo, J. Favier, M. Lubbers, J. Quist, G. Bot, T. de Jong, J. Frijters and J. de Wit (Agricultural University) and T. Gielen (Imag Institute Wageningen) for their contributions made in the course of the experiments. Credit is due to Mrs. Josie Zeevat-van Homelen, who took care of the preparations for publication of this manuscript.

References

8.1 R.J. Brewer, C.W. Bruce, J.L. Mater: Appl. Opt. **21**, 4092–4100 (1982)
8.2 D. Neljubov: Beitr. Bot. Zentralbl. **10**, 128–138 (1901)
8.3 R.T. Gane: Nature **134**, 1008 (1934)
8.4 R.E. Young, H.K. Pratt, J.B. Biale: Anal. Chem. **24**, 551–555 (1952)
8.5 S.F. Yang, N.E. Hoffmann: Annu. Rev. Plant Physiol. **35**, 155–189 (1984)
8.6 F.B. Abeles: *Ethylene in Plant Biology* (Academic, London 1973) pp. 302ff.
8.7 M.B. Jackson: "Ethylene and the Responses of Plants to Excess Water in Their Environment - A Review" in *Ethylene and Plant Development*, ed. by J.A. Roberts, G.A. Tucker (Butterworths, London 1985)
8.8 H. Mehlhorn, A.R. Wellburn: Nature **327**, 417–418 (1987)
8.9 C.H. Kao, S.F. Yang: Planta **155**, 261–266 (1982)
8.10 F. Harren, C. Sikkens, D. Bicanic: Proc. SPIE **701**, 525–531 (1987)
8.11 E.J. Woltering, F. Harren, H.A.M. Boerrigter: Plant Physiol. **88**, 506–510 (1988)
8.12 G. Scherer, K.K. Lehmann, W. Klemperer: J. Chem. Phys. **81**, 5319–5325 (1984)
8.13 R. Nichols: J. Hortic. Sci. **41**, 279–290 (1966)
8.14 J.D. Faragher, S. Mayak: J. Exp. Bot. **35**, 965–974 (1984)
8.15 C.J. Goh, A.H. Halevy, R. Engel, A.M. Kofranek: Sci. Hortic. **26**, 57–67 (1985)
8.16 M. Knee: J. Plant Growth Reg. **7**, 111–119 (1988)
8.17 E.J. Woltering, H. Harkema: Acta Hortic. **216**, 255–262 (1987)
8.18 V. Trippi, A. Paulin: Physiol. Plant. **60**, 221–226 (1984)
8.19 J.M.O. Eze, S. Mayak, J.E. Thomson, E.B. Dumbroff: Plant Physiol. **68**, 323–328 (1986)
8.20 S.P. Burg, M.J. Dijkman: Plant Physiol. **42**, 1648–1650 (1967)
8.21 E.J. Woltering, E.P. Sterling: Acta Hortic. **181**, 483–488 (1986)
8.22 A. Borochov, S. Mayak, R. Broun: J. Exp. Bot. **33**, 1202–1209 (1982)
8.23 S. Mayak, A. Borochov, T. Tirosh: J. Exp. Bot. **36**, 800–806 (1985)
8.24 I. Sylvestre, A. Paulin: Physiol. Plant. **70**, 530–536 (1987)
8.25 C.W.P.M. Blom: "Waarnemen en verklaren, een beschouwing over plantenoecologie", Inaugural address, Brakkenstein, Nijmegen (1985), The Netherlands
8.26 F.N. Ponnamperuma: "Effects of Flooding on Soils", in *Floodings of Plant Growth*, ed. by T.T. Kozlowski (Academic, London 1984) pp. 10–45

8.27 J.T. Lambers: "Energy metabolism in higher plants in different environments", Ph.D. Thesis, University of Groningen, The Netherlands (1979)
8.28 M.B. Jackson, M.C. Drew: "Effects of Floodings on Growth and Metabolism of Herbaceous Plants", in *Floodings and Plant Growth*, by T.T. Kozlowski (Academic, London 1984) pp.47–128
8.29 L.A.C.J. Voesenek, C.W.P.M. Blom: Plant, cell and environment (in press)
8.30 R.F. Horton: "Carbon Dioxide Flux and Ethylene Productions in Leaves", in *Ethylene and Plant Development*, ed. by J.A. Roberts, G.A. Tucker (Butterworths, London 1985) pp.37–46
8.31 G. Krause, E. Weis: Photosynth. Res. **5**, 139–157 (1984)
8.32 J.-M. Briantais, C. Venotte, G.H. Krause, E. Weis: "Chlorophyll *a* Fluorescence of Higher Plants: Chloroplasts and Leaves", in *Light Emission by Plants and Bacteria*, ed. by Govindjee, J. Amesz, D.C. Fork (Academic, New York 1986) pp.539–583
8.33 U. Schreiber, U. Schliwa, W. Bilger: Photosynth. Res. **10**, 51-62 (1986)
8.34 E. Weis, J.A. Berry: Biochim. Biophys. Acta **894**, 198–208 (1987)
8.35 U. Schreiber, W. Bilger: "Rapid Assessment of Stress Effects on Plant Leaves by Chlorophyll Fluorescence Measurements", in *Plant Response to Stress*, ed. by J.D. Tenhunen, F.M. Catarino, O.L. Lange, W.C. Oechel, NATO ASI Ser., Ser. G (Springer, Berlin, Heidelberg 1987) pp.27–53
8.36 S.E. Braslavsky: Photochem. Photobiol. **43** , 667–675 (1986)
8.37 G. Bults, B.A. Horwitz, S. Malkin, D.Cahen: Biochim. Biophys. Acta **679**, 452–465 (1982)
8.38 S. Malkin, D. Cahen: Photochem. Photobiol. **29**, 803–813 (1979)
8.39 U. Schreiber: Photosynth. Res. **9**, 261–272 (1986)
8.40 J.F.H. Snel: "Measurement and Analysis of Chlorophyll Fluorescence in Plants using Modulated Actinic Illumination", in *Applications of Chlorophyll Fluorescence*, ed. by H.K. Lichtenthaler (Kluwer, Dordrecht 1988) pp.93–98
8.41 J.F.H. Snel: "Regulation of Photosynthetic Electron Flow in Isolated Chloroplasts by Bicarbonate, Formate and Herbicides"; Ph. D. Thesis, Agricultural University Wageningen (1985)
8.42 P. Poulet, D. Cahen, S. Malkin: Biochim. Biophys. Acta **724**, 433–446 (1983)
8.43 V.S. Dorofeeva, V.P. Zharov, S.G. Zemchuzin, V.S. Letokhov, S.G. Montanari, L.M. Tumanova: Agrokhimiya **8**, 116-121 (1984)
8.44 D.D. Bicanic, H. Jalink, H. Sauren, P. Torfs, A. Roos, W. Kunze, F. Harren, J. Lloyd, L. Tuinstra: *Photoacoustic Investigations of Pesticides*, Proc. 3rd Int. Conf. Trends in Quantum Electronics TQE'88, Bucharest, 1988, to appear in J. de Phys. Roum. (1989)
8.45 F. Harren: "Photoacoustics – Refined and Applied to Biological Problems"; Ph. D. Thesis, Department of Laser and Molecular Physics, Catholic University, Nijmegen (1988)
8.46 D. Bicanic, H. Jalink, B. v. Veldhuyzen: submitted to Anal. Chem. (1989)
8.47 H. Jalink, D. Bicanic: submitted to Appl. Phys. Lett. (1989)
8.48 G.P.A. Bot: "Greenhouse Climate: From Physical Process to a Dynamic Model". Ph. D. Thesis, Department of Physics and Meteorology, Agricultural University, Wageningen (1985)
8.49 T. Takakura, K. Kurata, T. Honjo: Acta Hortic. **174**, 97–104 (1985)
8.50 D.D. Bicanic, A. Bizzarri, B.F.J. Zuidberg, G.P.A. Bot, T. de Jong, H.C.P. Wegh: "CO_2 Laser Based N_2O Photoacoustic Monitor" in *Photoacoustic and Photothermal Phenomena*, ed. by P. Hess, J. Pelzl, Springer Ser. Opt. Sci., Vol. 58 (Springer, Berlin, Heidelberg 1988) pp.143–145
8.51 H.G. Wolting, E.A.M. van Remortel, N. van Berkel: Acta Hortic. **174**, 351–358 (1986)
8.52 H.G. Wolting, E.A.M. van Remortel, N. van Berkel: Weekblad groenten en fruit, 27–29 November (1986)
8.53 A. van het Ooster, J.V. Klarenbeek: IMAG Institute of Agricultural Engineering, Wageningen, private communication (1988)
8.54 J.W.M. Budel: BAP: Bio Algeen Producten, Benelux, Arnhem, The Netherlands, private communication (1988)
8.55 B.M. Artyomov, E.M. Artyomov, V.P. Zharov, I.M. Nazarov, S.D. Fridman, V.B. Biryulin: "Photoacoustic study of ammonia produced by fertilized fields", Trudi ordena Trudovogo Krasnogo Znameni Instituta Prikladnoi Geofiziki imeni akademika E.K. Fedorova, Vol. 67, *Distancione sredstva i metodi izmerenia zagrzenii atmosferi i vibrosov*, ed. by V.I. Rozdestvenskoi (Hidrometeoizdata, Moscow 1986) pp.106–114
8.56 A.B. Antipov, B.M. Artyomov, E.M. Artyomov, V.B. Biryulin, V.P. Zharov, S.A. Kolesnikov, Yu.N. Ponomarjov: "Optoacoustic technique for gas analysis" in 5th Int. Topical Meeting on Photoacoustic and Photothermal Phenomena, Heidelberg, Tech. Digest. (1987) pp.186–187
8.57 B.F. Pain, J.V. Klarenbeek: IMAG Wageningen-Institute of Agricultual Engineering Research Report 88–2 (1988)
8.58 R.A. Rooth, A.J.L. Verhage, L.W. Wouters, L. van de Beeld: Proc. 4th Int. Conf. on Infrared Physics, Zurich (1988) pp.593–595

8.59 A.J.L. Verhage, R.A. Rooth: Photoacoustic measurement of atmospheric ammonia. Report on 3rd annual meeting of Dutch photoacoustic group (Kema N.V. Arnhem, The Netherlands 1988)
8.60 J. Mooi, A.J.A. van der Zalm: "Research on the effects of higher than ambient concentrations of SO_2 and NO_2 on vegetation under seminatural conditions", EEC contract EVN-672-NL (1986)
8.61 W.A.J. van Pul: "Investigation of the diurnal and seasonal fluxes of ozone and nitrogen oxides to and in a maize crop" Ph. D. Thesis, Department of Physics and Meteorology, Agricultural University Wageningen (1989) (To be published)
8.62 C.G. van Ginkel: "Oxidation of gaseous hydrocarbons by alkene utilizing bacteria"; Ph. D. Thesis, Department of Microbiology, Agricultural University Wageningen (1987)
8.63 E. Woltering, F. Harren, H. Boerrichter: Plant Physiol **88**, 506–510 (1988)
8.64 S. Hartmans, J.A.M. de Bout, T. Tramper, K.Ch.A.M. van der Luyben: Biotechnol. Lett. **7**, 383–388 (1985)
8.65 J. Sikkema, P. Smith: Department of Industrial Microbiology, Agricultural University, Wageningen, private communication (1989)
8.66 J.L. Beveridge, J. Dalziel, H.J. Duncan: Potato Res. **24**, 61-76 (1981)
8.67 D.D. Bicanic, B.F.J. Zuidberg, K. Hartman, A. van Es: submitted for publication in Appl. Spect.
8.68 L.W.A. van Hove, A.J. Koops, E.H. Adema, W.J. Vredenberg, G.A. Pieters: Atmos. Environ. **21**, 1799-1763 (1987)
8.69 L.W.A. van Hove: "The uptake of NH_3 and SO_2 by leaves", Report Project 78. Dutch Priority Program on Acidification, Nat. Inst. for Environ. Health, Bilthoven (1988)
8.70 L.W.A. van Hove, W.J.M. Tonk, G.A. Pieters, E.H. Adema, W.J. Vredenburg: in Atmos. Environ. **22**, 2515–2523 (1988)
8.71 H. Sauren, L.W.A. van Hove, W. Tonk, H. Jalink, D. Bicanic: "On the Adsorption Properties of Ammonia", in *Monitoring of Gaseous Pollutants by Tunable Diode Lasers*, ed. by R. Grisar, H. Preier, G. Schmidtke (D. Reidel Publishing Company, Dordrecht) (in press)
8.72 G.L. Loper, J.A. Gelbwachs, S.M. Beck: Can. J. Phys. **64**, 1124-1132 (1986)
8.73 H. Sauren, D. Bicanic, W. Hillen, H. Jalink, C. van Asselt, J. Quist, J. Reuss: submitted to Appl. Opt. (1989)
8.74 W. Tonk: Department of Plant Physiological Research, Agicultural University Wageningen, private communication (1988)
8.75 F.T. Schiet: "Laser photoacoustic spectroscopy and the psychophysics"; M.Sc. Thesis, Department of Physics and Meteorology, Agricultural University Wageningen (1986)
8.76 D. Bicanic, F.T. Schiet: 4th Int. Topical Meeting of Photoacoustic, Thermal and Related Sciences, Ville d'Esterel, Quebec, Tech. Digest paper MC9.1 (1985)
8.77 S.S. Stevens (ed.): *Psychophysics* (Wiley, New York 1975)
8.78 M.L. Lee, D.L. Smith, L.R. Freeman: Appl. Environ. Microbiol. **37**, 85–90 (1979)
8.79 A.C.J.M. van de Laar: M.Sc. Thesis, Wageningen University (1987)
8.80 R.H. Dainty, R.A. Edwards, C.M. Hibbard: J. Appl. Bacteriol. **59**, 303–309 (1985)
8.81 R.A. Edwards, R.H. Dainty: J. Sci. Food Agric. **38**, 57-66 (1987)
8.82 R.H. Dainty: AFRC Institute of Food Research, Langford, Bristol, private communication (1987)
8.83 D. Bicanic, B.F.J. Zuidberg, R. Friedhoff, F. Harren, D. van der Akker, A. Bizzarri: "Photoacoustic Monitoring of the Time Course of Ammonia Formation During the Spoilage of Inoculated Beef at Room Temperature" in *Photoacoustic and Photothermal Phenomena*, ed. by P. Hess, J. Pelzl, Springer Ser. Opt. Sci., Vol.58 (Springer, Berlin, Heidelberg 1988) pp.145–148
8.84 Y.R. Kolomiisky, V.P. Zharov: J. Photoacoust. **1**, 49–61 (1982)
8.85 H. Coufal, U. Möller, S. Schneider: "Photoacoustic cells for Measurements at Various Temperatures and Pressure – Design and Characterization" in *Photoacoustic Effect – Principles and Applications*, ed. by P. Luscher et al. (Vieweg, Braunschweig 1984) pp.350–354
8.86 M.M. Thompson, R.A. Palmer: Anal. Chem. **60**, 1027–1032 (1988)
8.87 D. Chisolm: *The Heat Pipe* (Mills and Boon, London 1971)
8.88 H. Scheingraber, C.R. Vidal: Rev. Sci. Instrum. **52**, 1010–1012 (1981)
8.89 S.A. Johnson, P.G. Cummins, S.A. Bone, P.B. Davies: SPIE Proc. 907, to be published
8.90 G.G. Shklover and S.E. Gusev: Inzh.-Fiz. Zh. **53**, 229-236 (1987)
8.91 Z.Sh. Semerkhanov, G.F. Smirnov, B.A. Afanesev: J. Eng. Phys. (USSR) **53**, 908–913 (1988)
8.92 T.H. Gieling, E. van Os, A. de Jager: Acta Hortic. **230**, 357–361
8.93 D.D. Bicanic, W.D. Kunze, H. Sauren, H. Jalink, M. Lubbers, E. Strauss: to appear in Air, Water and Soil Pollut., Vol. 45 (1989)
8.94 J. Murphy, J. Riley: Anal. Chim. Acta **27**, 31–36 (1962)
8.95 D. Bicanic, E. Strauss, J.P. Favier, M. Lubbers, G. Fleuren: submitted to Anal. Chem.

8.96 E. Strauss, D. Bicanic, J.P. Favier, M. Lubbers, A. Miklós: submitted to Appl. Opt.
8.97 P.L. Meyer, M.W. Sigrist: "Air Pollution Monitoring with a Mobile CO_2-Laser Photoacoustic System"; Ph. D. Thesis no. 8651, Institute of Quantum Electronics, Swiss Federal Institute of Technology ETH, Zurich (1988)
8.98 S. Bernegger, M.W. Sigrist: "CO Laser Photoacoustic Spectroscopy of Gases and Vapours for Trace Gas Analysis"; Ph. D. Thesis no. 8636, Institute of Quantum Electronics, Swiss Federal Institute of Technology ETH, Zurich (1988)
8.99 M.W. Sigrist: J. Appl. Phys. **60**, R83-R121 (1986)
8.100 G. Loper, J. Gelbwachs, S. Beck: Progress Reports, FY-1985, 1986. Aeorospace Corporation, El Segundo, CA
8.101 V.S. Zharov, V. Letokhov: *Laser Optoacoustic Spectroscopy*, Springer Ser. Opt. Sci., Vol. 37, (Springer, Berlin, Heidelberg 1986)

Subject Index

MIX
Papier aus verantwortungsvollen Quellen
Paper from responsible sources
FSC® C105338

If you have any concerns about our products,
you can contact us on
ProductSafety@springernature.com

In case Publisher is established outside the EU,
the EU authorized representative is:
Springer Nature Customer Service Center GmbH
Europaplatz 3, 69115 Heidelberg, Germany

Printed by Libri Plureos GmbH
in Hamburg, Germany